Lecture Notes in Mathematics

Edited by A. Dold and B. Eckmann

1149

Universal Algebra and Lattice Theory

Proceedings of a Conference held at Charleston, July 11 – 14, 1984

Edited by Stephen D. Comer

Springer-Verlag
Berlin Heidelberg New York Tokyo

Editor

Stephen D. Comer
Department of Mathematics and Computer Science, The Citadel
Charleston, S.C. 29409, USA

Mathematics Subject Classification (1980): 03G15, 06B05, 06B20, 06B25, 06C05, 06D10, 08A05, 08A35, 08A50, 08B10, 08B20

ISBN 3-540-15691-7 Springer-Verlag Berlin Heidelberg New York Tokyo
ISBN 0-387-15691-7 Springer-Verlag New York Heidelberg Berlin Tokyo

Library of Congress Cataloging in Publication Data. Main entry under title: Universal algebra and lattice theory. (Lecture notes in mathematics; 1149) 1. Algebra, Universal–Congresses. 2. Lattice theory–Congresses. I. Comer, Stephen D., 1941-. II. Series: Lecture notes in mathematics (Springer-Verlag); 1149. QA3.L28 no. 1149 [QA251] 510 s [512] 85-17329
ISBN 0-387-15691-7 (U.S.)

Printed in Germany

Printing and binding: Beltz Offsetdruck, Hemsbach/Bergstr.
2146/3140-543210

FOREWARD

The conference on Universal Algebra and Lattice Theory held at The Citadel in Charleston, South Carolina from July 11 to July 14, 1984 was attended by 71 mathematicians from seven countries. The purpose of the meeting was to bring together in North America for the first time since 1982 an international collection of active researchers representing current trends in the field. The scientific program consisted of eight invited lectures and 28 contributed talks.

The present volume contains research papers based on lectures presented at the conference as well as contributions from colleagues who were invited to submit manuscripts that contain significant new results. Several directions of current research are represented in the proceedings. In keeping with the tradition set by its origin, it is notable that there are a number of papers that deal with connections between lattice theory and universal algebra and other areas of mathematics such as geometry, graph theory, group theory and logic. The editor wishes to thank the numerous individuals who refereed the manuscripts submitted. The editor also wishes to thank the other members of the organizing committee, Bjarni Jonsson and Trevor Evans, for their valuable advice and guidance and to thank his colleagues at The Citadel for their practical support - in particular, Charles Cleaver, Leslie Crabtree, Margaret Francel, Jean-Marie Pages, and Russ Thompson.

Financial support for the conference from the National Science Foundation and The Citadel Development Foundation is gratefully acknowledged.

TABLE OF CONTENTS

UNIVERSAL TERMS FOR PSEUDO-COMPLEMENTED DISTRIBUTIVE LATTICES AND HEYTING ALGEBRAS

M. E. Adams and D. M. Clark
SUNY at New Paltz
New Paltz, New York 12561

A term $w(x_0,x_1,\ldots x_{k-1})$ in the language of an algebra $\underline{A}$ is called $\underline{A}$-universal if for every $b \in A$ there are $a_0,a_1,\ldots a_{k-1} \in A$ such that $b = w^{A}(a_0,a_1,\ldots a_{k-1})$. $w(x_0,x_1,\ldots x_{k-1})$ is universal for a class M of algebras if it is $\underline{A}$-universal for each $\underline{A}$ in M. We will give criteria to characterize universal terms for the variety of pseudo-complemented distributive lattices (p-algebras) and for the variety of Heyting algebras in terms of the validity of certain derived equations. Since both varieties have decidable equational theories our criteria are effective and, in many cases, quite efficient. In both varieties we find universal and nonuniversal terms occurring in substantial number and complexity.

The study of universal terms arose out of the work of Ore [12] who studied groups in which every member of the commutator subgroup is a commutator. Our notion of universal terms for an algebra comes from Isbell [7] and Mycielski [10], and has been extensively developed by Silberger as in, for example, [16]. McNulty and Pigozzi have extended the notion to varieties of algebras in quite a different way than ours in [9] and [13].

In case M is a variety of algebras, our notion of universality can be described in terms of the free algebras. $w(x_0,x_1,\ldots x_{k-1})$ is universal for a variety M if and only if there is a homomorphism of the M-free algebra on generators $G_0,G_1,\ldots G_{k-1}$ into the M-free algebra on one generator G_0 taking $w(G_0,G_1,\ldots G_{k-1})$ to G_0, if and only if $w(x_0,x_1,\ldots x_{k-1})$ is universal for the M-free algebra on one generator. In case the M-free algebra on one generator is finite, the latter condition provides an (admittedly inefficient) decision procedure for univeral terms. We will study universal terms in each variety by reinterpreting the problem in a dual category of structured

Boolean spaces. We have included a solution to the problem for Boolean algebras in order to illustrate this technique in an otherwise rather simple case. The solutions for p-algebras and for Heyting algebras follow the same pattern in successively more complex settings.

1. Boolean Algebras

In the next two sections we will use two variations of Stone's duality theorem for Boolean algebras. For our present needs, we formulate Stone's theorem for finite Boolean algebras as follows: To each finite set $\mathbf{X}$ we associate the Boolean algebra $\mathbf{B(X)}$ of all subsets of $\mathbf{X}$. Then $\mathbf{B}$ is an antiequivalence between the category of finite sets and the category of finite Boolean algebras. particular, each finite Boolean algebra is isomorphic to $\mathbf{B(X)}$ for somefinite set $\mathbf{X}$, and each homomorphism

$$f:B(\mathbf{X}) \to B(\mathbf{Y})$$

is induced by a map

$$\beta:\mathbf{Y} \to \mathbf{X}$$

by taking $f(N) = \beta^{-1}(N)$ for $N \in B(\mathbf{X})$.

THEOREM 1.1 A Boolean algebra term $w(x_0,x_1,\ldots x_{k-1})$ is universal if and only if neither it nor its negation is a tautology.

Proof. Let $\mathbf{B}(X_k)$ be the free Boolean algebra on generators $G_0,G_1,\ldots G_{k-1}$. Then X_1 is a set of two points, m and n, and $\mathbf{B}(X_1)$ is freely generated by $\{m\}$. Using Stone duality, we observe that

$$w(x_0,x_1,\ldots x_{k-1}) \text{ is universal}$$

if and only if

there is a homomorphism $f:\mathbf{B}(X_k) \to \mathbf{B}(X_1)$ taking $W = w(G_0,G_1,\ldots G_{k-1})$ to $\{m\}$

if and only if

there is a map $\beta:\mathbf{X}_1 \to \mathbf{X}_k$ such that $\beta^{-1}(W) = \{m\}$

if and only if

W is neither empty nor all of X_k

if and only if, since $G_0, G_1, \ldots G_{k-1}$ freely generate $B(\mathbf{X}_k)$,

neither $w(x_0, x_1, \ldots x_{k-1})$ nor $-\, w(x_0, x_1, \ldots x_{k-1})$ is a tautology. ∎

2. Pseudo-Complemented Distributive Lattices

A pseudo-complemented distributive lattice (p-algebra)

$$\mathbf{A} = \langle A, \wedge, \vee, {}^*, 0, 1\rangle$$

is a bounded distributive lattice $\langle A, \wedge, \vee, 0, 1\rangle$ augmented with a unary operation * (pseudo-complementation) such that for x and y in A,

$$x \wedge y = 0 \quad \text{if and only if} \quad y \leq x^*.$$

P-algebras form an equational class in which the finitely generated free algebras are finite (Lee [8]). Finite p-algebras can be described very concretely in terms of finite posets: Let $\mathbf{X} = (X, \leq)$ be a finite poset, $N \subseteq X$. Then

$$(N] = \{\, x \in X \mid x \leq y \text{ for some } y \in N \,\}$$
$$[N) = \{\, x \in X \mid y \leq x \text{ for some } y \in N \,\}$$
$$\mathrm{Min}(N) = \{\, x \in X \mid x \text{ is a minimal element of } (N]\}$$

A <u>p-map</u> is an order preserving map $\beta:\mathbf{X} \to \mathbf{Y}$ between finite posets such that $\mathrm{Min}\{\beta(x)\} = \beta\mathrm{Min}\{x\}$ for all $x \in X$. Note, in this case, that $\beta\mathrm{Min}(X) \subseteq \mathrm{Min}(Y)$.

A subset of a poset is <u>decreasing</u> if it contains every element below x whenever it contains x. The decreasing subsets of a finite poset **X** form a p-algebra

$$\mathbf{PA}(X) = \langle PA(\mathbf{X}), \wedge, \vee, {}^{*}, \emptyset, X \rangle$$

where $N^{*} = \{x \in X \mid Min\{x\} \cap N = \emptyset\}$. Notice that * acts like complementation on minimals:

$$Min(N^{*}) = Min(X)/Min(N)$$

It is shown in [1] that **PA** is an antiequivalence between the category of finite posets, with p-maps, and the category of finite p-algebras. In particular, every finite p-algebra is representable as **PA**(X) for some finite poset **X** and for any finite posets **X** and Y, $f:\mathbf{PA}(X) \to \mathbf{PA}(Y)$ is a homomorphism if and only if there is a p-map $\beta : Y \to \mathbf{X}$ such that

$$f(M) = \beta^{-1}(M) \quad \text{for all} \quad M \in PA(\mathbf{X}).$$

Let $\mathbf{Fp}_k = \mathbf{PA}(\mathbf{X}_k)$ be the free p-algebra on generators $G_0, G_1, \ldots G_{k-1}$ for each positive integer k. $\mathbf{Fp}_k$ is a finite p-algebra and $\mathbf{X}_k$ is a finite poset. We will focus on

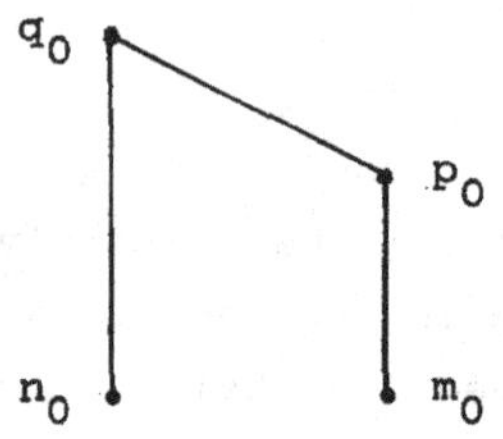

$G_0 = \{m_0\}$

$\mathbf{X}_1$

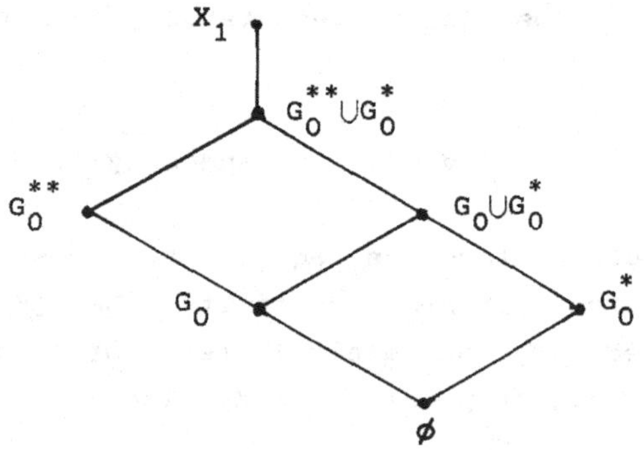

$\mathbf{Fp}_1 = \mathbf{PA}(\mathbf{X}_1)$

<u>LEMMA 2.1</u> A p-algebra term $w(x_0, x_1, \ldots x_{k-1})$ is universal if and only if there is a p-map $\beta : \mathbf{X}_1 \to \mathbf{X}_k$ such that $\beta^{-1}(w(G_0, G_1, \ldots G_{k-1}) = \{m_0\}$. ∎

<u>LEMMA 2.2</u>

(i) Let $p,q \in X_k$, $p \nleq q$ and öin{p} = Min{q}. Then some G_i contains q and not p.

(ii) Let $m,n \in \mathrm{Min}(X_k)$, $m \neq n$. Then some G_i separates m and n.

<u>Proof</u>. (i) Suppose not. Then for each G_i, $q \in G_i$ implies $p \in G_i$. This property is preserved under union, intersection and, since Min{p} = Min{q}, also under *. Thus every member of $\mathbf{Fp}_k$ containing q also contains p. But this cannot be since $p \notin (q] \in F_k$.

(ii) The proof is similar to (i). ∎

The 2^k different elements of F_k obtained as intersections of the generators and their pseudo-complements will determine the distinct minimal elements of X_k. To see this, let $\eta \in \{\text{blank}, {}^*\}^k$ be a k-tuple of blanks and *'s. We will write $\cap G^\eta$ as an abbreviation for the set

$$(G_0)^{\eta(0)} \cap (G_1)^{\eta(1)} \cap \ldots \cap (G_{k-1})^{\eta(k-1)}.$$

<u>LEMMA 2.3</u> For each $\eta \in \{\text{blank}, {}^*\}^k$, $\cap G^\eta = \{m\}$ where $m \in \mathrm{Min}(X_k)$. Conversely, for every $m \in \mathrm{Min}(X_k)$ there is an $\eta \in \{\text{blank}, {}^*\}^k$ such that $\cap G^\eta = \{m\}$.

<u>Proof</u>. By Lemma 2.2(ii), $\cap G^\eta$ contains at most one minimal element. It follows, by Lemma 2.2(i), that it contains at most one element, which must be minimal.. If it were empty, then the equation

$$(x_0)^{\eta(0)} \wedge (x_1)^{\eta(1)} \wedge \ldots \wedge (x_{k-1})^{\eta(k-1)} = 0$$

would hold in every p-algebra (and therefore every Boolean algebra), which is clearly not the case. This proves the first statement.

Conversely, if $m \in \mathrm{Min}(X_k)$ is given, then $m \in \cap G^\eta$ where $\eta(i)$ = blank if $m \in G_i$ and $\eta(i) = {}^*$ otherwise. ∎

Now let $w = w(x_0,x_1,\ldots x_{k-1})$ be a fixed p-algebra term, $W = w(G_0,G_1,\ldots G_{k-1})$. We can now refine Lemma 2.1:

LEMMA 2.4 w is universal if and only if

(i) $\mathrm{Min}(X_k)$ is not contained in W and

(ii) there are $m,p \in X_k$ such that $m \in \mathrm{Min}(X_k) \cap W$, $p \notin W$ and $(p] \cap \mathrm{Min}(X_k) = \{m\}$.

Proof. If w is universal, take β as in Lemma 2.1. Then $n = \beta(n_0) \in \mathrm{Min}(X_k)\backslash W$ and $p = \beta(p_0)$, $m = \beta(m_0)$ satisfy (ii) since β is a p-map.

Conversely, suppose $n \in \mathrm{Min}(X_k)\backslash W$ and m,p satisfy (ii). Let $Z_1 = (\{m,n,p,q\},\leq)$ be a copy of $\mathbf{X}_1$ with m_0,n_0,p_0,q_0 replaced, respectively, by m,n,p,q where $q \notin X_k$. Let $f:\mathbf{F}p_k \rightarrow \mathbf{PA}(Z_1)$ be the homomorphism where

$$f(G_i) = G_i \cap \{m,n,p\} \quad \text{for each } i < k.$$

Then there is a p-map $\beta:\mathbf{Z}_1 \rightarrow X_k$ such that

$$f(N) = \beta^{-1}(N) \quad \text{for} \quad N \in F_{pk}.$$

We show that β satisfies the conditions of Lemma 2.1. For $i<k$,

$$m \in G_i \quad \text{iff} \quad m \in f(G_i) = \beta^{-1}(G_i) \quad \text{iff} \quad \beta(m) \in G_i.$$

Since $m,\beta(m) \in \mathrm{Min}(X_k)$, we conclude from Lemma 2.2(ii) that $\beta(m) = m$. Similarly, $\beta(n) = n$.

We claim that $\beta(p) \geq p$. Suppose not. Then we could apply Lemma 2.2(i) to obtain a G_i containing $\beta(p)$ and not p. But this is not possible since $p \notin f(G_i) = \beta^{-1}(G_i)$ so $\beta(p) \notin G_i$. Thus $\beta(p) \geq p \notin W$ so $\beta(p) \notin W$.

Now $\beta^{-1}(W)$ is a decreasing subset of Z_1 containing m but neither n nor p, so it must be $\{m\}$. By Lemma 2.1, w is universal. ∎

We examine conditions (i) and (ii) of Lemma 2.4 independently.

LEMMA 2.5 (i) holds if and only if $W^* \neq \emptyset$.

Proof. $\mathrm{Min}(W^*) = \mathrm{Min}(X_k) \setminus \mathrm{Min}(W)$. ∎

LEMMA 2.6 (ii) holds if and only if for some $\eta \in \{\text{blank}, {}^*\}^k$,

$$(W \cap (\cap G^{\eta}))^{**} \nsubseteq W.$$

PROOF: Use Lemma 2.3 and the fact that for $m \in \mathrm{Min}(X_k)$, $p \in \{m\}^{**}$ if and only if $(p] \cap \mathrm{Min}(X_k) = \{m\}$. ∎

THEOREM 2.7 Let Σ_{pa} be the equational theory of p-algebras. Then a p-algebra term $w = w(x_0, x_1, \ldots x_{k-1})$ is universal if and only if

(i) $w^* = 0 \notin \Sigma_{pa}$ and

(ii) for some k-tuple $\eta \in \{\text{blank}, {}^*\}^k$,

$$(w \wedge (x_0)^{\eta(0)} \wedge \ldots \wedge (x_{k-1})^{\eta(k-1)})^{**} \leq w \notin \Sigma_{pa}.$$

Proof. Lemmas 2.4, 2.5, 2.6. ∎

To see that the conditions of Theorem 2.7 can be effectively checked, we use a result of Urquhart. Let B[k] denote the p-algebra obtained by adding a new unit to the Boolean algebra 2^k.

LEMMA 2.8 (Urquhart [17]) A k-variable p-algebra identity holds in all p-algebras if and only if it holds in $B[2^k]$. ∎

COROLLARY 2.9 A p-algebra term $w = w(x_0, x_1, \ldots x_{k-1})$ is universal if and only if for some $\eta \in \{\text{blank}, {}^*\}^k$, the formulas

(i) $w^* \neq 0$ and

(ii) $(w \wedge (x_0)^{\eta(0)} \wedge \ldots \wedge (x_{k-1})^{\eta(k-1)})^{**} \nleq w$

are both satisfiable in $B[2^k]$. ∎

3. Heyting Algebras

A Heyting algebra

$$A = \langle A, \wedge, \vee, \rightarrow, 0, 1\rangle$$

is a bounded distributive lattice $\langle A, \wedge, \vee, 0, 1\rangle$ augmented by a binary operation $\rightarrow$ (relative pseudocomplementation) such that for any x, y and z in A,

$$x \wedge z \leq y \quad \text{if and only if} \quad z \leq x \rightarrow y.$$

Heyting algebras also form an equational class (cf Balbes and Dwinger [2]). Each Heyting algebra has a natural reduct to a p-algebra which we obtain by defining $x^* = x\to 0$.

An extension to Heyting algebras of the Priestley duality for distributive lattices [14] is now part of the folklore (see, for example, Priestley [15]). We review this construction briefly, as it will be our primary tool for characterizing universal terms. A partially ordered topological space $\mathbf{X} = (X,\leq)$ is <u>totally order disconnected</u> if for every $x,y \in X$, $x \not\leq y$, there is a clopen decreasing set $N \subseteq X$ containing y but not x. For each $N \subseteq X$, we define $(N]$, $[N)$ and $\mathrm{Min}(N)$ as in the previous section. A compact totally order disconnected space is a <u>Heyting space</u> (<u>H-space</u>) if $[N)$ is clopen for every convex clopen set N. We will often use without specific reference the fact that every clopen subset N of an H-space contains an element which is minimal in N. An <u>H-map</u> is a continuous order preserving map $\beta:\mathbf{X} \to \mathbf{Y}$ between H-spaces such that $\beta((x]) = (\beta(x)]$ for each $x \in X$. As before we notice that in this case, $\beta\mathrm{Min}(X) \subseteq \mathrm{Min}(Y)$.

The clopen decreasing sets in an H-space $\mathbf{X}$ form a Heyting algebra

$$\mathbf{H}(X) = \langle H(\mathbf{X}),\cap,\cup,\to,\emptyset,X\rangle$$

where $M\to N = X \setminus [M\setminus N)$ for $M,N \in H(X)$. We write N^* for $N\to\emptyset$. As with p-algebras we observe that $\mathrm{Min}(N^*) = \mathrm{Min}(X) \setminus \mathrm{Min}(N)$ for $N \in H(\mathbf{X})$. The contravariant functor $\mathbf{H}$ is an antiequivalence between the category of H-spaces and the variety of Heyting algebras. In particular, $f:\mathbf{H}(X) \to \mathbf{H}(Y)$ is a homomorphism if and only if there is an H-map $\beta:\mathbf{Y} \to \mathbf{X}$ such that

$$f(M) = \beta^{-1}(M) \quad \text{for all} \quad M \in H(X).$$

Let $\mathbf{FH}_k = \mathbf{H}(\mathbf{X}_k)$ be the free Heyting algebra on generators $G_0,G_1,\ldots G_{k-1}$ for each positive integer k. Then each $\mathbf{FH}_k$ is denumerably infinite and, in case $k>1$, is apparently very complex. A simple description of $\mathbf{FH}_1$ was found in Nishimura [11]:

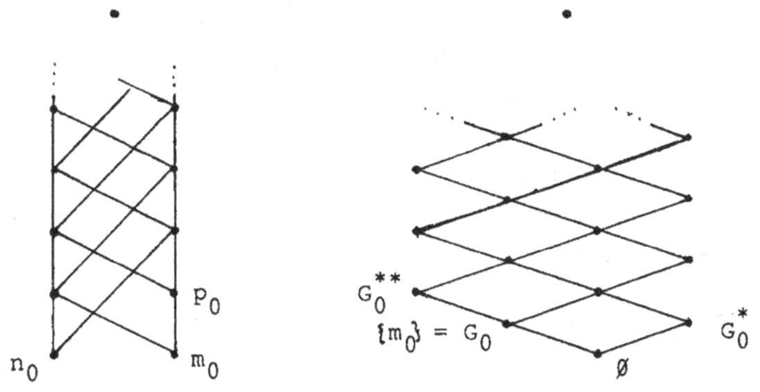

X_1 $\qquad$ $FH_1 = H(X_1)$

<u>LEMMA 3.1</u> A Heyting algebra term $w = w(x_0, x_1, \ldots x_{k-1})$ is universal if and only if there is an H-map $\beta: \mathbf{X}_1 \to \mathbf{X}_k$ such that $\beta^{-1}(w(G_0, G_1, \ldots, G_{k-1})) = \{m_0\}$. ∎

<u>LEMMA 3.2</u>

(i) If $m, p, q \in X_k$ are distinct, $(p] = \{m, p\}$ and $(q] = \{m, q\}$ then some G_i separates p and q.

(ii) If $m, n \in \mathrm{Min}(X_k)$, $m \neq n$, then some G_i separates m and n.

(iii) If $\mathrm{Min}(X_k) \cap (p] = \{m\}$ and $p \neq m$, then some G_i separates m and p.

<u>Proof</u>. (i) Suppose not. We will contradict the total order disconnectedness of X_k by showing that no member of $\mathbf{FH}_k$ would separate p and q. If $M, N \in F_k$ do not separate p and q, neither do $M \cap N$ and $M \cup N$. Suppose p and q were separated by $M \to N = X_k \setminus [M \setminus N)$:

$$p \in M \to N \qquad \text{and} \qquad q \notin M \to N.$$

Then

$$p \notin [M \setminus N) \qquad \text{and} \qquad q \in [M \setminus N).$$

Since $m \leq p$ we have $m \in M \to N$, $m \notin M \setminus N$. But $(q] = \{m, q\}$ contains something in $M \setminus N$, so we conclude that $q \in M \setminus N$. This is not possible since $p \notin M \setminus N$ and M and N do not separate p and q.

(ii) and (iii) follow arguments similar to (i). ∎

<u>LEMMA 3.3</u> For each $\eta \in \{\text{blank}, {}^*\}^k$, $\bigcap G^\eta = \{m\}$ where $m \in in(X_k)$. Conversely, for every $m \in Min(X_k)$ there is an $\eta \in \{\text{blank}, {}^*\}^k$ such that $\bigcap G^\eta = \{m\}$.

<u>Proof</u>. By Lemma 3.2(ii), $\bigcap G^\eta$ contains at most one minimal element m. By Lemma 3.2(iii), $\bigcap G^\eta \subseteq \{m\}$. If $\bigcap G^\eta$ were empty, then, again, the equation

$$(x_0)^{\eta(0)} \wedge (x_1)^{\eta(1)} \wedge \ldots \wedge (x_{k-1})^{\eta(k-1)} = 0$$

would hold in every Heyting (and therefore every Boolean) algebra, which is clearly not the case.

Conversely, let $m \in Min(X_k)$. Then $m \in \bigcap G^\eta$ where $\eta(i)$ is blank if $m \in G_i$, and $\eta(i)$ is $*$ otherwise. ∎

Let $w = w(x_0,x_1,\ldots x_{k-1})$ be a fixed Heyting algebra term, $W = w(G_0,G_1,\ldots G_{k-1})$. We can now refine Lemma 3.1:

<u>LEMMA 3.4</u> w is universal if and only if

(i) $Min(X_k) \not\subseteq W$ and

(ii) there are $m,p \in X_k$ such that $m \in Min(X_k) \cap W$, $p \notin W$ and $(p] = \{m,p\}$.

<u>Proof</u>. If w is universal, take β as in Lemma 3.1. Then $n = \beta(n_0) \in Min(X_k)\backslash W$ and $p = \beta(p_0)$, $m = \beta(m_0)$ safisfy (ii). Conversely, let Z_1 be a copy of X_1 where m_0,n_0,p_0 are replaced, respectively, by m,n,p, and $Z_1 \cap X_k = \{m,n,p\}$. Let $f:\mathbf{FH}_k \to \mathbf{H}(Z_1)$ be the homomorphism where

$$f(G_i) = G_i \cap \{m,n,p\} \quad \text{for } i<k.$$

Then there is an H-map $\beta:Z_1 \to X_k$ such that

$$f(N) = \beta^{-1}(N) \quad \text{for } N \in F_{Hk}.$$

We claim that β fixes m,n and p. Notice that for $i<k$,

$$m \in G_i \text{ iff } m \in f(G_i) = \beta^{-1}(G_i) \text{ iff } \beta(m) \in G_i.$$

Since m and $\beta(m)$ are minimal, we conclude from Lemma 3.2(ii) that $\beta(m) = m$. Similarly, $\beta(n) = n$. Now suppose $\beta(p) = q \neq p$. Since β is an H-map, $(q] = \{m,q\}$. By Lemma 3.2(i) there is a G_i that separates p and q. But

$$p \in G_i \quad \text{iff} \quad p \in f(G_i) = \beta^{-1}(G_i) \quad \text{iff} \quad \beta(p) = q \in G_i.$$

Thus $\beta(p) = p$.

Now $\beta^{-1}(W)$ is a decreasing subset of Z_1 containing m but neither n nor p, so it must be $\{m\}$. By Lemma 3.1, w is universal. ∎

We examine conditions (i) and (ii) of Lemma 3.4 independently.

LEMMA 3.5 (i) holds if and only if $W^* \neq \emptyset$.

Proof. $\mathrm{Min}(W^*) = \mathrm{Min}(X_k) \setminus \mathrm{Min}(W)$. ∎

LEMMA 3.6 (ii) holds if and only if for some $\eta \in \{\text{blank}, {}^*\}^k$, $\cap G^\eta \subseteq W$ but

$$(W \longrightarrow \cap G^\eta) \cap (\cap G^\eta)^{**} \not\subseteq W.$$

Proof. If (ii) holds, choose η so that $\cap G^\eta = \{m\}$. Then $\cap G^\eta \subseteq W$, $p \in (W \longrightarrow \cap G^\eta) \cap (\cap G^\eta)^{**}$ but $p \notin W$.

Conversely, let $\cap G^\eta = \{m\}$. Then m is a minimal element of X_k in W. Let p be a minimal element of the nonempty clopen set $(W \longrightarrow \{m\}) \cap \{m\}^{**} \setminus W$. Then $p \notin W$ and, since $p \in \{m\}^{**}$, $(p]$ contains no minimal other than m. To see that p covers m, suppose $m < q \leq p$. Then $q \notin W$ since $p \in W \rightarrow \{m\}$. By the minimality of p, $q = p$. ∎

THEOREM 3.7 Let Σ_H be the equational theory of Heyting algebras. Then a Heyting algebra term $w = w(x_0,x_1,\ldots x_{k-1})$ is universal if and only if

(i) $w^* = 0 \notin \Sigma_H$ and

(ii) for some k-tuple $\eta \in \{\text{blank}, {}^*\}^k$, $\wedge x^\eta \leq w \in \Sigma$ but

$$(w \longrightarrow \wedge x^\eta) \wedge (\wedge x^\eta)^{**} \leq w \notin \Sigma_H$$

where $\wedge x^{\eta} = (x_0)^{\eta(0)} \wedge (x_1)^{\eta(1)} \wedge \ldots \wedge (x_{k-1})^{\eta(k-1)}$.

Proof. Lemmas 3.4, 3.5, 3.6. ∎

Gentzen [5] gave an effective decision procedure for membership in Σ_H. Using this we obtain from Theorem 3.7 an effective proceedure to decide universality.

4. Universal Terms

In this section we will justify our original contention that both universal and nonuniversal terms occur in substantial number and complexity. Theorems 2.7 and 3.7 not only give us simple criteria to check universality but, perhaps more importantly, they tell us how to construct terms which are and are not universal.

Recall that a Heyting algebra becomes a p-algebra when we replace $\rightarrow$ by $*$ where $x^* = x \rightarrow 0$. A simple consequence of the definition of universality shows the relationship between our results for p-algebras and those for Heyting algebras.

LEMMA 4.1 A p-algebra term w is universal for p-algebras if and only if it is universal for Heyting algebras.

Proof. If w is universal for p-algebras, then, since each Heyting algebra can be made into a p-algebra, it is universal for Heyting algebras. Conversely, suppose w is universal for Heyting algebras. Now $\mathbf{Fp}_1$, being finite, is a reduct of a Heyting algebra where we define

$$M \rightarrow N = \bigcup \{Q \in \mathrm{Fp}_1 \mid Q \cap M \subseteq N\}$$

Consequently, w is universal for $\mathbf{Fp}_1$ and therefore for all p-algebras. ∎

This tells us that we can speak of universality of p-algebra terms without making reference to either variety. The same is true for the notion of equivalence of p-algebra terms. For, suppose u and v are p-algebra terms and that $u=v$ holds in every p-algebra.

Then it clearly holds in every Heyting algebra. Conversely, suppose u=v holds in every Heyting algebra and that the variables in u and v are among $x_0, x_1, \ldots x_{k-1}$. Then $\mathbf{Fp}_k$, being finite, becomes a Heyting algebra if we replace * by $\rightarrow$ as defined in the previous proof. Since u=v is satisfied by $\mathbf{Fp}_k$, it holds in all p-algebras.

For k>0, let d_k denote the cardinality of the free distributive lattice $\mathbf{FD}_k$ on k generators. The sequence d grows very fast: d = (1, 4, 18, 166, 7579, 7828352, ...) (See Berman and Kohler [3]).

THEOREM 4.3

(i) For k>0, there are at least d_{k-1} inequivalent k-variable p-algebra terms which are not universal.

(ii) There are infinitely many inequivalent 1-variable Heyting algebra terms which are not universal.

Proof. (i) Let $K = \{N \in Fp_k \mid G_0 \vee G_0^* \leq N\}$ be the filter on $\mathbf{Fp}_k$ determined by $G_0 \vee G_0^*$. If $w(G_0, G_1, \ldots G_{k-1})$ is in K, we have

$$w(G_0, G_1, \ldots G_{k-1})^* \leq (G_0 \vee G_0^*)^* = \emptyset.$$

By Theorem 2.7, $w(x_0, x_1, \ldots x_{k-1})$ is not universal. To prove the theorem, we will construct a map from K onto FD_k.

Let **A** be the p-algebra

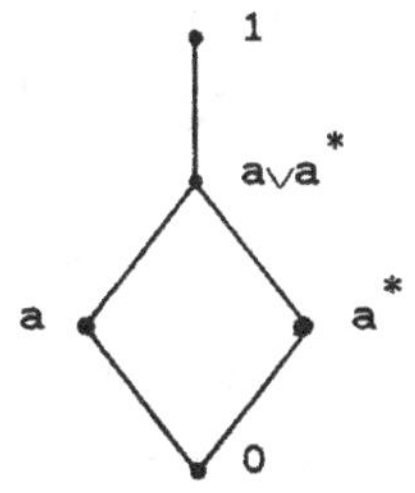

where the interval $[a \vee a^*, 1]$ is $\mathbf{FD}_{k-1}$. Let f be the homomorphism from $\mathbf{Fp}_k$ onto **A** taking $G_1, G_2, \ldots G_{k-1}$ to the free generators of $\mathbf{FD}_{k-1}$ and taking G_0 to a. Then for any choice of $t \in FD_{k-1}$ we have

$$t = f(N) \quad \text{for some} \quad N \in Fp_k,$$

$$N \vee G_0 \vee G_0^* \in K \quad \text{and}$$
$$f(N \vee G_0 \vee G_0^*) = t \vee a \vee a^* = t.$$

Thus f takes K onto DF_{k-1}. ∎

(ii) The only elements $N \in FH_1$ for which $N^* \neq \emptyset$ are $\emptyset$, G_0, G_0^*, G_0^{**}. By Theorem 3.7 the remaining elements are not universal. ∎

We can also give many examples of complex terms which are universal. For each N in Fp_k or in FH_k, let $N\downarrow$ denote the ideal in Fp_k or in FH_k, respectively, consisting of all elements contained in N. The filter $N\uparrow$ is defined similarly.

LEMMA 4.4 Let $k>0$.

(i) For $G_0 \in Fp_k$, $|G_0\downarrow| \geq |Fp_{k-1}|$.

(ii) For $G_0 \in FH_k$, $|G_0\downarrow| \geq |FH_{k-1}|$.

Proof. (i) Let f be the homomorphism from $\mathbf{Fp}_k$ onto $\mathbf{Fp}_{k-1}$ which takes $G_1, G_2, \ldots G_{k-1}$ to the generators of $\mathbf{Fp}_{k-1}$ and takes G_0 to X_{k-1} (the unit in $\mathbf{Fp}_{k-1}$). Then f maps $G_0\downarrow$ onto Fp_{k-1}. Indeed, if $M \in Fp_{k-1}$, then there is an $N \in Fp_k$ such that $f(N) = M$. But then

$$N \cap G_0 \in G_0\downarrow \quad \text{and} \quad f(N \cap G_0) = M \cap X_{k-1} = M.$$

The proof of (ii) is identical. ∎

Referring to either $\mathbf{Fp}_k$ or $\mathbf{FH}_k$ we define, for each element $W \in G_0\downarrow$,

$$W' = (G_0^* \cap G_1 \cap \ldots \cap G_{k-1}) \cup W.$$

LEMMA 4.5 If $W, V \in G_0\downarrow$ and $W \neq V$, then $W' \neq V'$.

Proof. Suppose $W' = V'$:

$$(G_0^* \cap G_1 \cap \ldots \cap G_{k-1}) \cup W = (G_0^* \cap G_1 \cap \ldots \cap G_{k-1}) \cup V$$

Intersecting both sides with G_0 and distributing, we obtain
$\emptyset \cup W = \emptyset \cup V$ and $W = V$. ∎

LEMMA 4.6 Let $k>0$, $W = w(G_0,G_1,\ldots G_{k-1}) \in G_0\downarrow$, $W' = w'(G_0,G_1,\ldots G_{k-1})$. Then w' is universal.

Proof. Since $W \subseteq G_0$, $w(0,x_1,x_2\ldots x_{k-1}) = 0$ holds in the variety. Then

$$w'(\emptyset,G_1,G_1,\ldots G_1) = G_1 \cup w(\emptyset,G_1,G_1,\ldots G_1) = G_1$$

and therefore $w'(0,x,x,\ldots x) = x$ holds in the variety. ∎

THEOREM 4.7

(i) For $k>1$, there are at least $|Fp_{k-1}|$ inequivalent k-variable p-algebra terms which are universal.

(ii) There are infinitely many inequivalent 2-variable Heyting algebra terms which are universal.

Proof. Use Lemmas 4.4, 4.5, 4.6 and the fact that FH_1 is infinite. ∎

REFERENCES

[1] M. E. Adams, Implicational classes of pseudo-complemented distributive lattices, J. London Math. Soc. (2) 13 (1976), 381-384.

[2] R. Balbes and P. Dwinger, Distributive Lattices, Univ. Missouri Press, Columbia, Miss., 1974.

[3] J. Berman and P. Kohler, Cardinalities of finite distributive lattices, Mitt. Math. Sem. Giessen 121 (1976), 103-124.

[4] B. A. Davey and D. Duffus, Exponentiation and duality, Ordered Sets, NATO Advanced Study Institute Series 83, D. Reidel Publishing Co., Dordrecht, Holland, 1982, 43-95.

[5] G. Gentzen, Untersuchungen uber das logische Schliessen, Math. Z., vol. 39 (1934), 176-210.

[6] G. Gratzer, Lattice Theory: First Concepts and Distributive Lattices, Freeman, San Francisco, California, 1971.

[7] J. R. Isbell, On the problem of universal terms, Bull. de l'Academie Polonaise des Sciences XIV (1966).

[8] K. B. Lee, Equational classes of distributive pseudo-complemented lattices, Canad. J. Math. 22 (1970), 881-891.

[9] G. McNulty, Decidable properties of finite sets of equations, Journal Symb. Logic, 41 (1976), 589-604.

[10] J. Mycielski, Can one solve equations in groups?, Amer. Math. Monthly 84 (1977), 723-726.

[11] I. Nishimura, On formulas of one variable in intuitionistic propositional calculus, J. Symbolic Logic 25(1960), 327-331.

[12] O. Ore, Some remarks on commutators, Proc. Amer. Math. Soc. 2 (1951), 307-314.

[13] D. Pigozzi, The universality of the variety of quasigroups, Journal Australian Math. Soc. (Series A), XXI (1976), 194-219.

[14] H. A. Priestley, Representation of distributive lattices by means of ordered Stone spaces, Bull. London Math. Soc. 2(1970), 186-190.

[15] H. A. Priestley, Ordered sets and duality for distributive lattices, Proc. Conf on Ordered Sets and their Applications, Lyon, 1982, North Holland Series Ann. Discrete Math.

[16] D. M. Silberger, When is a term point universal?, Algebra Universalis, 10 (1980), 135-154.

[17] A. Urquhart, Free distributive pseudo-complemented lattices, Algebra Univeralis 3(1973), 13-15.

[18] A. Urquhart, Free Heyting algebras, Algebra Universalis 3(1973) 94-97.

CLONES OF OPERATIONS ON RELATIONS

H. Andreka,* S. D. Comer**, and I. Nemeti*

* Math. Inst. Hungarian Acad. Sci., Budapest 1364, Pf. 127, Hungary

** The Citadel, Charleston, S.C. 29409, U.S.A.

It is well known that every Boolean polynomial can be written as a polynomial using only the Sheffer stroke operation $\uparrow$ which is defined by

$$x \uparrow y = \bar{x} + \bar{y}.$$

This fact is expressed by saying that the clone of any Boolean algebra is one-generated (by { $\uparrow$ }). This paper deals with clones of Boolean algebras with additional operations. We show that, for such algebras, finitely generated clones are one-generated, give specific sets of generators for clones of relation algebras, and show that, in at least one case, the generating set given is best possible. Our observations were motivated by questions posed by Bjarni Jonsson in [1]. The original proof of Theorem 2 has been simplified thanks to a suggestion of Roger Maddux.

1. INTRODUCTION. The first result applies to most of the natural clones which extend the classical Boolean clone.

THEOREM 1. *Every finitely generated clone that contains the Sheffer stroke operation is one-generated.*

In the particular case of relation algebras we show

THEOREM 2. *The clone of every relation algebra is generated by*
(a) *the Sheffer stroke and one binary operation*

$$\tau(x,y) = (x;y^{\smile}) + (1' - (1;(x+y);1)),$$

and
(b) *one ternary operation*

$$\delta(x,y,z) = ((x \uparrow y) - (1;(y \oplus z);1)) + \tau(x,y) \cdot (1;(y \oplus z);1)$$

where $\oplus$ *denotes symmetric difference.*

Is it possible to do better than Theorem 2 ? The following shows that 2(a) cannot be improved.

THEOREM 3. For every set V with at least 6 elements there is a proper relation algebra on V in which relative composition is not contained in the clone generated by the Sheffer stroke operation together with all unary terms of the algebra.

It is not known whether 2(b) can be improved to a single binary operation.

2. THE PROOFS

Proof of Theorem 1. It is shown in [1] that every finitely generated clone that contains $\uparrow$ can be generated by $\{\uparrow,\kappa\}$ for some n-ary term $\kappa(x_1,\ldots,x_n)$. Now, let

$$\rho(y,z,w,x_1,\ldots,x_n) = \bar{y} + \bar{z} + \bar{w}\cdot\kappa(x_1,\ldots,x_n)$$

where $\bar{y}$ denotes $-y$, etc. Then $\rho(y,z,z,x_1,\ldots,x_n) = y\uparrow z$ and $\rho(1,1,\emptyset,x_1,\ldots,x_n) = \kappa(x_1,\ldots,x_n)$. Thus, $\text{clone}\{\rho\} = \text{clone}\{\uparrow,\kappa\}$ as desired.

Theorem 2 produces special generating sets for the clone $C_c(\mathfrak{A})$ of all operations definable from the basic operations of the relation algebra $\mathfrak{A} = \langle A,\uparrow,;,{}^{\smile},1'\rangle$.

Proof of Theorem 2. To show that the set $\{\uparrow,\tau\}$, given in (a), generates the clone $C_c(\mathfrak{A})$ of a relation algebra $\mathfrak{A}$ it suffices to verify that the identities (i)-(iii) below hold in every relation algebra.

(i) $1' = \tau(\emptyset,\emptyset)$

(ii) $x^{\smile} = \tau(1',x)$

(iii) $x;y = \tau(x,y^{\smile}) - \tau(x+y^{\smile},\emptyset)$.

It is easily seen that (i) holds in $\mathfrak{A}$. Since $1;(1'+x);1 \geq 1;1';1 = 1$ in every relation algebra $\mathfrak{A}$, $\tau(1',x) = 1';x^{\smile} + (1'-(1;(1'+x);1)) = x^{\smile}$ so (ii) holds. To see that (iii) holds in $\mathfrak{A}$ observe that

$$\begin{aligned}\tau(x,y^{\smile}) - \tau(x+y^{\smile},\emptyset) &= [(x;y) + (1'-1;(x+y^{\smile});1)] - [1' - 1;(x+y^{\smile});1]\\ &= (x;y)\cdot-(1'-1;(x+y^{\smile});1)\\ &= (x;y)\cdot(\emptyset' + 1;(x+y^{\smile});1)\\ &= x;y\end{aligned}$$

because $x;y \leq 1;(x+y^{\smile});1 \leq \emptyset' + 1;(x+y^{\smile});1$. It follows that $C_c(\mathfrak{A})$ is generated by $\{\uparrow,\tau\}$.

To show that the operation δ, in (b), generates the clone of a relation algebra $\mathfrak{A}$ it suffices to verify that the identities (iv) and (v) below hold in $\mathfrak{A}$.

(iv) $x \dagger y = \delta(x,y,y)$

(v) $\tau(x,y) = \delta(x,y,-y)$.

Since $y \oplus y = 0$ for every y in $\mathfrak{A}$, $1;(y \oplus y);1 = 0$ from which $\delta(x,y,y) = x \dagger y$ easily follows. So, (iv) holds in $\mathfrak{A}$. Condition (v) follows from the definition of δ because $1;(y \oplus \bar{y});1 = 1;(y+\bar{y});1 = 1;1;1 = 1$ holds in every relation algebra $\mathfrak{A}$. This completes the proof of Theorem 2.

Proof of Theorem 3. The proof will be given in two steps. First, a proper relation algebra on a 6 element set with the desired properties will be constructed and then the general case will be treated.

Let $H = \{ 0,\ldots,5 \}$, $I_H = \{ (i,i) : i \in H \}$, $\mathfrak{A}_\emptyset = \langle Sb({}^2H), \cup, \cap, \sim, \emptyset, {}^2H \rangle$ denote the Boolean algebra of all subsets of ${}^2H$, and let $\mathfrak{A}[H] = \langle \mathfrak{A}_\emptyset, |, {}^\vee, I_H \rangle$ denote the (full) proper algebra of relations on H with $\mathfrak{A}_\emptyset$ as its Boolean part. To prove the theorem for $|V| = |H| = 6$, it suffices to find a collection $\mathfrak{B}$ of binary relations on H such that $\mathfrak{B}$ is a Boolean subalgebra of $\mathfrak{A}_\emptyset$, $\mathfrak{B}$ is closed under all unary relation algebra terms, but $\mathfrak{B}$ is not a subuniverse of the relation algebra $\mathfrak{A}[H]$. Consider the following relations on H:

$x = \{ (0,1),(1,0),(2,3),(3,2),(4,5),(5,4) \}$

$y = \{ (0,2),(2,0),(1,4),(4,1),(3,5),(5,3) \}$

$z = \{ (0,5),(5,0),(1,3),(3,1),(2,4),(4,2) \}$

$u = \{ (0,4),(1,2),(2,5),(3,0),(4,3),(5,1) \}$

$v = \{ (0,3),(1,5),(2,1),(3,4),(4,0),(5,2) \}$

$w = u \cup v$

Let $\mathfrak{B}$ denote the Boolean subalgebra of $\mathfrak{A}_\emptyset$ generated by $\mathcal{P} = \{ I,x,y,z,w \}$. Since $\mathcal{P}$ partitions ${}^2H$, an elements of $\mathfrak{B}$ is just a union of a set of generators. To see that $\mathfrak{B}$ is the desired model it suffices to show

(i). $x|y \notin B$, and

(ii). for every $a \in \mathfrak{B}$ the subalgebra of $\mathfrak{A}[H]$ generated by a, $\mathfrak{Sg}(a)$, is contained in $\mathfrak{B}$.

Note that property (ii) implies that $\mathfrak{B}$ is closed under every unary relational term. To assist in the verification of (i) and (ii), it is convenient to know how the relative composition (partial) operation $|$ works on the atoms of $\mathfrak{B}$. The "multiplication" table for $|$ is given below in Table 1.

From Table 1, $x|y = u$; so (i) holds since $u \notin \mathfrak{B}$. For $a \in \mathfrak{B}$ observe that $\mathfrak{Sg}\{ a \} = \mathfrak{Sg}\{ \sim a \}$ and $\mathfrak{Sg}\{ a \} = \mathfrak{Sg}\{ a \sim I \}$. Hence it suffices to verify the conclusion of (ii) for elements a which are the union of at most two atoms in the set $\{ x,y,z,w \}$. For $a = \emptyset$, $\mathfrak{Sg}\{ a \}$ is the minimal subalgebra of $\mathfrak{A}[H]$ which is contained in $\mathfrak{B}$. If $a \in \{ x,y,z,w \}$, let $\hat{a}$ denote $\sim a \cap \sim I$. It is easily seen that the atoms of $\mathfrak{Sg}\{ a \}$ are I, a, and $\hat{a}$ because $a;\hat{a} = \hat{a} = \hat{a};a$ and $\hat{a};\hat{a} = 1$. Hence $\mathfrak{Sg}\{ a \} \subseteq \mathfrak{B}$ when a is an atom of $\mathfrak{B}$. Now, suppose that a is the union of two atoms in the set $\{x,y,z,w\}$. As before let $\hat{a} = \sim a \cap \sim I$. Either $a \geq w$ or $\hat{a} \geq w$. Without loss of generality we assume that $\hat{a} \geq w$ and set $a' = \hat{a} \sim w$. The atoms of $\mathfrak{Sg}\{a\}$ are I, a, a' and w as can be seen from the "multiplication" table of | for these relations (Table 2).

	I	x	y	z	w
I	I	x	y	z	w
x	x	I	u	v	y∪z
y	y	v	I	u	x∪z
z	z	u	v	I	x∪y
w	w	y∪z	x∪z	x∪y	I∪w

Table 1

	I	a	a'	w
I	I	a	a'	w
a	a	I∪w	w	a∪a'
a'	a'	w	I	a
w	w	a∪a'	a	I∪w

Table 2

It follows that $\mathfrak{Sg}\{a\} \subseteq \mathfrak{L}$ when a is a join of two atoms of $\mathfrak{L}$. Hence (ii) holds which completes the proof of Theorem 3 when $|V| = 6$.

Now suppose $|V| \geq 6$ and $H = \{0,\ldots,5\} \subseteq V$. Let $G = V \sim H$ and $N = \{I_G, {}^2G \sim I_G, G \times H, H \times G\}$. Then, recalling notation from the first case, $\mathcal{P} \cup N$ partitions ${}^2V$ and, for every $a,b \in N$, $\{a^{\vee}, a;b\} \subseteq N \cup \{0, {}^2G\}$. Let $\mathfrak{L}'$ be the Boolean algebra generated by $\mathcal{P} \cup N$. As in the $|V| = 6$ case, it suffices to show that (i) and (ii) hold for the modified $\mathfrak{L}'$.

It is clear that $x;y \notin \mathfrak{L}'$, because, for example, every element of $\mathfrak{L}$ is symmetric. Thus, (i) holds for $\mathfrak{L}'$.

To show (ii) for $\mathfrak{L}'$ we shall use the fact that (ii) holds for $\mathfrak{L}$ and that the domain and range of every $b \in \mathfrak{L}$ is H. Suppose $a \in \mathfrak{L}'$. Then $a = b + \Sigma X$ for some $b \in \mathfrak{L}$ and some $X \subseteq N$. Let $\mathfrak{S}$ be the relation subalgebra of $\mathfrak{A}[H]$ generated by {b}. Then $\mathfrak{S} \subseteq \mathfrak{L}$ since (ii) holds for $\mathfrak{L}$. Let $\mathfrak{S}'$ be the Boolean algebra generated by $\mathfrak{S} \cup N$. Since $a \in \mathfrak{S}' \subseteq \mathfrak{L}'$, in order to establish (ii) for $\mathfrak{L}'$, it is enough to show that $\mathfrak{S}'$ contains I_V and is closed under $^\smile$ and $;$. It is clear that $I_V = I_H \cup I_G$ belongs to $\mathfrak{S}'$. Now, suppose $a \in \mathfrak{S}'$ has the form $a = b + \Sigma X$ for some $b \in \mathfrak{S}$ and some $X \subseteq N$. Then $a^\smile = b^\smile + \Sigma\{ t^\smile : t \in X \}$ belongs to $\mathfrak{S}'$ since $b^\smile \in \mathfrak{S}$ and N is closed under $^\smile$. Hence $\mathfrak{S}'$ is closed under $^\smile$. Now, suppose $a,c \in \mathfrak{S}'$ where $a = b + \Sigma X$ and $c = d + \Sigma Y$ with $b,d \in \mathfrak{S}$ and $X,Y \subseteq N$. Then

$$a;c = b;d + b;\Sigma Y + \Sigma X;d + \Sigma X;\Sigma Y.$$

Since $b;d \in \mathfrak{S}$, $b;\Sigma Y \in \{ \emptyset, H \times G \}$, $\Sigma X;d \in \{ \emptyset, G \times H \}$, and $\Sigma X;\Sigma Y \in N \cup \{ \emptyset, {}^2G \}$, it follows that $a;c \in \mathfrak{S}'$. Thus, $\mathfrak{S}'$ is a relation subalgebra of $\mathfrak{R}[V]$ and it follows that (ii) holds for $\mathfrak{L}'$ which completes the proof of Theorem 3.

ACKNOWLEDGEMENT. The research of the second author was supported by The Citadel Development Foundation.

REFERENCES

1. Jonsson, B., *The theory of binary relations. A first draft.* Manuscript. July 1984.
2. Jonsson, B. and Tarski, A., *Boolean algebras with operators II.*, Amer. J. Math. 74(1952), 127-162.

SEPARATION CONDITIONS ON CONVEXITY LATTICES

M. K. Bennett
Department of Mathematics and Statistics
University of Massachusetts
Amherst, MA 01003

1. INTRODUCTION. If V is a vector space over an ordered division ring D, the convex subsets of V form a complete atomic algebraic lattice denoted Co(V). We will use [12] and [6] as standard references for the properties of convex sets and lattices respectively. For X a convex subset of V, the principal ideal Co(X) of all convex subsets of X inherits many of the properties of Co(V). In [2], [4] and [5] the present author and G. Birkhoff began an investigation of a class of 'convexity lattices' which includes (but is not limited to) the lattices Co(V) and Co(X) described above. There the following lattice-theoretic concepts were introduced.

For natural number n, Co($\underline{n}$) is the lattice of points and segments generated by n collinear points in V. The Hasse diagrams for Co($\underline{3}$) and Co($\underline{4}$) are given in Figures 1 and 2 below.

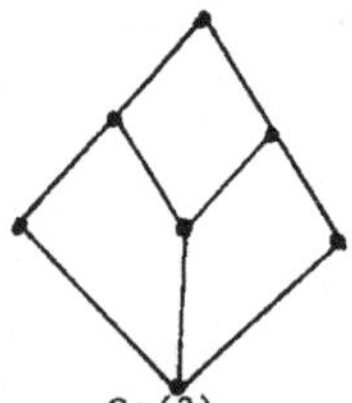
Co($\underline{3}$)
Figure 1

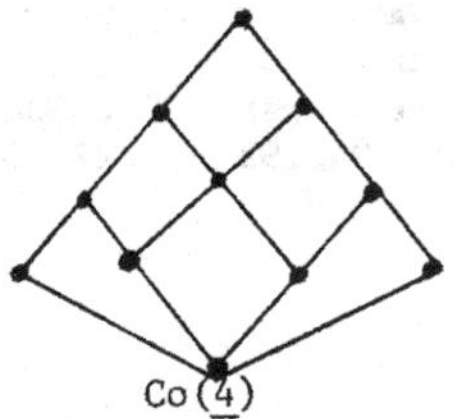
Co($\underline{4}$)
Figure 2

DEFINITION. An atomic lattice L is said to be biatomic when given a and b nonzero elements of L and p an atom under $a \vee b$, there are atoms $a_1 \leq a$ and $b_1 \leq b$ with $p \leq a_1 \vee b_1$.

DEFINITION. A complete algebraic biatomic lattice is a convexity lattice when

(CL1) Given p, q, r distinct atoms of L, then $\langle p,q,r\rangle$ (the sublattice generated by p, q and r) is isomorphic to $\underline{2}^3$ (the atoms are considered non collinear) or $Co(\underline{3})$ (one atom is 'between' the other two),

(CL2) If p, q, r and s are distinct atoms and if $\langle p,q,r\rangle$ and $\langle q,r,s\rangle$ are both isomorphic to $Co(\underline{3})$, then $\langle p,q,r,s\rangle$ is isomorphic to $Co(\underline{4})$ (i.e. if two triples of the four atoms are collinear, all four are collinear).

Axioms (CL1) and (CL2) describe properties of linear betweenness. For distinct atoms p, q and r in $Co(X)$, $p \leq q \vee r$ means that point p is strictly between points q and r (written $(q\,p\,r)\beta$); and if $\langle p,q,r,s\rangle$ is isomorphic to $Co(\underline{4})$, these four points are on a (closed) segment bounded by two of them.

Various conditions have been used by synthetic geometers to axiomatize planar betweenness. (See [15, p.351] for a discussion of these.) Pasch, and later Hilbert [8] postulated that "a line in the plane of a triangle which intersects one side of that triangle intersects one of the other two sides." (See Fig. 3a from [8], p.5).

Two weaker assumptions were used by Peano and others. The first was called the triangle transversal axiom by Veblen [15] and the Pasch axiom by Szmielew [14]: If a, b and c are non-collinear then $(acd)\beta$ and $(bec)\beta$ imply there is an f such that $(def)\beta$ and $(afb)\beta$. The second statement, which we call the Peano axiom is: If a, b and c are non-collinear then $(acd)\beta$ and $(afb)\beta$ imply there is an e such that $(bec)\beta$ and $(def)\beta$. (See Fig. 3b where a, b, c and d are given in both statements above; the existence of e implies that of f in the triangle transversal axiom, and the existence of f implies that of e in the Peano axiom.)

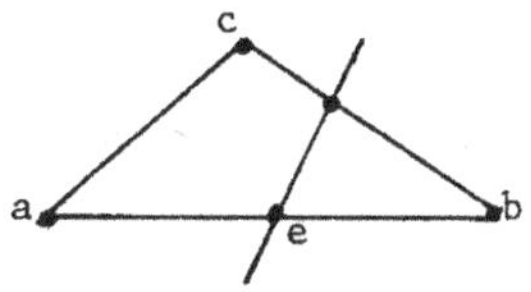

Figure 3a

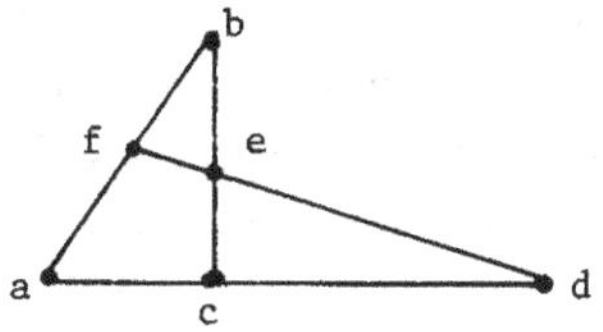

Figure 3b

The lattice theoretic version of the triangle transversal axiom holds in any biatomic lattice (in particular in any convexity lattice).

THEOREM 1. *For atoms a, b, c, d, e in a biatomic lattice L, $c \leq a \vee d$ and $e \leq b \vee c$ imply there is an atom $f \leq a \vee b$ such that $e \leq d \vee f$.*

Proof: $e \leq b \vee c \leq b \vee a \vee d$ implies $e \leq f \vee d$ for some atom $f \leq b \vee a$.

The lattice-theoretic version of the Peano axiom is given below.

DEFINITION. A convexity lattice is a Peano (convexity) lattice when (PC) given a, b, c, d, f distinct atoms, with $f \le a \vee b$ and $c \le a \vee d$, then $(b \vee c) \wedge (d \vee f) \neq 0$.

Not all Peano lattices are of the form Co(X). The lattices Co($\underline{n}$) are Peano lattices, and in [5] further examples of finite Peano lattices are presented.

The main result of this paper gives necessary and sufficient conditions for a Peano lattice to be Co(V) for some V. We first present a series of conditions for convexity lattices which are equivalent to the Peano condition.

2. SEPARATION BY HYPERPLANES. If L is Co(V), the affine flats of V (translates of linear subspaces) constitute the core of modular elements of L, M(L), i.e. those a in L such that xMa holds for all x in L. (Recall that xMa means $b \vee (x \wedge a) = (b \vee x) \wedge a$ whenever $b \le a$). These flats also form a complete atomic algebraic lattice (see [4]) denoted A(V). Moreover its interval sublattices of the form $[p,1]_{A(V)}$ for p an atom are projective geometries isomorphic to the (modular) lattice of (linear) subspaces of V.

It is a classic result that any hyperplane h in D^n separates the vectors in D^n into three equivalence classes, those in h and those on either "side" of h. This condition can be defined in any convexity lattice as well. The lattice-theoretic analogue of a hyperplane is a coatom of M(L); hence we recall some previous results (3.2-3.6 of [3]). We will assign letters to known theorems, and number the theorems proved here for the first time.

THEOREM A. If L is a convexity lattice, M(L) is closed under arbitrary meets. $(M(L),\bar{\vee},\wedge)$ is a complete atomic algebraic lattice where

$$a \mathbin{\bar{\vee}} b = \bigwedge\{c \in M(L) : a \vee b \le c\} .$$

Furthermore for a in L the following are equivalent. (i) a is modular; (ii) pMa for all atoms p; (iii) if q and r are distinct atoms under a, $q \le p \vee r$ implies $p \le a$.

In D^n points p and q are in different equivalence classes defined by a hyperplane h exactly when $(prq)\beta$ for some r in h. This motivates the following concept.

DEFINITION. A convexity lattice L has the separation property when each coatom h of $M(L)$ gives rise to an equivalence relation E_h on the atoms of L where

$$p\ E_h\ q \text{ if and only if } \begin{cases} p \vee q \leq h & \text{or} \\ (p \vee q) \wedge h = 0 & . \end{cases}$$

(In any convexity lattice, every E_h is reflexive and symmetric.)

In this section we will show that when $M(L)$ is coatomic for convexity lattice L, the Peano condition is equivalent to the separation property. We first need to introduce the lattice-theoretic equivalent of the Pasch axiom.

In view of (CL1) a plane in a convexity lattice is a join of atoms $x \bar{\vee} y \bar{\vee} z$ where $\langle x,y,z\rangle \cong \underline{2}^3$. Thus we can state the planar condition used by Pasch and Hilbert as follows.

DEFINITION. A convexity lattice L satisfies the Pasch condition when, given atoms x, y, z, p, q of L with $\langle x,y,z\rangle \cong \underline{2}^3$, $p \leq x \vee y$ and $q \leq x \bar{\vee} y \bar{\vee} z$, then either (i) $(p \bar{\vee} q) \wedge (x \vee y) > p$; (ii) $(p \bar{\vee} q) \wedge (x \vee z) \neq 0$; or (iii) $(p \bar{\vee} q) \wedge (y \vee z) \neq 0$.

THEOREM B. ([4], Theorem 12). A convexity lattice is a Peano lattice if and only if it satisfies the Pasch condition.

THEOREM C. ([4], Theorem 7). Let L be a Peano lattice, S a set of atoms in L, and r an atom under $\bar{\vee}S$. Then $r \leq p \bar{\vee} q$ for some atoms $p, q \leq \vee S$.

LEMMA. Any Peano lattice has the separation property.

Proof: In what follows, we will use the symbols $h_0,\ldots,h_9$ to represent atoms under the coatom h. Let x, y and z be atoms of L, none of which is under the coatom h. Let $(x \vee y) \wedge h = h_0$. We must show that either $(x \vee z) \wedge h \neq 0$ or $(y \vee z) \wedge h \neq 0$. Since h is a coatom, $h \bar{\vee} y = 1$, hence $z \leq h \bar{\vee} y$. Thus $z \leq r \bar{\vee} s$ (Theorem C) where r and s are atoms under $h \vee y$, and we may assume $r \leq h_1 \vee y$ and $s \leq h_2 \vee y$. By the Peano condition (since $h_0 \leq x \vee y$ and $r \leq h_1 \vee y$), $(x \vee r) \wedge (h_1 \vee h_0) = h_3$ and similarly $(x \vee s) \wedge (h_0 \vee h_2) = h_4$.

If $z \leq r \vee s$ then $z \leq y \vee h_1 \vee h_2$, so $z \leq y \vee h_5$, and by (PC) ($z \leq y \vee h_5$, $h_0 \leq y \vee x$) we have $(x \vee z) \wedge (h_0 \vee h_5) = h_6$ and we are done.

Thus we can assume without loss of generality that $s \leq r \vee z$. Then $h_3, h_4 \leq x \vee r \vee z$ and by Theorem B either $(h_3 \bar{\vee} h_4) \wedge (x \vee z) \neq 0$ or $(h_3 \bar{\vee} h_4) \wedge (r \vee z) \neq 0$. In the former case we are done. Otherwise let $h_7 = (h_3 \bar{\vee} h_4) \wedge (r \vee z)$. If $h_7 \leq r \vee s$ then $h_7 \leq y \vee h_1 \vee h_2$ so $h_7 \leq y \vee h_8$ where $h_8 \leq h_1 \vee h_2$ and $y \leq h$, a contradiction. If $h_7 \leq s \vee z$ then $h_7 \leq h_2 \vee y \vee z$, so $h_7 \leq h_2 \vee h_9$ where $h_9 \leq y \vee z$. Thus $h \vee (y \vee z) \neq 0$ and L has the separation property.

In any convexity lattice L, if $M(L)$ is coatomic, the separation property is inherited by the principal ideals $[0,m]_L$ where $m \in M(L)$. In the rest of this section L will be assumed to be a convexity lattice such that $M(L)$ is coatomic.

LEMMA. The following are equivalent:

(1) $[0,n]_L$ has the separation property for all $n \in M(L)$.

(2) L has the separation property.

Proof: That (1) implies (2) is trivial. To show the reverse implication assume L has the separation property and let $n \in M(L)$ cover $m \in M(L)$. Then there is a coatom $h \in M(L)$ with $m \leq h$ and $n \not\leq h$.

Let n_1, n_2, n_3 be distinct atoms under n and let $(n_1 \vee n_2) \wedge m = m_1$ while $(n_2 \vee n_3) \wedge m = (n_1 \vee n_3) \wedge m = 0$. Since $m \leq n \wedge h \leq n$, and since $n \not\leq h$ implies $n \wedge h \neq n$, we have $m = n \wedge h$. Thus if $n_1 \leq h$, $n_1 \leq m$ and $(n_1 \vee n_3) \wedge m \neq 0$. Similarly $n_2, n_3 \not\leq h$. Since $n_1 \not{E}_h n_2$ we may assume (by the separation property) that there is an atom $h_1 \leq n_1 \vee n_3$, $h_1 \leq h$. But then $h_1 \leq n_1 \vee n_3 \leq n$, so $h_1 \leq n \wedge h = m$. Thus $(n_1 \vee n_3) \wedge m \neq 0$ and $[0,n]_L$ has the separation property.

THEOREM 2. Let L be a convexity lattice. Then L is a Peano lattice if and only if L has the separation property.

Proof: Any Peano lattice has the separation property. If L is not a Peano lattice there are distinct atoms x, y, z, p, q such that $z \leq q \vee x$, $p \leq y \vee x$ and $(z \vee y) \wedge (p \vee q) = 0$. Let $\ell = p \bar{\vee} q$ and $m = x \bar{\vee} y \bar{\vee} z$. Then ℓ is a coatom of $[0,m]_L$ and $xE_\ell y$ fails. Now $zE_\ell x$ since $(z \bar{\vee} x) \wedge \ell = q$ and $q \not\leq z \vee x$. If $(z \vee y) \wedge \ell = r$, we have $r \leq q \vee x \vee y$, so $r \leq q \vee t$ for some atom $t \leq x \vee y$. But $\ell = q \bar{\vee} p = q \bar{\vee} r$ and $\ell \wedge (x \bar{\vee} y) = p$, hence $t = p$ and $r \leq q \vee p$. Thus $r \leq (z \vee y) \wedge (p \vee q)$, a contradiction. Hence $(z \vee y) \wedge \ell = 0$, so $zE_\ell y$ and E_ℓ is not an equivalence relation. Thus $[0,m]_L$ does not have the separation property, from which it follows that the separation property fails in L.

3. PRENOWITZ-JANOSCIAK CONGRUENCE. Prenowitz and Janosciak introduced a collection of equivalence relations in their 'join geometries' [11, p. 52] which lend themselves nicely to interpretation in convexity lattices. Given an affine subspace A in D^n,

they defined p and q to be congruent modulo A when there is a point r such that $(pra_1)\beta$ and $(qra_2)\beta$; $a_1, a_2 \in A$. (Recall that since β represents strict betweenness $r \neq p, q, a_1, a_2$.) Congruence modulo a single point simply means being on the same open ray determined by that point. The configuration in Figure 4 shows two points p and q which are congruent modulo the line A.

Interpreted in a convexity lattice we have:

DEFINITION. If p and q are atoms of L with m modular, let $p\ \theta_m q$ mean that either $p = q$; $p \vee q \leq m$; or $(p \vee m) \wedge (q \vee m) > m$.

(Because of biatomicity, the last condition says there are atoms $a_1, a_2 \leq m$ and r not under m with $r \leq (p \vee a_1) \wedge (q < a_2)$ as shown in Fig. 4).

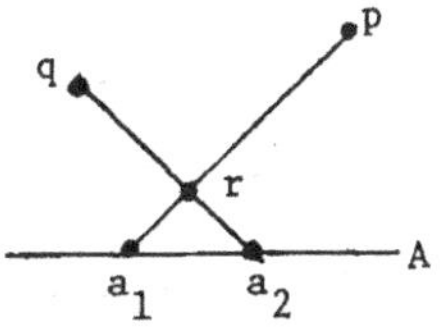

Figure 4

We shall first show that θ_m is an equivalence relation (it is always reflexive and symmetric) exactly when L is a Peano lattice. We shall then extend the θ_m to become join-congruences on L which will enable us to interpret M(L) as a collection of join-congruences on L.

LEMMA. For L a Peano lattice and m modular, θ_m is an equivalence relation.

Proof: Reflexivity and symmetry are obvious, and transitivity is clear except when $p\ \theta_m q$ and $q\ \theta_m r$ with $p \wedge m = q \wedge m = r \wedge m = 0$. Then there are atoms x and y not under m with $x \leq (p \vee m) \wedge (q \vee m)$ and $y \leq (q \vee m) \wedge (r \vee m)$. Then $x \leq (p \vee m_1) \wedge (q \vee m_2)$ and $y \leq (q \vee m_3) \wedge (r \vee m_4)$ with the m_i atoms under m.

The Peano condition gives an atom $z \leq (m_2 \vee y) \wedge (m_3 \vee x)$. If $z \leq m$ we have $y \leq m \bar{\vee} z \leq m$ and therefore $q \leq y \bar{\vee} m_3 \leq m$, a contradiction; hence $z \not\leq m$. But $z \leq m_3 \vee x \leq m_3 \vee p \vee m_1$, implies that $z \leq p \vee m_5$ for some atom $m_5 \leq m_3 \vee m_1 \leq m$. But similarly $z \leq m_2 \vee y$ implies $z \leq r \vee m_6$ for some atom $m_6 \leq m_2 \vee m_4 \leq m$. Thus $z \leq (r \vee m) \wedge (p \vee m)$ and $z \not\leq m$; hence $p\ \theta_m r$.

We shall next prove the converse of this lemma, which will imply the following theorem.

THEOREM 3. A convexity lattice is a Peano lattice if and only if θ_m is an equivalence relation for all modular m.

Proof: We take distinct atoms x, y, q, m_1, m_2 such that $x \leq q \vee m_1$ and $y \leq q \vee m_2$, and must show that $(y \vee m_1) \wedge (x \vee m_2) \neq 0$. For $m = m_1 \bar{\vee} m_2$ we have $q\ \theta_m x$ and $q\ \theta_m y$; hence $x\ \theta_m y$; so there are atoms s, m_3, m_4 with $m_i \leq m$ and $(x \vee m_3) \wedge (y \vee m_4) = s$. The proof now breaks down into cases according to whether $m_i \leq m_j \vee m_k$; $\{i,j,k\} \subseteq \{1,2,3,4\}$. We shall reproduce here only one typical case.

Let $m_3 \leq m_2 \vee m_4$; $m_4 \leq m_1 \vee m_3$. Since $s \leq m_1 \vee m_3 \vee y$, $s \leq m_3 \vee t$ for some $t \leq m_1 \vee y$. Then $t \leq x \vee m_3$ (by CL2) so $t \leq x \vee m_1 \vee m_2$. Thus there is an atom w with $t \leq m_1 \vee w$, $w \leq x \vee m_2$. Hence $w \leq (y \vee m_1) \wedge (x \vee m_2)$.

In light of the theorem above it is not surprising that θ_m and E_m are the same for the coatoms m of a Peano lattice.

THEOREM 4. For L a Peano lattice and m a coatom of $M(L)$, the relations θ_m and E_m are equivalent.

Proof: (In the following proof $m_1,\ldots,m_4$ will represent atoms under m.) If $p\ \theta_m q$ there is an atom x with $x \wedge m = 0$ and $x \leq (p \vee m) \wedge (q \vee m)$. Then $x \leq (p \vee m_1) \wedge (q \vee m_2)$. If $(p \vee q) \wedge m \neq 0$, then we have $m_3 \leq p \vee q$. By (PC) $(m_2 \vee m_3) \wedge (x \vee p) = m_4$ which implies $x \leq m_1 \bar{\vee} m_4 \leq m$, a contradiction.

Suppose $(p \vee q) \wedge m = 0$. Since m is a coatom of $M(L)$, $p \bar{\vee} m = 1$ so $q \leq p \bar{\vee} m$. By Theorem C, $q \leq r \bar{\vee} s$ where r and s are atoms under $p \vee m$. Thus we may assume $r \leq p \vee m_1$ and $s \leq p \vee m_2$. If $q \leq r \vee s$, then $q \leq p \vee m_3$ so $q \leq (p \vee m) \wedge (q \quad m)$ and $p\ \theta_m q$. If $s \leq q \vee r$, then $s \leq p \vee m_1 \vee q$ so $s \leq p \vee t$ for some atom $t \leq m_1 \vee q$. If $t \leq p \vee m_2$, then $t \leq (q \vee m) \wedge (p \vee m)$ and $p\ \theta_m q$. If $m_2 \leq p \vee t$, then $m_2 \leq p \vee m_1 \vee q$, so $m_2 \leq m_1 \vee m_3$ for $m_3 \leq p \vee q$, a contradiction.

To extend the θ_m to relations on all of L we must do more than simply substitute general lattice elements where the atoms appear in the definition of $\tilde{\theta}_m$ given above. For, to say that $a\ \tilde{\theta}_m b$ means that either $a = b$, $a \vee b \leq m$, or $(a \vee m) \wedge (b \vee m) > m$, can lead to the situation shown in Figure 5 where $a\ \tilde{\theta}_m b$, $b\ \tilde{\theta}_m c$, but not $a\ \tilde{\theta}_m c$ when m is a line and a, b and c are sets in $Co(\mathbb{R}^2)$.

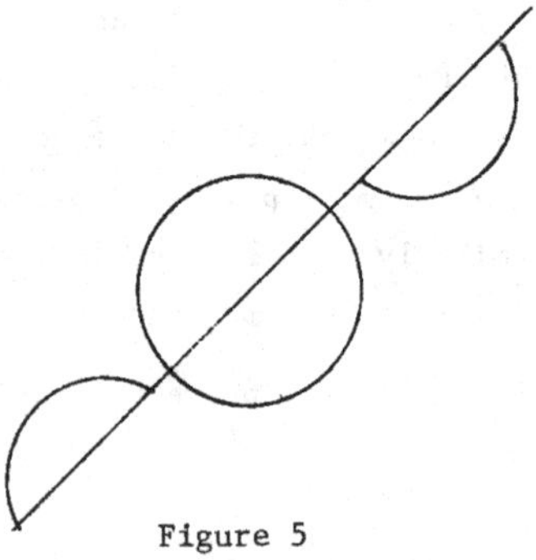

Figure 5

Hence we must make the following more complicated extension.

DEFINITION. Let L be a Peano lattice with $a,b \in L$ and m modular. Then $a\ \theta_m b$ if and only if (1) $a = b$, (2) $a \vee b \leq m$ or (3) for every atom $a_0 \leq a$ with $a_0 \wedge m = 0$, there is an atom $b_0 \leq b$ such that $a_0\ \theta_m b_0$, and for every atom

$b_1 \leq b$ with $b_1 \wedge m = 0$ there is an atom $a_1 \leq a$ such that $a_1 \, \theta_m b_1$.

REMARK. We note that $0 \, \theta_m a$ if and only if $a \leq m$, that θ_0 is equality, and that θ_1 is the relation with one equivalence class.

THEOREM 5. If m is a modular element of Peano lattice L, then θ_m is an equivalence relation. Moreover $a \, \theta_m b$ and $c \in L$ imply $(a \vee c)\theta_m(b \vee c)$.

Proof: θ_m is always reflexive and symmetric hence we first assume $a_0 \theta_m b_0$ and $b \, \theta_m c$. For atom $a_0 \leq a$ with $a_0 \wedge m = 0$, there is an atom $b_0 \leq b$ with $a \, \theta_m b$ (whence $b_0 \wedge m = 0$). Thus there is an atom $c_0 \leq c$ with $b_0 \, \theta_m c_0$ and by the lemma before Theorem 3 $a_0 \, \theta_m c_0$. Similarly starting with $c_1 \leq c$ $(c_1 \wedge m = 0)$ we obtain $a_1 \leq a$ with $c_1 \, \theta_m a_1$, so $a \, \theta_m c$ and θ_m is transitive.

To show that θ_m is a join-congruence we suppose $a \, \theta_m b$ and $c \in L$. The proof now breaks down into cases:

(1) $a,b \leq m$: If $c \leq m$, we are done. Otherwise there is an atom $p \leq a \vee c$ with $p \wedge m = 0$. Then $p \leq a_1 \vee c_1$ where a_1 and c_1 are atoms under a and c respectively. Now $p \leq (p \vee m) \wedge (c_1 \vee m)$ (since $a_1 \leq a \leq m$) and $p \nleq m$ so $p \, \theta_m c_1$. But $c_1 \leq c \leq b \vee c$, hence given $p \leq a \vee c$ with $p \nleq m$ there is an atom $c_1 \leq b \vee c$ with $p \, \theta_m c_1$. Similarly, given $q \leq b \vee c$ with $q \wedge m = 0$ we can find an atom $c_2 \leq a \vee c$ with $q \, \theta_m c_2$. Thus $a \vee c \, \theta_m b \vee c$.

(2) $a = b$: Clearly $a \vee c = b \vee c$; hence $a \vee c \, \theta_m b \vee c$.

(3) $a \nleq m$; $b \nleq m$: If p is an atom under $a \vee c$ with $p \wedge m = 0$, then $p \leq a_1 \vee c_1$ with a_1 and c_1 atoms under a and c respectively. If $a_1 \leq m$ then $c_1 \nleq m$ (otherwise $p \leq m$). Thus $(p \vee m) \wedge (c_1 \vee m) > m$, hence $p \, \theta_m c_1$ with $c_1 \leq b \vee c$.

If $a_1 \nleq m$ then there is a $b_1 \leq b$ (b_1 an atom) with $a_1 \, \theta_m b_1$. Thus there is an atom $x \nleq m$ with $x \leq (a_1 \vee m) \wedge (b_1 \vee m)$ and there are atoms $m_1, m_2 \leq m$ with $x \leq (a_1 \vee m_1) \wedge (b_1 \vee m_2)$. By the Peano condition $(m_1 \vee p) \wedge (c_1 \vee x) = y$ for some atom y. Since $y \leq c_1 \vee x \leq c_1 \vee b_1 \vee m_2$ we have $y \leq m_2 \vee z$ with z an atom under $c_1 \vee b$. If $y \nleq m$ then since $y \leq (m \vee z) \wedge (m \vee p)$ we have $p \, \theta_m z$ where z is an atom under $b \vee c$. If $y \leq m$ then $p \leq m_1 \bar{\vee} y$ implies $p \leq m$, a contradiction. Thus given $p \leq a \vee c$, $p \nleq m$, there is an atom $z \leq b \vee c$ with $p \, \theta_m z$. The converse holds similarly, so $a \vee c \, \theta_m b \vee c$.

The join-congruences θ_m induce an order-monomorphism from $M(L)$ to the lattice of equivalence relations on L as follows.

THEOREM 6. Let m, n be modular elements in a Peano lattice. Then $m < n$ if and only if $\theta_m < \theta_n$.

Proof: First, let $m < n$. If $p\ \theta_m q$ where p and q are atoms, then $p,q \leq m$ implies $p,q \leq n$ and $p\ \theta_n q$. If $p,q \not\leq m$ and x is an atom with $x \wedge m = 0$ and $x \leq (p \vee m) \wedge (q \vee m)$, then $x \leq (p \vee m_1) \wedge (q \vee m_2)$ (m_i are atoms under m). If $x \leq n$ then $p \leq x \bar{\vee} m_1 \leq n$. Likewise $q \leq n$ and $p\ \theta_n q$. If $x \not\leq n$, then $x \leq (p \vee n) \wedge (q \vee n)$ so $p\ \theta_n q$.

Now for a and b arbitrary elements of L with $a\ \theta_m b$, if $a = 0$ then $b \leq m \leq n$, so $a\ \theta_n b$. Otherwise we have:

(1) $a \leq m$, $b \leq m$, in which case $a \vee b \leq n$ and $a\ \theta_n b$.

(2) $a = b$, in which case $a\ \theta_n b$, or

(3) For a_1 an atom under a with $a_1 \wedge m = 0$, there is an atom $b_1 \leq b$ with $a_1\ \theta_m b_1$; and for b_2 an atom under b, with $b_2 \wedge m = 0$, there is an atom $a_2 \leq a$ with $b_2\ \theta_m a_2$. But for any atom $a_1 \leq a$ with $a_1 \not\leq n$, $a_1 \not\leq m$, so we have b_1 as above, and since a_1 and b_1 are atoms, $a_1\ \theta_m b_1$ implies $a_1\ \theta_n b_1$. A similar argument holds for atoms under b, hence $a\ \theta_m b$ implies $a\ \theta_n b$.

Finally, if $m \neq n$, we take p an atom under n but not m. Thus $0\ \theta_n p$ holds but $0\ \theta_m p$ fails so $\theta_m \neq \theta_n$.

Conversely if $\theta_m < \theta_n$ then for p an atom under m we have $0\ \theta_m p$, hence $0\ \theta_n p$ and $p \leq n$. Thus $m \leq n$. If $m = n$ then $\theta_m = \theta_n$, hence $m < n$.

Using a result from [5], we conclude this section by listing the various equivalents to the Peano condition.

THEOREM 7. Let L be a convexity lattice. The following are equivalent.

(1) L is a Peano lattice.

(2) L satisfies the Pasch condition.

(3) θ_m is an equivalence relation on the atoms of L whenever $m \in M(L)$.

(4) For $a,b \in L$ and p an atom, $p \wedge x = p \wedge y = x \wedge y = 0$ implies $(p \vee x) \wedge y = 0$ or $(p \vee y) \wedge x = 0$.

Furthermore, any of the above implies

(5) L has the separation condition.

Proof: The equivalence of (2) and (3) with (1) are Theorems B and 3 respectively. The equivalence of (1) and (4) is [5, Theorem 3], and (1) implies (5) is the first lemma preceding Theorem 2 above.

THEOREM 8. If L *is a convexity lattice and* $M(L)$ *is coatomic, the five statements above are equivalent.*

4. CHARACTERIZATION OF $Co(V)$. In this section we give necessary and sufficient conditions for a Peano lattice to be $Co(V)$ for some V. Here some lattice theoretic properties of affine and projective spaces are needed, as well as the notion of the distributive cover of an element.

If $L = Co(X)$, then the (join)-distributive elements $a \in L$ such that $(x \vee y) \wedge a = (x \wedge a) \vee (y \wedge a)$ for all $x,y \in L$ constitute the lattice $D(L)$ whose elements correspond to the faces of X. For any a in L if we denote by $D(a)$ the smallest face of X containing a, then the interior points of X are exactly those atoms p in L with $D(p) = 1$. The elements of $M(L)$ are the intersections of X with the affine subspaces of V. Furthermore if $D(p) = 1$, then $[p,1]_{M(L)}$ is isomorphic to the linear subspaces of V', a vector space over D whose dimension is that of the affine closure of X.

THEOREM D. ([3], Theorem 3.2). If L *is a convexity lattice and* $D(L) = \{a \in L:(x \vee y) \wedge a = (x \wedge a) \vee (y \wedge a)$ *for all* $x,y \in L\}$, *then* $D(L)$ *is closed under arbitrary meets. Furthermore* $(D(L),\tilde{\vee},\wedge)$ *is a complete atomic algebraic lattice where* $a \tilde{\vee} b = \bigwedge\{c \in D(L):a \vee b \leq c\}$.

DEFINITION. For a in convexity lattice L, the distributive cover of a, $D(a)$, is $\bigwedge\{b \in D(L):a \leq b\}$.

The notion of the distributive cover of an atom is of central importance in what follows, and it is useful to have this characterization of it.

THEOREM 9. Let p *be an atom of Peano lattice* L. *Then* q, *an atom of* L *distinct from* p, *is in* $D(p)$ *if and only if there is an atom* $r \neq p,q$ *with* $p \leq q \vee r$.

Proof: If such an r exists, $(q \vee r) \wedge D(p) = (q \wedge D(p)) \vee (r \wedge D(p))$ and since $p \leq (q \vee r) \wedge D(p)$, $p \leq (q \wedge D(p)) \vee (r \wedge D(p))$ so $q \wedge D(p) \neq 0$ and $q \leq D(p)$. Conversely, let $R = \{q|q$ an atom of L such that there is an atom r with $p \neq q,r;p \leq q \vee r\} \cup \{p\}$. We will show the following: (A) R is the set of atoms under an element of L: (B) $\bigvee R \in D(L)$: (C) $\bigvee R = D(p)$.

(A): Let $q_1,q_2 \in R$ with $s \leq q_1 \vee q_2$, s an atom. Then $p \leq (q_1 \vee r_1) \wedge (q_2 \vee r_2)$. By (PC) there is an atom w with $w \leq (r_2 \vee s) \wedge (q_1 \vee p)$. By CL2 we have $p \leq w \vee r_1$ since $p \leq q_1 \vee r_1$ and $w \leq q_1 \vee p$. Hence $p \leq s \vee r_1 \vee r_2$, so $p \leq s \vee r_3$ with r_3 an atom under $r_1 \vee r_2$, which implies $s \in R$.

(B): For t and u atoms of L we must show $(t \vee u) \wedge (\vee R) \leq (t \wedge (\vee R)) \vee (u \wedge (\vee R))$. If x is an atom in R with $x \leq t \vee u$, then if $x = t$ or u, $x \leq (t \wedge (\vee R)) \vee (u \wedge (\vee R))$. Otherwise, if t and u are in R, $t = t \wedge (\vee R)$ and $u = u \wedge (\vee R)$ so $x \leq (t \wedge (\vee R)) \vee (u \wedge (\vee R))$. If $t \in R$ and $u \notin R$ then there is an atom $x' \neq x,p$ with $p \leq x \vee x'$ and an atom $t' \neq t_1,p$ with $p \leq t \vee t'$. Hence $p \leq t \vee u \vee x'$ so $p \leq u \vee y$ for some atom $y \leq t \vee x'$. If $y = p$, then $x = (x' \bar{\vee} p) \wedge (t \bar{\vee} u) = (x' \bar{\vee} y) \wedge (t \vee u) = t$, a contradiction. Thus $y \neq p$ and since $p \leq u \vee y$, $p \neq u,y$ we have $u \in R$.

If $t \notin R$ and $u \notin R$, then $p \leq x \vee x' \leq t \vee u \vee x'$, hence $p \leq t \vee w$ for some atom $w \leq x' \vee u$. If $w = p$ then $u = (x' \bar{\vee} w) \wedge (t \bar{\vee} x) = (x' \bar{\vee} p) \wedge (t \bar{\vee} x) = x$ a contradiction. Thus $w \neq p$ and hence $p \leq t \vee w$ implies $t \in R$.

(C): In the proof above we showed that each atom in R is under $D(p)$. Hence $p \leq \vee R \leq D(p)$ and since $\vee R \in D(L)$, $\vee R = D(p)$

The lattices $M(L)$ are not biatomic; however we can introduce the notion of a biatomic pair in $M(L)$ as seen below. The following lemma will be used to show that for L a Peano lattice with p an atom, if $D(p) = 1$ then $[p,1]_{M(L)}$ is modular.

LEMMA. Let L be a Peano lattice with a, b in $M(L)$ and p an atom such that $p \leq a \wedge b$, and $D(p) = D(a \vee b) \leq a \bar{\vee} b$. Then if p is an atom under $a \bar{\vee} b$, there are atoms $a_1 \leq a$ and $b_1 \geq b$ with $p \leq a_1 \bar{\vee} b_1$. (We shall call this last condition $\bar{B}(a,b)$.)

Proof: For x an atom under $a \bar{\vee} b$, there are atoms $y,z \leq a \vee b$ with $x \leq y \bar{\vee} z$ by Theorem C. Now there are atoms $a_1,a_2 \leq a$ and $b_1,b_2 \leq b$ with $y \leq a_1 \vee b_1$ and $z \leq a_2 \vee b_2$. If $x \leq y \vee z$ then $x \leq a \vee b$, so $x \leq a_0 \vee b_0$ where a_0 and b_0 are atoms under a and b respectively, hence $x \leq a_0 \bar{\vee} b_0$ and $\bar{B}(a,b)$ holds.

If $z \leq x \vee y$, since $p \leq a \wedge b$ and $a \bar{\vee} b \leq D(a \wedge b) = D(p)$ we have an atom $a_3 \leq a$ with a_1, a_3 and p distinct and $p \leq a_1 \vee a_3$. Similarly $p \leq b_1 \vee b_3$, $p \leq a_2 \vee a_4$ and $p \leq b_2 \vee b_4$ with the $a_i \leq a$ and the $b_i \leq b$. All atoms (x,y,z,a_i,b_i) are under $a_1 \bar{\vee} a_2 \bar{\vee} b_1 \bar{\vee} b_2 = a_1 \bar{\vee} b_1 \bar{\vee} p$ (since $M(L)$ has the exchange property); hence by the separation property in $[0,a_1 \bar{\vee} a_2 \bar{\vee} b_1 \bar{\vee} p]_L$ $E = E_{a_1 \bar{\vee} a_2 \bar{\vee} p}$ is an equivalence relation. Since $(b_1 \vee b_3) \wedge (a_1 \bar{\vee} a_2 \bar{\vee} p) \geq p \neq 0$, $b_1 \mathrel{E} b_3$ fails. Hence (by the transitivity of E) either $b_1 \mathrel{E} x$ fails or $b_2 \mathrel{E} x$ fails, i.e. either $(b_1 \vee x) \wedge (a_1 \bar{\vee} a_2 \bar{\vee} p) \neq 0$ or $(b_3 \vee x) \wedge (a_1 \bar{\vee} a_2 \bar{\vee} p) \neq 0$. In the former case, since $a_1 \bar{\vee} a_2 \bar{\vee} p \leq a$ we have an atom $a_5 \leq a$ with $a_5 \leq b_1 \vee x$; hence $x \leq b_1 \bar{\vee} a_5$ and $\bar{B}(a,b)$ holds. Otherwise, $x \leq b_3 \bar{\vee} a_6$ for some atom $a_6 \leq a$ and again $\bar{B}(a,b)$ holds.

THEOREM 10. Let L be a Peano lattice with p an atom of L and $D(p) = 1$. Then $[p,m]$ is modular for all $m \in M(L)$.

Proof: $[p,1]_{M(L)}$ has what Maeda [9] calls the "weak covering property" i.e. the join of a pair of atoms covers both of them, hence by his 'Remark' on p. 77 [9], $[p,1]_{M(L)}$ is modular if it is biatomic. By the lemma above, for $a,b \in [p,1]_{M(L)}$ $\bar{B}(a,b)$ holds, i.e. a and b form a biatomic pair <u>in</u> <u>M(L)</u>. But for x an atom of L (and therefore $p \mathbin{\bar\vee} x$ an atom of $[p,1]_{M(L)}$) we have $p \mathbin{\bar\vee} x \leq a \mathbin{\bar\vee} b$ implies $x \leq a \mathbin{\bar\vee} b$, so $x \leq a_0 \mathbin{\bar\vee} b_0$ (a_0 an atom under a and b_0 an atom under b). Hence $p \mathbin{\bar\vee} x \leq p \mathbin{\bar\vee} a_0 \mathbin{\bar\vee} b_0 = (p \mathbin{\bar\vee} a_0) \mathbin{\bar\vee} (p \mathbin{\bar\vee} b_0)$, hence $[p,1]_{M(L)}$ is biatomic, therefore modular.

Wyler [16] and Sasaki [13] characterized the lattice of flats of an <u>incidence space</u> (i.e. a space satisfying Hilbert's incidence axioms [8]) as follows:

THEOREM E. A complete atomic algebraic lattice L is the lattice of flats of an incidence space when

(1) aMb implies bMa for all $a,b \in L$, (equivalently L has the exchange property) and

(2) $a \wedge b \neq 0$ implies aMb.

We shall call lattices which satisfy all the conditions in Theorem E <u>Hilbert lattices</u>. Affine geometries were characterized by the present author in [1] by:

THEOREM F. Let L be a lattice of height ≥ 4. Then $L \cong A(V)$ for V a vector space over a (not necessarily ordered) division ring if and only if

(1) L is a Hilbert lattice;

(2) L has no sublattice isomorphic to K_6 (whose Hasse diagram is shown in Figure 6) and

(3) Each coatom of L has a complementary coatom.

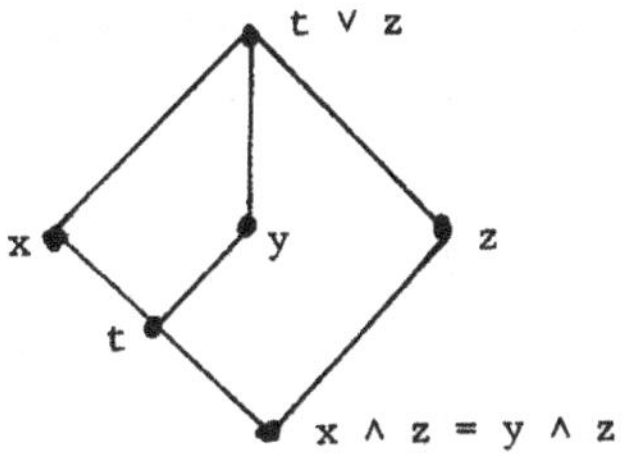

Figure 6

It was shown earlier [4, Theorem 8] that when L is a Peano lattice, M(L) has the exchange property. It therefore follows immediately from Theorem 10 that:

THEOREM 11. If L is a Peano lattice and $D(p) = 1$ for each atom p, then M(L) is a Hilbert lattice.

Making use of the results in [10] Theorem 10 can be extended as follows:

THEOREM 12. Let L be a Peano lattice with p any atom of L. Then $[p,1]_{M(L)}$ is modular if and only if it is biatomic.

Proof: Biatomicity implies modularity by ([9], p.77, Remark). Conversely, if $[p,1]_{M(L)}$ is modular, it is dual modular, hence for x any element of $[p,1]_{M(L)}$ and q any atom of L, x and $p \bar{\vee} q$ form a dual modular pair in $[p,1]_{M(1)}$. By ([9], p.77, Lemma 2) x and a form a biatomic pair in $[p,1]_{M(L)}$ whenever a is a finite element (finite join of atoms).

Now take $a,b \in [p,1]_{M(L)}$ and let $p \bar{\vee} q \leq a \bar{\vee} b$, q an atom of L. Then $p \bar{\vee} q \leq p \bar{\vee} a_1 \bar{\vee} \ldots \bar{\vee} a_n \bar{\vee} b_1 \bar{\vee} \ldots \bar{\vee} b_k$, with a_i, b_i atoms under a and b respectively. By the remarks above $p \bar{\vee} q \leq (p \bar{\vee} a_0) \bar{\vee} (p \bar{\vee} b_0)$ with $p \bar{\vee} a_0$ an atom of $[p,1]_{M(L)}$ under $p \bar{\vee} a_1 \bar{\vee} \ldots \bar{\vee} a_n \leq a$ and $p \bar{\vee} b_0$ an atom of $[p,1]_{M(L)}$ under $p \bar{\vee} b_1 \bar{\vee} \ldots \bar{\vee} b_k \leq b$. Hence $[p,1]_{M(L)}$ is biatomic.

Since 0 and 1 are contained in any $D(L)$, and since $a \leq b$ clearly implies that $D(a) \lesssim D(b)$, the condition $D(p) = 1$ for every atom p is equivalent to saying that $D(L)$ is as small as possible, i.e. that $D(L) = \{0,1\}$. This condition provides the final connection between the Peano condition and the lattice theoretic analogue of the statement "between every pair of distinct points there is a third."

DEFINITION. A convexity lattice has the divisibility property when, given distinct atoms p and q, there is an atom $r \leq p \vee q$, $r \neq p,q$.

The svnthetic analogue of Theorem 13 appeared in [15, Theorem 6].

THEOREM 13. Let L be a Peano lattice in which the height of $M(L)$ is greater than 2 and $D(L) = \{0,1\}$. Then L has the divisibility property.

Proof: Let p and q be atoms of L, and choose r an atom with $r \wedge (p \bar{\vee} q) = 0$. Then $r \leq p \vee s$ for some atom s distinct from p and r (since $D(r) = 1$). Since $D(s) = 1$ there is an atom t with $s \leq q \vee t$; q,s and t distinct. Hence $r \leq p \vee s \leq p \vee q \vee t$, so $r \leq t \vee u$ for some atom $u \leq p \vee q$. Since $t \neq s$ we can show $u \neq p$, and since $r \neq s$ we obtain $u \neq q$, hence L has the divisibility property.

If L were $Co(X)$ for X an ellipse in $\mathbf{R}^2$, then the lines a and b shown in Figure 7 have nonempty intersection, while $a \bar{\vee} b$ is the ellipse and $a \vee b$ is the quadrilateral whose diagonals are a and b. In $CO(V)$ such a situation is impossible; there if two lines intersect both their affine and their convex join is the plane.

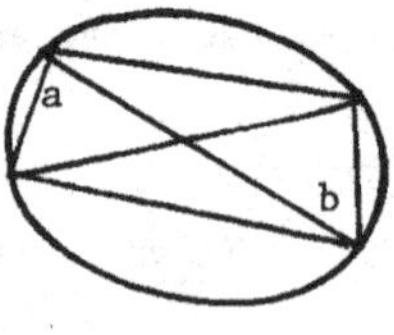

Figure 7

This motivates the condition in Theorem 14 under which $M(L)$ is the lattice of flats of an incidence space which satisfies the uniqueness part of the parallel axiom.

THEOREM 14. Let L be a Peano lattice such that for $a,b \in M(L)$ with $a \wedge b \neq 0$, $a \bar{\vee} b = a \vee b$ and $D(L) = \{0,1\}$. Then $M(L)$ is a Hilbert lattice which has no sublattice isomorphic to K_6 (shown in Figure 6) for x, y and z of height 2 in $M(L)$.

Proof: If $a \bar{\vee} b = a \vee b$ whenever $a \wedge b \neq 0$, then $\bar{B}(a,b)$ holds whenever $a \wedge b \neq 0$, and it follows from Theorem 12 that $M(L)$ is a Hilbert lattice.

For x, y, z of height 2 in $M(L)$ with $t = x \wedge y$ and $x \bar{\vee} y = y \bar{\vee} z = x \bar{\vee} z = t \bar{\vee} z$, we must show that $z \wedge x \neq 0$ or $y \wedge x \neq 0$. Given atoms $z_1, z_2 \leq z$, we may assume $z_i \leq x \bar{\vee} y = x \vee y$; hence there are atoms x_1, x_2, y_1, y_2 ($x_i \leq x$, $y_i \leq y$) with $z_i \leq x_i \vee y_i$. Since $t \leq x \wedge y$, $t \leq (x_1 \bar{\vee} x_2) \wedge (y_1 \bar{\vee} y_2)$ one of the following three cases can be assumed to hold:

(1) $x_1 \leq t \vee x_2$, $y_1 \leq t \vee y_2$;

(2) $t \leq y_1 \vee y_2$, $x_2 \leq t \vee x_1$;

(3) $t \leq x_1 \vee x_2$, $t \leq y_1 \vee y_2$.

In each case, Theorem B is used to show $z \wedge x \neq 0$ or $z \wedge y \neq 0$. If (1) we obtain $(z_1 \bar{\vee} z_2) \wedge (t \vee y_2) \neq 0$ or $(z_1 \bar{\vee} z_2) \wedge (t \vee x_2) \neq 0$, and if (2) or (3) we have $(z_1 \bar{\vee} z_2) \wedge (t \vee x_1) \neq 0$ or $(z_1 \bar{\vee} z_2) \wedge (t \vee y_1) \neq 0$.

It follows from Maeda ([10] 18.6, 19.9) that a line (element of height 2) has a parallel through any point not on it if there is an element in $M(L)$ with which it fails to form a modular pair. From this the existence part of the parallel axiom can be shown. In this case $M(L)$ is the lattice of flats of an affine space, and the present author [3, Theorem 4.6] has shown that this implies that L is $Co(V)$ for some vector space V over an ordered division ring D. Thus:

THEOREM 15. Let L be a Peano lattice whose modular core has height greater than 3. Then L is isomorphic to some $Co(V)$ if and only if

(1) For all $a,b \in M(L)$ with $a \wedge b \neq 0$, $a \bar{\vee} b = a \vee b$,

(2) $D(L) = \{0,1\}$,

(3) $M(M(L))$ contains only 0, 1 and the atoms of L.

ACKNOWLEDGEMENT: The author wishes to thank Garrett Birkhoff for several helpful ideas and discussions.

BIBLIOGRAPHY

[1] Bennett, M. K. Affine Geometry: A Lattice Characterization. Proc. AMS 88 (1983), 21-26.

[2] Bennett, M. K. Lattices of Convex Sets. Trans. AMS 234 (1977), 279-288.

[3] Bennett, M. K. On Generating Affine Geometries. Alg. Univ. 4 (1974), 207-219.

[4] Bennett, M. K. and G. Birkhoff. Convexity Lattices. To appear in Alg. Univ.

[5] Bennett, M. K. and G. Birkhoff. A Peano Axiom for Convexity Lattices. To appear in Bull. Calcutta Math. Soc.

[6] Birkhoff, G. "Lattice Theory", 3rd ed. Providence, AMS, 1967.

[7] Gorn, S. On Incidence Geometries, Bull. AMS 46 (1940), 158-167.

[8] Hilbert, D. "Foundations of Geometry", (transl. by E. J. Townsend), Open Court, La Salle, Ill. 1902.

[9] Maeda, S. On Finite-Modular Atomistic Lattices. Alg. Univ. 12 (1981), 76-80.

[10] Maeda, F. and S. Maeda. "Theory of Symmetric Lattices." Springer, New York, 1971.

[11] Prenowitz, W., J. Jantosciak. "Join Geometries", Springer, Undergrad. Texts in Math. New York, 1979.

[12] Rockafellar, T. "Convex Analysis," Princeton University Press, 1970.

[13] Sasaki, U. Lattice Theoretic Characterization of Geometries Satisfying 'Axiome der Verknüpfung.' Hiroshima J. Ser. A, 16 (1953), 417-423.

[14] Szmielew, W. The Role of the Pasch Axiom in the Foundations of Euclidean Geometry. In Proc. Tarski Symp., Proc. Symp. Pure Math. XXV, Providence, AMS 1974, 123-132.

[15] Veblen, O. A New System of Axioms for Geometry. Trans. AMS 4 (1903), 343-384.

[16] Wyler, O. Incidence Geometry. Duke Math. J. 20 (1953), 601-610.

SOME INDEPENDENCE RESULTS IN THE CO-ORDINIZATION OF ARGUESIAN LATTICES

by Alan Day [(1)]
Lakehead University
Thunder Bay, Canada P7B 5E1

0. INTRODUCTION. In [1], Douglas Pickering and this author showed that every n-frame ($n \geq 3$) in an Arguesian lattice induces a ring (with 1) in the classical von Staudt - von Neumann way. This ring, however, depended on a particular orientation of the given n-frame. In this paper it is shown that this supposed dependence on the orientation is baseless in that the rings resulting from different orientations are isomorphic. Another result of classical geometry is that the ring depends only on the three chosen points, 0, 1, and ∞. This we shall also see is true modulo the proper lattice reformulation.

1. PRELIMINARIES. We will use [1] as a basic reference, and recall here only the essential information needed in the sequel. All lattices, L, are assumed to be Arguesian, and if **a** and **b** are triangles in L, **a** and **b** are called centrally perspective in the *weak sense* if $(a_0 + b_0)(a_1 + b_1) \leq (a_0 + a_2)(a_1 + a_2) + (b_0 + b_2)(b_1 + b_2)$. As proven in [2] and mentioned in [1.2.1], in an Arguesian lattice two triangles are axially perspective iff they are centrally perspective in the weak sense.

---

(1) This research was supported by NSERC Operating Grant A-8190.

Let L have a spanning 3-frame, $\langle x, y, z, t\rangle$. We define (1) $h := x + y$, the line at infinity, (2) $w := (z + t)(x + y)$, (3) $A_h := \{p.\ p + h = 1 \ \&\ ph = 0\}$, the affine plane, (4) $D_{\langle x,y,z,t\rangle} := \{a.\ a + w = z + t \ \&\ aw = 0\} = \{a \text{ in } A_h.\ a \leq z + t\}$, the diagonal line, (5) $v := (y + z)(x + t)$, and (6) $u := (x + z)(y + t)$.

We will usually write A and D for the sets in (3) and (4) if no possible confusion arises. For any b in D, we define (1) $b_0 := (y + z)(x + b)$, the y-intercept of b, (2) $b_1 := (y + t)(x + b)$, the affine slope point of b, and (3) $b_\infty := h(z + b_1)$, the slope point at infinity of b. On D, there is the (planar) ternary ring operation $T.D^3 \longrightarrow D$ defined by.

$$T(a,b,c) = (z + t)(x + (y + a)(b_\infty + c_0))$$

This ternary operation in turn defines an addition and multiplication on D by

$$a+b := T(a,t,b) = (z + t)(x + (y + a)(w + b_0))$$

$$a\times b := T(a,b,z) = (z + t)(x + (y + a)(z + b_1))$$

With these definitions, one of the main results of [1] is

THEOREM 1. $(D, +, z, \times, t)$ is a(n) (associative) ring with zero, z, and unit, t.

In proving this theorem, many ancillary results were produced. A list of those required here is as follows.

LEMMA 2. The following properties hold for T, +, and $\times$.

(1) $T(a,b,c) = a\times b + c$

(2) $a-b = (z + t)(y + (x + a)(w + b_0))$

$= (z + t)(x + (y + z)(w + (y + b)(x + a)))$

(3) $a-b+c = (z + t)(x+ (y + a)(w + (y + b)(x + c)))$

LEMMA 3. There is a bijection from D^2 onto A given by $(a,b) \mapsto (y + a)(x + b)$. Moreover, $(y + a)(w + b_0) = (y + a)(x + a+b)$, and $(x + a)(w + b_0) = (y + a-b)(x + b)$.

THEOREM 4. For any (x', y') such that $\langle x', y', z, t\rangle$ is a spanning 3-frame satisfying $w = (x' + y')(z + t)$, and $x', y' \leq h$, $T(a,b,c) = T'(a,b,c)$ where T' is obtained from T by substituting x' and y' for x and y respectively. Furthermore we have.

(1) $a+b = (z + t)[(v' + w)(y' + a) + (v' + b)(y' + w)]$

(2) $a \times b = (z + t)[(v' + z)(x' + a) + (v' + b)(x' + w)]$ where $v' := (y' + z)(x' + t)$.

THEOREM 5. For any a, b, c, d, e, f in D,

$$h[(y + a)(x + b) + (y + c)(x + d)] = h[(y + a+e)(x + b+f) + (y + c+e)(x + d+f)]$$

2. INDEPENDENCE OF THE 3-FRAMES. A pair, (x', y'), is called *admissible* if $\langle x', y', z, t\rangle$ is a spanning 3-frame of L, and $w = (x' + y')(z + t)$. The pair is called *h-admissible* if $x', y' \leq h$, and called *A-admissible* if x', y' belong to A. Note that (x', y') is (h- or A-) admissible iff (y', x') is. We repeat part of (1.4) for continuity.

THEOREM 1. For any h-admissible pair, (x', y'), and T', the associated ternary operation on D for the frame $\langle x', y', z, t\rangle$, $T = T'$.

LEMMA 2. Let (x', y') be an admissible pair, and take c in D. Then for $x_c := h(x' + c)$, and $y_c := h(y' + c)$, (x_c, y_c) is h-admissible with $x' + c = x_c + c$, and $y' + c = y_c + c$. Moreover if (x', y') is A-admissible, then $x' + x_c = x' + c$, $y' + y_c = y' + c$, and $x'x_c = y'y_c = 0$ hold as well.

<u>Proof</u>. Easy but tedious calculations with the modular law.

This lemma is extremely useful in combination with (2.1) in that we may assume without loss of generality, for whichever c is appropriate, that $x = x_c$ and $y = y_c$.

LEMMA 3. For any A-admissible pair, (x', y'), $T = T'$.

Proof. We break T up into its usual components.

Re addition. Take a, b in D, and, using c:= z, assume that $x + x' = x + z = x' + z$, $y + y' = y + z = y + z$, and $xx' = yy' = 0$. This makes the triangles, $\langle b, w, y'\rangle$ and $\langle x, (y + a)(w + b_0), y\rangle$, triply centrally perspective at b_0. Since L is Arguesian, we obtain the formula.

$$a{+}b = (z + t)[h(y' + b) + h'(y + a)]$$

By the symmetry of primed and non-primed elements, and the commutativity of + and +′, we get $a{+}'b = b{+}'a = a{+}b$.

Re multiplication. Take a, b in D, and, choosing c:= t, assume that $x + x' = x + t = x' + t$, $y + y' = y + t = y' + t$, and $xx' = yy' = 0$. The triangles, $\langle z, b, x\rangle$ and $\langle (y' + z)(x' + a), h'(v' + b), x'\rangle$, are centrally perspective in the weak sense, so we have by (1.4), $a\mathbf{x}'b = (z + t)[(x + z)(x' + a) + (x + b)(x' + w)] = a\mathbf{x}b$, using (1.4) again with the h-admissible pair $(x'', y'') = (h(x' + a), x)$.

We now need to consider various 3-frames derived from the given one.

LEMMA 4. Let $(D, +', t, \mathbf{x}', z)$ be the ring for the 3-frame $\langle x, y, t, z\rangle$. Then the map $a \longmapsto 1-a$ is a ring isomorphism from $(D, +, z, \mathbf{x}, t)$ onto $(D, +', t, \mathbf{x}', z)$.

Proof. Easy calculations show that $w + (y + t)(x + b) = w + (y + z)(x + b-t)$. Therefore $a{+}'b = a{+}b-t$.

For multiplication, $a{+}b-a\mathbf{x}b = T(a, t-b, b) = (z + t)[x + (y + a)((t-b)_\infty + b_0)]$, and $a\mathbf{x}'b=(z + t)[x + (y + a)(t + b_0)]$. Since $(t-b)_\infty=h(z + (t-b)_1)= h(t + b_0)$ by (1.5), we obtain that $a\mathbf{x}'b = a{+}b-a\mathbf{x}b$.

The indicated map is easily seen to be the desired isomorphism.

LEMMA 5. Let $(D', +', z, \times', v)$ be the ring associated with the 3-frame $\langle x, w, z, v\rangle$. Then the one-step projectivity from $[0, z + v]$ to $[0, z + t]$, determined by x, provides an isomorphism between this ring and $(D, +, z, \times, t)$.

<u>Proof</u>. For a in D we have $a' = (y + z)(x + a)$ in D', and $a'+'b' = (y + z)[x + [w + a'][y + (w + z)(x + b')]] = (y + z)[x + [w + a'][y + b]] =(y + z)[x + a+b] =(a+b)'$.
For multiplication, the triangles, $\langle x, (w + a')(z + (w + v)(x + b)), z\rangle$ and $\langle a, w, y\rangle$, are centrally perspective at a'. Therefore $(z + t)(x + a'\times'b') \le (x + z)(y + a) + h(z + (w + v)(x + b)) = (x + z)(y + a) + h(u + b)$ by (1.5). By (1.4), we obtain $a'\times'b' =(a\times b)'$.

LEMMA 6. Let $(D', +', y, \times', t)$ be the ring for the 3-frame $\langle x, z, y, t\rangle$. Then the one-step projectivity from $[0, z+t]$ to $[0, y+t]$, determined by $p:= (y + z)(w + u)$, provides a ring isomorphism between this ring and $(D, +, z, \times, t)$.

<u>Proof</u>. Easy calculations show that the pair, (p,u), is $A = A_{x+y}$-admissible for the given 3-frame. Similarly (p, w) is A_{x+z}-admissible for the new 3-frame $\langle x, z, y, t\rangle$. We have then, for $a' = (y + t)(p + a)$, etc..

$$a'+'b' = (y + t)(p + (w + a')(u + (w + y)(p + b'))),$$

$$a+b = (b+t)+(a-t) = (z + t)(p + (u + b+t)(w + (u + z)(p + a-t))), \text{ and}$$

$$a-t = (z + t)(p + (u + z)(w + (u + t)(p + a))).$$

Since $u + b+t = u + h(p + b) = u + h(p + b')$, we obtain $a'+b' = (a+b)'$.
For multiplication, we have $a'\times'b' = (y + t)(p + (w + a')(y + (w + t)(p + b'))) = (y + t)(p + (w + a')(y + b))$. Thus we must show that $a\times b = (z + t)(p + (y + b)(w + a'))$. The triangles, $\langle p, (y + b)(w + a'), y\rangle$ and $\langle a, w, (y + t)(x + a)\rangle$, are centrally perspective at a'. Axial perspectivity implies that $(z + t)(p + (y + b)(w + a')) = (z + t)[(y + z)(x + a) + (y + b)(w + (y + t)(x + a))] = a\times b$ by (2.3).

LEMMA 7. Let (D′, +′, y, ×′, x) be the ring associated with the 3-frame, <z, t, y, x>. Then the one-step projectivity from [0, z + t] to [0, y + x], determined by v, provides a ring isomorphism from this ring to (D, +, z, ×, t).

<u>Proof</u>. Let e:= (y + t)(w + v). Easy calculations show that (v, e) [and (e, v)] are A-admissible pairs. This gives us a+t = (z + t)(e + h(v + a)), and therefore for a in D, a′= h(v + a) = h(e + a+t). Now (a+b)′ = h(v + a+b) = h[v + (a+t)−t+b] =h[v + [e + a+t][w + (e + t)(v + b)]] by (1.2(3)). Therefore (a+b)′= h[v + [e + a′][w + (e + y)(v + b′)]] = a′+′b′.

Lemmas (2.4), (2.6), and (2.7) tell us that any of the 4! permutations of the letters {x, y, z, t} will produce, from the corresponding 3-frame, an isometric ring. This is our first independence result.

THEOREM 8. The ring of a 3-frame in an Arguesian lattice is, up to isomorphism, independent of the orientation of the 3-frame.

We now attack the general admissibility problem. Let (x′, y′) be an admissible pair, and assume that x + z = x′ + z, and y + z = y′ + z. Since z is a complement of both h and h′:= x′ + y′, z determines a one-step projectivity from [0, h] to [0, h′], that takes y, x, and w onto y′, x′, and w respectively. We need then that this projectivity provides a ring isomorphism between the relevant rings, D and D′, <u>and</u> that the three-step projectivity from [0, z + t] to [0, h] to [0, h′] to [0, z + t], determined by v, z, and v′:= (y′ + z)(x′ + t) = (y + z)(x′ + t), is the identity. We handle the second problem first.

LEMMA 9. Let (x′, y′) be an admissible pair with x + z = x′ + z, and y + z = y′ + z. Then for all p in [0, z + t], p = (z + t)[v + h(z + h′(v′ + p))].

<u>Proof</u>. Easy modular calculations show that the above statement is equivalent to the equality, z + h(v + p) = z + h′(v′ + p) for all p in [0, z + t]. The triangles, <p, v, v′> and <w,

x, x'>, are doubly centrally perspective at $t(p + w)$, and the third point of axial perspectivity of these triangles is $(x + x')(v + v') \le (x + z)(y + z) = z$. Therefore $z + h(v + p) = z + h'(v' + p)$ as desired.

We are now left with the first problem of the isomorphic rings. By using (2.1) and (2.7), we obtain the following equivalent statement.

LEMMA 10. For c in D, $z' := c_0$, and $t' := (y + t)(w + c_0)$, $\langle x, y, z', t'\rangle$ is a spanning 3-frame. If $(D', +', z', \times', t')$ is the associated ring, then the one-step projectivity from $[0, z + t]$ to $[0, z' + t']$, determined by y, provides an isomorphism from this ring onto $(D, +, z, \times, t)$.

Proof. For a in D, let $a' := (z' + t')(y + a) = (y + a)(w + c_0) = (y + a)(x + a{+}c)$. Then

$$
\begin{aligned}
a'{+'}b' &= (w + c_0)(x + (y + a')(w + (y + z')(x + b'))) \\
&= (w + c_0)(x + (y + a)(w + (y + z)(x + b{+}c))) \\
&= (w + c_0)(x + a{+}b{+}c) \\
&= (y + [(a{+}b{+}c) - c])(x + a{+}b{+}c) \\
&= (y + a{+}b)(x + a{+}b{+}c).
\end{aligned}
$$

Therefore $(z + t)(y + a'{+'}b') = a{+}b$.

$$
\begin{aligned}
a'{\times'}b' &= (w + c_0)(x + (y + a')(z' + (y + t')(x + b'))) \\
&= (w + c_0)(x + (y + a)(c_0 + (y + t)(x + b{+}c))) \\
&= (w + c_0)(x + (y + a)(c_0 + b_\infty)), \text{ since } b{+}c = T(t, b, c), \\
&= (w + c_0)(x + T(a,b,c)) \\
&= (y + a{\times}b)(x + T(a,b,c)).
\end{aligned}
$$

Therefore $(z + t)(y + a'{\times'}b') = a{\times}b$.

We can now state the other principal result of this section.

THEOREM 11. Let L be an Arguesian lattice with spanning 3-frame, $\langle x, y, z, t\rangle$. Then for any admissible pair, (x', y'), $T' = T$.

3. GENERAL INDEPENDENCE, THE OPEN PROBLEM. Let L be an Arguesian lattice with a spanning 3-frame, $\langle x, y, z, t\rangle$. It was noted in [1], Appendix 1, that any pair of elements, (x^*, y^*), that satisfied $x^*(z + t) = y^*(z + t) = 0$ and $(x^* + y^*)(z + t) = w$, did indeed produce a bona fide ternary operator, T^*, on D. We call such a pair *permissible for D.* and let (x^*, y^*) be a permissible pair for D.

LEMMA 1. If (x^*, y^*) is a permissible pair for D, then so is $(x^{\star}, y^{\star})$ where $x^{\star} := x^*(y^* + z + t)$ and $y^{\star} := y^*(x^* + z + t)$. Moreover $T^{\star} = T^*$.

<u>Proof</u>. That the $\star$-pair is permissible for D is an easy modular calculation. Since $x^{\star} \leq x^*$ and $y^{\star} \leq y^*$, $T^{\star} \leq T^*$ as functions from D^3 into D. Since the functions' values are comparable complements of w, they must be equal.

Therefore we may assume, without loss of generality, that the given permissible pair also satisfies the conditions $x^* + y^* + z = x^* + y^* + t = x^* + z + t = y^* + z + t$. These conditions also imply that $x^* + w = y^* + w = x^* + y^*$.
For these special permissible pairs it is an easy calculation to show that $\langle x^*, y^*, z + x^*y^*, t + x^*y^*\rangle$ is a 3-frame spanning the interval $[x^*y^*, z + t + x^*]$. Moreover the interval $[0, z + t]$ transposes up to the interval $[x^*y^*, z + t + x^*y^*]$ since $(x^*y^*)(z + t) = 0$. This transposition takes w to $w + x^*y^*$ and therefore D to $D^* := \{ q.\ q + w + x^*y^* = z + t + x^*y^* \ \&\ q(w + x^*y^*) = x^*y^*\}$. Further easy calculations show that for $a^* := a + x^*y^*$ etc., $x^*y^* + T^*(a,b,c) = T(a^*,b^*,c^*)$ in D^*. This produces the following result.

THEOREM 2. Let L be an Arguesian lattice, and take z, t, w in L satisfying $z + t = z + w = t + w$, and $zt = zw = tw = 0$. If $D = \{a \text{ in } L.\ a + w = z + t\ \&\ aw = 0\}$, and (x, y) is a permissible pair, then (D, +, z, **x**, t) is a ring.

We would have liked to prove that, starting with a 3-frame ⟨x, y, z, t⟩ etc., and choosing any permissible pair (x*, y*), we would produce the same ring. This we have not been able to do and so must leave it as an open problem.

4. REFERENCES.

[1] A. Day and D. Pickering, The coordinatization of Arguesian lattices, Trans.Amer.Math.Soc.**278**, 507-522, 1983.

[2] B. Jónsson and G. Monk, Representation of primary Arguesian lattices, Pacific J.Math. **30**, 95-139, 1969.

1980 *Mathematical Subject Classification* : Primary 06C05; Secondary 08Bxx, 16A30, 51A30, 51C05.

UNARY OPERATIONS ON COMPLETELY DISTRIBUTIVE COMPLETE LATTICES

Philip Dwinger
University of Illinois at Chicago
Chicago, Illinois 60680

1. INTRODUCTION AND PRELIMINARIES

If L is a complete lattice then $a+b$, ab, ΣS and ΠS denote sums (least upper bounds) and products (greatest lower bounds) for $a,b \in L$ and $S \subseteq L$. Occasionally, we will write $\Sigma^{L}S$ and $\Pi^{L}S$ instead of ΣS and ΠS. Likewise, the symbols 0_L and 1_L will also be used rather than 0 and 1 to denote the zero - and one element of L. If $S \subseteq L$, then S is said to be a Σ-subset (Π-subset) of L, if, whenever sums (products) in S exist, they are the same as in L. S is closed under sums (products) if S is a Σ-subset (Π-subset) and if sums (products) always exist in S. A sublattice of L is always meant to be a non-void subset of L which is closed under (finite and infinite) sums and products. L is completely distributive if it satisfies

$$(1.1) \qquad \prod_{s\in S} \sum_{t\in T} x_{st} = \sum_{\varphi\in T^S} \prod_{s\in S} x_{s,\varphi(s)}$$

It is known that (1.1) holds if and only if the dual of (1.1) holds [1]. The class of completely distributive lattices will be denoted by $\mathcal{D}_c$. If $L \in \mathcal{D}_c$, then L is a complete ring of sets if the elements of L are subsets of some set X with set-theoretic operations as lattice operations. For convenience we will always assume that $\emptyset$

I am indebted to J. Berman who has read a first draft of this paper and for many of his valuable suggestions.

and $X \in L$. The class of complete rings of sets will be denoted by $\mathfrak{D}_r$. A lattice L is dense in itself if for $a, b \in L$, $a < b$ there exists $c \in L$ such that $a < c < b$. The subclass of $\mathfrak{D}_c$ consisting of dense in itself lattices is denoted by $\mathfrak{D}_s$. Finally, the class of partially ordered sets is denoted by $\mathfrak{P}$. The symbols, $\mathfrak{D}_c$, $\mathfrak{D}_r$ and $\mathfrak{D}_s$ will also be used to denote the corresponding categories with complete homomorphisms as morphisms. Likewise, the symbol $\mathfrak{P}$ will also stand for the corresponding category with order preserving maps as morphisms. It is a classical result of Raney that if $L \in \mathfrak{D}_c$ then L is a subdirect product of complete chains [18]. If $L \in \mathfrak{D}_c$, $a \in L$ then a is completely join irreducible if for $S \subseteq L$, $a = \Sigma S$ implies $a \in S$. Completely meet irreducible is defined dually. The sets of completely join irreducible elements and of completely meet irreducible elements are denoted by $\mathfrak{J}(L)$ and $\mathfrak{M}(L)$ respectively. Note $0 \notin \mathfrak{J}(L)$ and $1 \notin \mathfrak{M}(L)$. Recall, that if $a \in L$, then $a \in \mathfrak{J}(L) \iff a$ has an immediate predecessor a_- and $a \in \mathfrak{M}(L) \iff a$ has an immediate successor a_+. [1] If $P \in \mathfrak{P}$ and $S \subseteq P$, then S is called join dense (meet dense) in P if every element of P is the sum (product) of elements of S. Also recall that if $L \in \mathfrak{D}_c$, then $L \in \mathfrak{D}_r \iff \mathfrak{J}(L)$ is join dense in $L \iff \mathfrak{M}(L)$ is meet dense in $L \iff$ for $a, b \in L$, $a \nleq b$ there exists $s \in \mathfrak{J}(L)$ such that $s \leq a$, $s \nleq b \iff$ for $a, b \in L$, $a \nleq b$, there exists $s \in \mathfrak{M}(L)$ such that $s \ngeq a$, $s \geq b$ [1]. Finally, $L \in \mathfrak{D}_s \iff \mathfrak{J}(L) = \emptyset \iff \mathfrak{M}(L) = \emptyset$ [10]. If $a \in L$, $L \in \mathfrak{D}_c$, then a is a node if $L = (a] \cup [a)$. A non trivial node is a node a, $a \neq 0, 1$. If L is a complete lattice and if ${}^c : L \to L$ is a closure operator on L then L^c will stand for the set of "closed" elements of L. Similarly, if ${}^k : L \to L$ is a kernel (interior) operator on L then L^k stands for the set of "open" elements of L. Recall that L^c (L^k) is closed under products (sums) and therefore L^c (L^k) is a complete lattice and also recall that the maps ^{c} and ^{k} preserve sums and

and products respectively. For lattice-theoretic and categorical concepts used in this paper and not defined we refer to [1] and [16]. The following lemma is immediate.

LEMMA 1.2. Let $L \in \mathcal{D}_r$ and let $\mathcal{J}(L) \subseteq S \subseteq L$. Then S is a Π-subset of L and dually. In particular, $\mathcal{J}(L)$ is a Π-subset and $\mathfrak{m}(L)$ is a Σ-subset of L.

The following two unary operations on a complete lattice L were introduced in [9]. For $a \in L$, let

$$(1.3) \qquad \begin{aligned} a^u &= \Pi\{s : s \not\leq a,\ s \in L\} \\ a_v &= \Sigma\{s : s \not\geq a,\ s \in L\}. \end{aligned}$$

The pair $\langle {}^u, {}_v \rangle$ is a Galois connection. Indeed it can be easily checked that the maps ^{u} and $_v$ are order preserving and $a^u \geq b \Leftrightarrow a \geq b_v$. For details on Galois connections refer for example to [15]. We recall that $(a_v)^u \geq a$, $(a^u)_v \leq a$, $((a^u)_v)^u = a^u$, $((a_v)^u)_v = a_v$ and also that the maps ^{u} and $_v$ preserve (finite and infinite) products and sums respectively. The sets $\{a^u : a \in L\}$ and $\{a_v : a \in L\}$ will be denoted by L^u and L_v respectively. The maps ^{u} and $_v$ play an important role in Raney's work [19]. (Also cf. Bandelt [4] and [6].) The maps

$$(1.4) \qquad a \mapsto a^c = (a_v)^u \quad \text{and} \quad a \mapsto a^k = (a^u)_v$$

are a closure and a kernel operation on L respectively, and we have [15]

$$(1.5) \qquad \begin{aligned} &{}^c \text{ is the identity map} \Leftrightarrow {}_v \text{ is one-one} \Leftrightarrow {}^u \text{ is onto;} \\ &{}^k \text{ is the identity map} \Leftrightarrow {}_v \text{ is onto} \Leftrightarrow {}^u \text{ is one-one.} \end{aligned}$$

We will occasionally write for $a \in L$, a^{u_L} and a_{v_L} rather than a^u and a_v respectively. It is easy to check that if $L \in \mathcal{D}_c$, $a \in \mathcal{J}(L)$.

then $a_v \in \mathfrak{M}(L)$, $a_v \nleq a$ and $(a_v)^u = a$. Similarly, if $a \in \mathfrak{M}(L)$, then $a^u \in \mathfrak{J}(L)$, $a^u \nleq a$ and $(a^u)_v = a$. Thus the map $_v$ restricted to $\mathfrak{J}(L)$ is an order isomorphism between $\mathfrak{J}(L)$ and $\mathfrak{M}(L)$ whose inverse is the map ^{u} restricted to $\mathfrak{M}(L)$. Thus the map ^{u} and $_v$ on L, $L \in \mathfrak{S}_c$ are extensions of the isomorphisms between $\mathfrak{J}(L)$ and $\mathfrak{M}(L)$. It is also known (see section 4) that if $L \in \mathfrak{S}_r$ then the structure of L is determined by the structure of $\mathfrak{J}(L)$. It may therefore be expected that the special properties of the maps ^{u} and $_v$ may serve to characterize subclasses of $\mathfrak{S}_c$ and $\mathfrak{S}_r$. In this paper a first attempt is made to obtain such a characterization. Section 3 is devoted to the characterization of the subclasses of Stone algebras and relative Stone algebras of $\mathfrak{S}_c$. In the subsequent sections we focus our attention on the characterization of several subclasses of $\mathfrak{S}_r$. Note that if $L \in \mathfrak{S}_r$ and $a \in \mathfrak{J}(L)$, then $a^c = a$. Also, if $0 \notin \mathfrak{M}(L)$ then $0^c = 0$. Therefore we have for $L \in \mathfrak{S}_r$,

(1.6)
$$\mathfrak{J}(L) \cup \{1\} \subseteq L^c \subseteq L \quad \text{if} \quad 0 \in \mathfrak{M}(L)$$
$$\mathfrak{J}(L) \cup \{0\} \subseteq \{1\} \subseteq L^c \subseteq L \quad \text{if} \quad 0 \notin \mathfrak{M}(L), \text{ and dually.}$$

Note that (1.6) also holds if $L \in \mathfrak{S}_c$ and not necessarily $L \in \mathfrak{S}_r$, but we will only use (1.6) for the case of $L \in \mathfrak{S}_r$. Because the closure operators (and kernel operators) on a complete lattice can be partially ordered in a natural way, we will say that ^{c} is <u>maximal</u> if equality holds on the left side of (1.6) and <u>minimal</u> if equality holds on the right side, and dually for ^{k}. Sections 5 and 6 will deal with the case that ^{c} is maximal for $L \in \mathfrak{S}_r$. We will show that for $L \in \mathfrak{S}_r$, the condition that ^{c} is maximal is equivalent to each of the following conditions: ^{k} is maximal; $\mathfrak{J}(L) \cup \{0\} \cup \{1\}$ is a complete lattice; $\mathfrak{M}(L) \cup \{0\} \cup \{1\}$ is a complete lattice. In section 6 we will in particular consider the case that $\mathfrak{J}(L)$ (and thus $\mathfrak{M}(L)$) is a complete lattice. In the subsequent sections we deal with the cases that ^{c} and ^{k} are minimal for $L \in \mathfrak{S}_r$. Unlike in the previous case, these two conditions are in general not equivalent, but they are equivalent in the finite case. Whereas the finite members of $\mathcal{D}_r$ for which ^{c} and thus ^{k}, are minimal can be fully characterized (cf. Bandelt [6]), the infinite case is more complex. It is known [14] that the categories $\mathcal{P}$ and $\mathcal{D}_r$ are dually equivalent. The properties of this dual equivalence will be stated in section 4. Because of this dual relation between $\mathcal{P}$ and $\mathcal{D}_r$, our examination will lead to the characterization of those partially ordered sets for which the corresponding complete rings of sets have the property that ^{c} is minimal and dually, that ^{k} is minimal. In particular we will focus our attention on those partially ordered sets for which the corresponding complete rings of sets have the property that both ^{c} and ^{k} are minimal.

2. PROPERTIES OF THE UNARY OPERATIONS ^{u} AND $_v$.

For the proofs of the following three lemmas we refer to [19] (also cf. [4]).

LEMMA 2.1. If L is a complete lattice and $a \in L$, then

$$\Sigma\, \{s^u : s \ngeq a,\ s \in L\} = \Sigma\, \{s : s_v \ngeq a,\ s \in L\} \quad \text{and}$$
$$\Pi\, \{s_v : s \leq a,\ s \in L\} = \Pi\, \{s : s^u \ngeq a,\ s \in L\}.$$

LEMMA 2.2. If L is a complete lattice then the following are equivalent:

(i) $L \in \mathcal{D}_c$;

(ii) $a = \Sigma\, \{s^u : s \ngeq a,\ s \in L\}$ for all $a \in L$;

(iii) $a = \Pi\, \{s_v : s \nleq a,\ s \in L\}$ for all $a \in L$.

LEMMA 2.3. If $L \in \mathcal{D}_c$ and $a \in L$ then

$$a = \Sigma\, \{\Pi\, \{s_v : s \nleq t,\ s \in L\} : t \ngeq a,\ t \in L\} =$$
$$\Pi\, \{\Sigma\, \{s^u : s \ngeq t,\ s \in L\} : t \nleq a,\ t \in L\}.$$

The following lemma is immediate.

LEMMA 2.4. Let $L \in \mathcal{D}_c$. Then $0_v = 0$, $0^u = 0_+$ if $0 \in \mathfrak{M}(L)$, $0^u = 0$ if $0 \notin \mathfrak{M}(L)$, $1^u = 1$, $1_v = 1_-$ if $1 \in \mathfrak{J}(L)$, $1_v = 1$ if $1 \notin \mathfrak{J}(L)$.

LEMMA 2.5. Let $L \in \mathcal{D}_c$, $a \in L$. Then (i) $a \nleq a_v \Leftrightarrow a \in \mathfrak{J}(L)$; (ii) $a \ngeq a^u \Leftrightarrow a \in \mathfrak{M}(L)$; (iii) $a \in \mathfrak{J}(L) \Rightarrow (a_v)^u = a$ and $a_v \in \mathfrak{M}(L)$; (iv) $a \in \mathfrak{M}(L) \Rightarrow (a^u)_v = a$ and $a^u \in \mathfrak{J}(L)$; (v) $a = a_v \Leftrightarrow a$ is a node and $a \notin \mathfrak{J}(L)$; (vi) $a = a^u \Leftrightarrow a$ is a node and $a \notin \mathfrak{M}(L)$; (vii) $a_v < a \Leftrightarrow a$ is a node and $a \in \mathfrak{J}(L)$; (viii) $a^u > a \Leftrightarrow a$ is a node and $a \in \mathfrak{M}(L)$.

<u>Proof</u>. (i) If $a \in \mathfrak{J}(L)$ then it is immediate that $a \nleq a_v$. If $a \notin \mathfrak{J}(L)$, then there exists $S \subseteq L$ such that $a = \Sigma S$ and $a \nleq s$ and thus $s \leq a_v$ for each $s \in S$. Thus $a \leq a_v$. (iii) By (i) we have $(a_v)^u \leq a$. Also $(a_v)^u \geq a$, hence $(a_v)^u = a$. If $a_v \notin \mathfrak{M}(L)$ then there exists $S \subseteq L$ such $a_v = \Pi S$ and $s \nleq a_v$ and thus $s \geq a$ for each $s \in S$. It follows $a \leq a_v$. Contradiction.
(v) $a = a_v \Rightarrow$ (by (i)) $a \notin \mathfrak{J}(L)$. Also, $a = a_v$ and $a \nleq b \Rightarrow b \leq a_v = a$. Thus a is a node. Conversely, if a is a node and $a \notin \mathfrak{J}(L)$ then $a_v = \Sigma\{s : s < a,\ s \in L\} = a$. (vii) $a_v < a \Rightarrow$ (by (i)) $a \in \mathfrak{J}(L)$. Again $a_v < a$ and $a \nleq b \Rightarrow b \leq a_v < a$ and thus a is a node. Conversely, if a is a node and $a \in \mathfrak{J}(L)$ then by (i) $a \nleq a_v$ and thus $a_v < a$. (ii), (iv), (vi) and (viii) by dual arguments.

It will be useful to introduce some additional notation. If $L \in \mathcal{D}_c$ and $a, b \in L$, $a \leq b$ then we will write $b^{u[a)} = \Pi\{s : s \nleq b,\ s \geq a,\ s \in L\}$. Thus ${}^{u[a)}$ denotes the operation ^{u} applied to the lattice $[a)$. It is easy to check that

$$(2.6) \qquad b^{u[a)} = b^u + a.$$

If $L \in \mathcal{D}_c$ then L is of course pseudocomplemented (in fact, L is a Heyting algebra) and for $a \in L$, $a^* = \Sigma\{s : as = 0,\ s \in L\}$. The relation between the operations * and ^{u} is provided by the following lemma.

LEMMA 2.7. Let $L \in \mathcal{S}_c$, $a \in L$. Then $a^* = \Pi\{s : 0 < s^u \leq a,\ s \in L\}$.

Proof. By Lemmas 2.1 and 2.2, $a\ \Pi\{s : 0 < s^u \leq a,\ s \in L\} = 0$. It remains to show that if $ba = 0$ and $0 < s^u \leq a$ then $b \leq s$. But $s^u \nleq 0 = ba$ thus $s^u \nleq b$ implying $b \leq s$.

The following two lemmas deal with the special case that $L \in \mathcal{S}_r$.

LEMMA 2.8. Let $L \in \mathcal{S}_r$, $a \in L$. Then (i) $a^u = \Pi\{s : s \nleq a,\ s \in \mathfrak{M}(L)\}$; (ii) $a^u = \Pi\{s : s \nleq a,\ s \in \mathcal{J}(L)\}$; (iii) $a_v = \Sigma\{s : s \ngeq a,\ s \in \mathcal{J}(L)\}$; (iv) $a_v = \Sigma\{s : s \ngeq a,\ s \in \mathfrak{M}(L)\}$.

Proof. (i) By (1.3) it suffices to show that $a^u \geq \Pi\{s : s \nleq a,\ s \in \mathfrak{M}(L)\}$. But this follows from the fact that if $s_1 \nleq a$, $s_2 \in \mathfrak{M}(L)$, $s_2 \geq s_1$, then $s_2 \geq a$. (ii) Since ^{u} preserves products, we have $a^u = \Pi\{s^u : s \geq a,\ s \in \mathfrak{M}(L)\}$. But $s \geq a$, $s \in \mathfrak{M}(L) \Longrightarrow$ (by Lemma 2.5) $s^u \in \mathcal{J}(L)$ and $s^u \nleq a$. Conversely, $s \nleq a$, $s \in \mathcal{J}(L) \Longrightarrow$ (by Lemma 2.5) $s = (s_v)^u$, $s_v \in \mathfrak{M}(L)$ and also $s_v \geq a$. This proves (ii). (iii) and (iv) by dual arguments.

The following result is analogous to a well known result for distributive lattices [1]. Recall that in this paper the term "sublattice of a complete lattice L" is defined as a non-void subset of L closed under finite and infinite sums and products.

LEMMA 2.9. Let $L \in \mathcal{S}_r$ and let L_1 be a sublattice of L. If $a \in \mathcal{J}(L_1)$ then there exists $b \in \mathcal{J}(L)$ such that $a \geq b$ and $[b) \cap L_1 = [a) \cap L_1$.

Proof. By Lemma 2.5, $a_{v_{L_1}} \in \mathfrak{m}(L_1)$ and $a_{v_{L_1}} \ngeq a$. Thus there exists $b \in \mathcal{J}(L)$, $a \geq b$ and $a_{v_{L_1}} \ngeq b$. Obviously $[a) \cap L_1 \subseteq [b) \cap L_1$. Conversely $s \geq b$, $s \in L_1 \Longrightarrow s \nleq a_{v_{L_1}} \Longrightarrow s \geq a$, sompleting the proof.

3. CHARACTERIZATION OF SOME SUBCLASSES OF $\mathcal{B}_c$.

The following theorems characterize the class $\mathcal{B}_s$ and some subclasses of $\mathcal{B}_s$. Refer to [10] and [11] for details on $\mathcal{B}_s$.

THEOREM 3.1. Let $L \in \mathcal{B}_c$. The following are equivalent: (i) $L \in \mathcal{B}_s$; (ii) $a \leq a_v$ for all $a \in L$; (iii) $a^u \leq a$ for all $a \in L$.

Proof. Recall $L \in \mathcal{B}_s \Longleftrightarrow \mathcal{J}(L) = \emptyset \Longleftrightarrow \mathfrak{m}(L) = \emptyset$. Apply Lemma 2.5.

THEOREM 3.2. Let $L \in \mathcal{B}_c$. The following are equivalent: (i) $L \in \mathcal{B}_s$ and L is a chain; (ii) $a = a_v$ for all $a \in L$; (iii) $a = a^u$ for all $a \in L$.

Proof. Immediate from Theorem 3.1 and Lemma 2.5.

THEOREM 3.3. Let $L \in \mathcal{B}_c$. The following are equivalent: (i) $L \in \mathcal{B}_s$ and L has no non trivial nodes; (ii) $a < a_v$ for $a \neq 0,1$ for $a \in L$ and $1_v = 1$; (iii) $a^u < a$ for $a \neq 0,1$ for $a \in L$ and $0^u = 0$.

Proof. Immediate from Lemmas 2.4 and 2.5 and from Theorem 3.1.

THEOREM 3.4. Let $L \in \mathcal{B}_c$. The following are equivalent: (i) $a \ngeq a^u$, for all $a \in L$, $a \neq 1$; (ii) L is an ordinal; (iii) $a^u > a$ for all $a \in L$, $a \neq 1$.

Proof. (i) $\Rightarrow$ (ii) Apply Lemma 2.5. $a \in \mathfrak{M}(L)$ for $a \neq L$. Thus a has an immediate successor for $a \neq 1$. L is complete, thus L is an ordinal. (ii) $\Rightarrow$ (iii) If $a \neq 1$ then $a \in \mathfrak{M}(L)$, hence $a \nleq a^u$, but L is a chain thus $a^u > a$. (iii) $\Rightarrow$ (i) trivial.

THEOREM 3.5. Let $L \in \mathfrak{D}_c$. The following are equivalent: (i) L is a Stone algebra; (ii) is $s,t \in L$, $s^u, t^u \neq 0$, $s^u t^u = 0$ then $s + t = 1$.

Proof. (i) $\Rightarrow$ (ii). We have $s^u \neq 0$, $(s^u)(s^u)^* = 0 \Rightarrow s^u \nleq (s^u)^*$ $\Rightarrow (s^u)^* \leq s$. Again, $t^u \neq 0$, $(s^u)^*(s^u)^{**} = 0 \Rightarrow$ either $t^u \nleq (s^u)^*$ or $t^u \nleq (s^u)^{**}$. But $s^u t^u = 0 \Rightarrow t^u \leq (s^u)^* \Rightarrow t^u \nleq (s^u)^{**} \Rightarrow (s^u)^{**} \leq t$. It follows that $s + t = 1$. (ii) $\Rightarrow$ (i) If $a \in L$ then by Lemma 2.7, $a^* + a^{**} = \Pi\,\{s+t : 0 < s^u \leq a,\ 0 < t^u \leq a^*,\ s,t \in L\}$. But $0 < s^u \leq a$ and $0 < t^u \leq a^* \Rightarrow s^u t^u = 0 \Rightarrow s + t = 1$. Hence $a^* + a^{**} = 1$.

THEOREM 3.6. Let $L \in \mathfrak{D}_c$. The following are equivalent: (i) L is a relative Stone algebra; (ii) if $s,t \in L$, $s^u \nleq t^u$, $t^u \nleq s^u$ then $s + t = 1$.

Proof. (i) $\Rightarrow$ (ii) By hypothesis $[s^u t^u)$ is a Stone algebra. Note $s^u \nleq t^u \Rightarrow s \geq s^u t^u \Rightarrow s \in [s^u t^u)$. By Lemma 2.6, $s^{u[s^u t^u)} = s^u \neq s^u t^u$. Similarly $t^u \neq s^u t^u$. By Theorem 3.5, $s + t = 1$. (ii) $\Rightarrow$ (i). It suffices to show that $[a)$ is a Stone algebra (cf. [1]) for $a \in L$. Suppose $s,t \geq a$, $s^{u[a)}, t^{u[a)} \neq a$, $s^{u[a)} t^{u[a)} = a$. By Lemma 2.6, $s^u t^u + a = a$. Hence $s^u t^u \leq a$. Also $s^{u[a)} \neq a \Rightarrow s^u + a \neq a \Rightarrow s^u \nleq a$. Similarly $t^u \nleq a$. Thus $s^u \nleq t^u$, $t^u \nleq s^u$. Apply Theorem 3.5.

If $L \in \mathfrak{D}_r$, and thus in particular if L is a finite distributive lattice, the condition that L is a Stone algebra (relative Stone algebra) can be expressed as a condition on $\mathfrak{J}(L)$ (and thus $\mathfrak{M}(L)$). The proofs of the following two theorems are straightforward and will therefore be omitted.

THEOREM 3.7. Let $L \in \mathcal{D}_r$. The following are equivalent: L is a Stone algebra; (ii) if $s,t \in \mathcal{J}(L)$, $st = 0$ then $s_v + t_v = 1$; (iii) if $s,t \in \mathfrak{m}(L)$, $s^u t^u = 0$ then $s + t = 1$.

THEOREM 3.8. Let $L \in \mathcal{D}_r$. The following are equivalent: (i) L is a relative Stone algebra; (ii) if $s,t \in \mathcal{J}(L)$, $s \nleq t$, $t \nleq s$ then $s_v + t_v = 1$; (iii) if $s,t \in \mathfrak{m}(L)$, $s^u \nleq t^u$, $t^u \nleq s^u$ then $s + t = 1$.

4. DUAL EQUIVALENCE BETWEEN THE CATEGORIES $\mathcal{P}$ and $\mathcal{D}_r$.

It is known [14] that the categories $\mathcal{P}$ and $\mathcal{D}_r$ are dually equivalent. In this section we will briefly describe this dual equivalence and its properties. Recall first that if P is a partially ordered set then a subset of A of P is a semi-ideal of P if $a \in A$, $b \in P$, $b \leq a$ implies that $b \in A$. The relevant functors are $\mathcal{S}:\mathcal{P} \longrightarrow \mathcal{D}_r$ and $\mathcal{J}:\mathcal{D}_r \longrightarrow \mathcal{P}$ and are defined as follows. If P is a partially ordered set, let $\mathcal{S}(P)$ be the complete ring of semi-ideals of P. If $h:P_1 \longrightarrow P_2$ is a morphism in $\mathcal{P}$ then $\mathcal{S}(h):\mathcal{S}(P_2) \longrightarrow \mathcal{S}'(P_1)$ is defined by $\mathcal{S}(h)(S) = h^{-1}[S]$ for $S \in \mathcal{S}(P_2)$. For $L \in \mathcal{D}_r$, let $\mathcal{J}(L)$ be defined, as before as the partially ordered set of completely join irreducible elements of L. If $h:L_1 \longrightarrow L_2$ is a morphism in $\mathcal{D}_r$, then $\mathcal{J}(h):\mathcal{J}(L_2) \longrightarrow \mathcal{J}(L_1)$ is defined by $\mathcal{J}(h)(a) = \Pi\{s:h(s) \geq a, s \in L_1\}$ for $a \in \mathcal{J}(L_2)$. Using Lemma 2.9 one can easily prove the following lemma.

LEMMA 4.1. Let $h:L_1 \longrightarrow L_2$ be a morphism in $\mathcal{D}_r$. Then, (i) h is one-one $\Longleftrightarrow$ $\mathcal{J}(h)$ is onto; (ii) h is onto $\Longleftrightarrow$ $\mathcal{J}(h)$ is an order embedding.

The property that $\mathcal{P}$ and $\mathcal{D}_r$ are dually equivalent will be utilized in subsequent sections. However in this section we will give some examples which will illustrate how this dual equivalence can also be used to provide some easy proofs of results in other areas.

First we note that products in $\mathcal{P}$ are direct products and coproducts (sums) in $\mathcal{P}$ are disjoint unions as can be easily seen. The following result which is known for finite distributive lattices easily follows in particular by using Lemma 4.1.

THEOREM 4.2. Suppose $L \in \mathcal{D}_r$. Then, (i) if L is the direct product of $\{L_i\}_{i \in I}$, $L_i \in \mathcal{D}_r$ for $i \in I$, then $\mathcal{J}(L)$ is the coproduct (sum) of $\mathcal{J}\{L_i\}_{i \in I}$; (ii) if $\mathcal{J}(L)$ is the coproduct (sum) of disjoint partially ordered sets $\{P_i\}_{i \in I}$ then L is isomorphic to the complete ring of semi-ideals of the product of $\{\delta(P_i)\}$, $i \in I$.

It is known [11], [17] that the free completely distributive complete lattice exists and is a complete ring of sets. This result also easily follows.

THEOREM 4.3. [11],[17]. The free completely distributive complete $\mathcal{F}_{\mathcal{D}_c}(S)$ on the set S of free generators exists and is a complete ring of sets.

<u>Proof</u>. We first show that $\mathcal{F}_{\mathcal{D}_r}(S)$ exists. Let for each $s \in S$, L_s denote the three-elements chain $\{0, s, 1\}$, $0 < s < 1$. Then L_s is the free object on one generator s in $\mathcal{D}_r$ and $\mathcal{J}(L_s) = \{s, 1\}$. By the dual equivalence of $\mathcal{P}$ and $\mathcal{D}_r$ it follows that $\mathcal{F}_{\mathcal{D}_r}(S)$ is the complete ring of semi-ideals of the power algebra 2^S. But it is known [1] that if $L \in \mathcal{D}_c$, then L is a complete homomorphic image of a complete ring of sets. A standard argument shows that $\mathcal{F}_{\mathcal{D}_c}(S) = \mathcal{F}_{\mathcal{D}_r}(S)$.

5. THE CASE THAT ^{c} IS MAXIMAL (AND ^{k} IS MAXIMAL) FOR $L \in \mathcal{D}_r$.

Recall that for $L \in \mathcal{D}_r$, L^c always satisfies the inequalities (1.6), and dually for L^k. The objective of this section and of the next one is to examine the case that ^{c} maximal which means that equality holds on the left side (1.6). We will, in particular, see that the conditions that ^{c} is maximal and that ^{k} is maximal are equivalent.

LEMMA 5.1. Let $L \in \mathcal{D}_r$. The following are equivalent: (i) $L_v \subseteq \mathfrak{m}(L) \cup \{0\} \cup \{1\}$; (ii) $\mathfrak{m}(L) \cup \{1\}$ is closed under sums of non-void subsets; (iii) $L^c = \mathcal{J}(L) \cup \{1\}$ if $0 \in \mathfrak{m}(L)$ and $L^c = \mathcal{J}(L) \cup \{0\} \cup \{1\}$ if $0 \notin \mathfrak{m}(L)$; (iv) $\mathcal{J}(L) \cup \{0\}$ is closed under products of non-void subsets; (v) $L^k = \mathfrak{m}(L) \cup \{0\}$ if $1 \in \mathcal{J}(L)$ and $L^k = \mathfrak{m}(L) \cup \{0\} \cup \{1\}$ if $1 \notin \mathcal{J}(L)$; (vi) $L^u \subseteq \mathcal{J}(L) \cup \{0\} \cup \{1\}$.

<u>Proof</u>. (i) $\Longrightarrow$ (ii). Let $S \subseteq \mathfrak{m}(L) \cup \{1\}$, $S \neq \emptyset$. If $1 \in S$ then $\Sigma S = 1 \in \mathfrak{m}(L) \cup \{1\}$. Thus we may assume that $1 \notin S$ and thus $S \subseteq \mathfrak{m}(L)$ and therefore $s = (s^u)_v$ for each $s \in S$. It follows that $\Sigma S = \Sigma\{(s^u)_v : s \in S\} = (\Sigma\{s^u : s \in S\})_v \in \mathfrak{m}(L) \cup \{0\} \cup \{L\}$. But if $\Sigma S = 0$ then $0 \in S$, since $S \neq \emptyset$ and thus $0 \in \mathfrak{m}(L)$. Hence $\Sigma S \in \mathfrak{m}(L) \cup \{1\}$. (ii) $\Longrightarrow$ (iii) By (1.6) it suffices to show that $L^c \subseteq \mathcal{J}(L) \cup \{1\}$ if $0 \in \mathfrak{m}(L)$ and $L^c \subseteq \mathcal{J}(L) \cup \{0\} \cup \{1\}$ if $0 \notin \mathfrak{m}(L)$. We first show that if $a \in L$, $a \neq 0$ then $a^c \in \mathcal{J}(L) \cup \{1\}$. Since $a \neq 0$, we have $a = \Sigma S$ for some $S \subseteq \mathcal{J}(L)$, $S \neq \emptyset$. Hence $a_v = \Sigma\{s_v : s \in S\}$. But $\{s_v : s \in S\} \neq \emptyset$ thus $a_v \in \mathfrak{m}(L) \cup \{1\}$. If $a_v \in \mathfrak{m}(L)$ then $a^c = (a_v)^u \in \mathcal{J}(L)$ and if $a_v = 1$ then by Lemma 2.4, $a^c = 1$. It follows that for $a \neq 0$, $a^c \in \mathcal{J}(L) \cup \{1\}$. It remains to show that if $0 \in \mathfrak{m}(L)$ then $0^c \in \mathcal{J}(L) \cup \{1\}$ and if $0 \notin \mathfrak{m}(L)$ then $0^c \in \mathcal{J}(L) \cup \{0\} \cup \{1\}$. By Lemma 2.4, if $0 \in \mathfrak{m}(L)$ then $0^c = 0_+ \in \mathcal{J}(L)$ and if $0 \notin \mathfrak{m}(L)$ then $0^c = 0$. (iii) $\Longrightarrow$ (iv) Recall that L^c is closed under products. Suppose $\emptyset \neq S \subseteq \mathcal{J}(L) \cup \{0\}$. If $0 \in S$ then $\Pi S = 0 \in \mathcal{J}(L) \cup \{0\}$. If $0 \notin S$ then $S \subseteq \mathcal{J}(L)$ and

thus by (1.6), $S \subseteq L^c$. It follows that $\Pi S \in L^c$ and therefore by hypothesis, $\Pi S \in \mathcal{J}(L) \cup \{0\} \cup \{1\}$. But if $\Pi S = 1$ then $1 \in S$ since $S \neq \emptyset$ and thus $1 \in \mathcal{J}(L)$. It follows that $\Pi S \in \mathcal{J}(L) \cup \{0\}$.
(iv) $\Rightarrow$ (v) By an argument dual to the proof of (ii) $\Rightarrow$ (iii).
(v) $\Rightarrow$ (vi) Suppose $a \in L$. Since $1^u = 1$, we may assume that $a \neq 1$ thus $a^k < 1$. By hypothesis, $a^k \in \mathfrak{m}(L) \cup \{0\}$. If $a^k \in \mathfrak{m}(L)$ then $a^u = ((a^u)_v)^u = (a^k)^u \in \mathcal{J}(L)$. If $a^k = 0$ then $a^u = (0^k)^u = 0^y$. But if $0 \in \mathfrak{m}(L)$ then by Lemma 2.4, $0^u \in \mathcal{J}(L)$ and if $0 \notin \mathfrak{m}(L)$ then $0^u = 0$. (vi) $\Rightarrow$ (i) If $a \in L$, then by hypothesis, $(a_v)^u \in \mathcal{J}(L) \cup \{0\} \cup \{1\}$. If $(a_v)^u \in \mathcal{J}(L)$ then $a_v = ((a_v)^u)_v \in \mathfrak{m}(L)$. If $(a_v)^u = 0$, then by Lemma 2.4, $a_v = ((a_v)^u)_v = 0$ and $(a_v)^u = 1$, $1 \notin \mathcal{J}(L)$ then $a_v = 1$ and finally, if $a_v = 1$, $1 \in \mathcal{J}(L)$, then $a_v = 1_v = 1_- \in \mathfrak{m}(L)$.

We are now able to characterize those complete rings of sets for which ^{c} and ^{k} are maximal.

THEOREM 5.2. If $L \in \mathcal{S}_r$, then the following are equivalent: (i) ^{c} is maximal; (ii) ^{k} is maximal; (iii) $\mathcal{J}(L) \cup \{0\} \cup \{1\}$ is a complete lattice; (iv) $\mathfrak{m}(L) \cup \{0\} \cup \{1\}$ is a complete lattice. Moreover if these conditions are satisfied then products in $\mathcal{J}(L) \cup \{0\} \cup \{1\}$ and sums in $\mathfrak{m}(L) \cup \{0\} \cup \{1\}$ are the same as in L.

Proof. (i) $\Rightarrow$ (iii) Immediate from Lemma 5.1. (iii) $\Rightarrow$ (i) By Lemma 5.1, it suffices to show that $\mathcal{J}(L) \cup \{0\}$ is closed under products of non-void subsets. Suppose $\emptyset \neq S \subseteq \mathcal{J}(L) \cup \{0\}$. Let $a = \Pi S$. By hypothesis, $a \in \mathcal{J}(L) \cup \{0\} \cup \{1\}$. If $a = 1$, then $S = \{1\}$ since $S \neq \emptyset$. But $S \subseteq \mathcal{J}(L) \cup \{0\}$, thus $a = 1 \in \mathcal{J}(L) \cup \{0\}$. If $a \neq 1$, then of course $a \in \mathcal{J}(L) \cup \{0\}$. (i) $\Leftrightarrow$ (ii) follows from Lemma 5.1, and (ii) $\Leftrightarrow$ (iv) follows from an argument dual to the proof of (i) $\Leftrightarrow$ (iii). The "Moreover" part is immediate from Lemma 1.2.

6. THE CASE THAT $\mathcal{J}(L)$ AND $\mathcal{M}(L)$ ARE COMPLETE LATTICES FOR $L \in \mathcal{S}_r$.

We note that for $L \in \mathcal{S}_r$ the condition that $\mathcal{J}(L) \cup \{0\} \cup \{1\}$ is a complete lattice (cf. Theorem 5.2), does not imply that $\mathcal{J}(L)$ is a complete lattice or even a lattice. Example: the 4-elements Boolean algebra. In this section we consider the case that $\mathcal{J}(L)$ and thus $\mathcal{M}(L)$, is a complete lattice.

THEOREM 6.1. If $L \in \mathcal{S}_r$ then the following are equivalent: (i) $\mathcal{J}(L)$ is a complete lattice; (ii) $\mathcal{M}(L)$ is a complete lattice; (iii) $L^u = \mathcal{J}(L)$; (iv) $L^c = \mathcal{J}(L)$; (v) $L_v = \mathcal{M}(L)$; (vi) $L^k = \mathcal{M}(L)$.

Proof. (i) $\Leftrightarrow$ (ii) Immediate, since $\mathcal{J}(L)$ and $\mathcal{M}(L)$ are order isomorphic. (ii) $\Rightarrow$ (iii) $a \in \mathcal{J}(L)$ then $a = (a_v)^u \in L^u$ and thus $\mathcal{J}(L) \subseteq L^u$. $\mathcal{J}(L)$ is a complete lattice, so $\mathcal{J}(L) \cup \{0\} \cup \{1\}$ is a complete lattice and thus by Lemma 5.2, ^{c} is maximal. By Lemma 5.1, $\mathcal{J}(L) \subseteq L^u \subseteq \mathcal{J}(L) \cup \{0\} \cup \{1\}$. Now $0 \leq 0_{\mathcal{J}(L)}$. But $0 \notin \mathcal{J}(L) \Rightarrow 0 < 0_{\mathcal{J}(L)}$. Also, $a \in L$, $0 < a \Rightarrow$ there exists $b \in \mathcal{J}(L)$ such that $0 < b \leq a \Rightarrow 0_{\mathcal{J}(L)} \leq b \leq 0 \in \mathcal{M}(L)$, $0_+ = 0_{\mathcal{J}(L)} \Rightarrow 0^u = 0_{\mathcal{J}(L)} \Rightarrow 0_{\mathcal{J}(L)} \leq a^u$ for $a \in \mathcal{J}(L) \Rightarrow 0 \notin L^u$. Again, $1_{\mathcal{J}(L)} < 1 \Rightarrow$ there exists $a \in \mathcal{J}(L)$, $a \nleq 1_{\mathcal{J}(L)}$. Contradiction. So $1_{\mathcal{J}(L)} = 1$ and thus $1 \in \mathcal{J}(L)$. It follows that $\mathcal{J}(L) \subseteq L^u \subseteq \mathcal{J}(L)$ implying $L^u = \mathcal{J}(L)$. (iii) $\Rightarrow$ (iv) $a \in L^c \Rightarrow a = a^c = (a_v)^u \in L^u \Rightarrow L^c \subseteq L^u = \mathcal{J}(L)$. By (1.6), $\mathcal{J}(L) \subseteq L^c$ hence $L^c = \mathcal{J}(L)$. (iv) $\Rightarrow$ (i) Immediate. (ii) $\Rightarrow$ (v) $\Rightarrow$ (vi) $\Rightarrow$ (ii) by dual arguments.

We have seen in Theorem 6.1 that if $L \in \mathcal{S}_r$ and if $\mathcal{J}(L)$ is a complete lattice then the maps $^u : L \to \mathcal{J}(L)$, $^c : L \to \mathcal{J}(L)$, $_v : L \to \mathcal{M}(L)$ and $^k : L \to \mathcal{M}(L)$ are onto maps. Moreover ^{u} and ^{k} preserve products and $_v$ and ^{c} preserve sums. We will now state under what conditions all these maps are complete homomorphism. First, we need some lemmas.

LEMMA 6.2. Let $L \in \mathcal{D}_r$ and suppose $\mathcal{J}(L)$ (and thus $\mathfrak{m}(L)$ is a complete lattice. If $a \in L$ then (i) $\Sigma^{\mathcal{J}(L)}\{s : s \leq a, s \in \mathcal{J}(L)\} = \Pi^{L}\{s : s \geq a, s \in \mathcal{J}(L)\}$; (ii) $\Pi^{\mathfrak{m}(L)}\{s : s \geq a, s \in \mathfrak{m}(L)\} = \Sigma^{(L)}\{s : s \leq a, s \in \mathfrak{m}(L)\}$.

Proof. (i) By Lemma 1.2 the left side is contained in the right side. The reverse inequality follows by observing that $a \leq \Sigma^{\mathcal{J}(L)}\{s : s \leq a, s \in \mathcal{J}(L)\} \in \mathcal{J}(L)$. (ii) By a dual argument.

LEMMA 6.3. Let $L \in \mathcal{D}_r$ and suppose $\mathcal{J}(L)$ (and thus $\mathfrak{m}(L)$) is a complete lattice. Then if $a \in L$ (i) $a^c = \Sigma^{\mathcal{J}(L)}\{s : s \leq a, s \in \mathcal{J}(L)\}$; (ii) $a^k = \Pi^{\mathfrak{m}(L)}\{s : s \geq a, s \in \mathfrak{m}(L)\}$.

Proof. Apply Theorem 6.1 and Theorem 6.2.

THEOREM 6.4. Let $L \in \mathcal{D}_r$ and suppose $\mathcal{J}(L)$ (and thus $\mathfrak{m}(L)$) is a complete lattice. The following are equivalent: (i) $\mathcal{J}(L) \in \mathcal{D}_c$; (ii) $\mathfrak{m}(L) \in \mathcal{D}_c$; (iii) ${}^c : L \to \mathcal{J}(L)$ is a complete homomorphism; (iv) ${}_v : L \to \mathfrak{m}(L)$ is a complete homomorphism; (v) ${}^k : L \to \mathfrak{m}(L)$ is a complete homomorphism; (vi) ${}^u : L \to \mathcal{J}(L)$ is a complete homomorphism.

Proof. (i) $\Longleftrightarrow$ (ii) Immediate, since ^{u} and ${}_v$ are isomomorphisms (i) $\Rightarrow$ (iii) We only need to chow that ^{c} preserves product. Therefore, it suffices to show that for $A \subseteq L$, $\Pi A^c \leq (\Pi A)^c$. We have by Lemma 6.3 and by hypothesis, $\Pi A^c = \Sigma^{\mathcal{J}(L)}\{\Pi\{\varphi(a) : a \in A\} : \varphi \in (\mathcal{J}(L))^A\}$ where for every $\varphi \in (\mathcal{J}(L))^A$, $\varphi(a) \leq a$ for all $a \in A$ and $(\Pi A)^c = \Sigma^{\mathcal{J}(L)}\{s : s \leq \Pi A, s \in \mathcal{J}(L)\}$. It suffices to show that for $\varphi_o \in (\mathcal{J}(L))^A$, $\Pi\{\varphi_o(a) : a \in A\} \leq (\Pi A)^c$. But $\varphi_o(a) \leq a$ for all $a \in A$, hence $\Pi\{\varphi_o(a) : a \in A\} \leq \Pi A$. But by Lemma 1.2, $\mathcal{J}(L)$ is closed under products and thus $\Pi\{\varphi_o(a) : a \in A\} \in \mathcal{J}(L)$. Since ^{c} preserves products, the desired inequality follows. (iii) $\Rightarrow$ (iv) We have for $a \in L$, $(a^c)_v = a_v$. But ${}^c : L \to \mathcal{J}(L)$ is a complete homomorphism and ${}_v : \mathcal{J}(L) \to \mathfrak{m}(L)$ is an isomorphism and thus ${}_v : L \to \mathfrak{m}(L)$ is a complete homomorphism. (iv) $\Rightarrow$ (i) Immediate. (ii) $\Rightarrow$ (v) $\Rightarrow$ (vi) $\Rightarrow$ (ii) by dual arguments.

The following theorem provides more information on the structure of L, $L \in \mathcal{D}_r$, if $\mathcal{J}(L)$ (and thus $\mathcal{M}(L)) \in \mathcal{D}_c$.

THEOREM 6.5. Let $L \in \mathcal{D}_r$ and suppose $\mathcal{J}(L)$ (and thus $\mathcal{M}(L)) \in \mathcal{D}_c$ then $\mathcal{J}(L) \cap \mathcal{M}(L) = \mathcal{M}(\mathcal{J}(L)) = \mathcal{J}(\mathcal{M}(L))$.

<u>Proof</u>. We show $\mathcal{J}(L) \cap \mathcal{M}(L) = \mathcal{M}(\mathcal{J}(L))$. Suppose $a \in \mathcal{J}(L) \cap \mathcal{M}(L)$. If $a = \Pi^{\mathcal{J}(L)} S$ for $S \subseteq \mathcal{J}(L)$, then by Lemma 1.2, $a = \Pi S$. But $a \in \mathcal{J}(L)$, so $a \in S$. Hence $a \in \mathcal{M}(\mathcal{J}(L))$. Conversely, suppose $a \in \mathcal{M}(\mathcal{J}(L))$. We only need to show that $a \in \mathcal{M}(L)$. Suppose $a = \Pi S$ for $S \subseteq L$. Thus $a^c = (\Pi S)^c$. By Theorems 6.1 and 6.4, $a = a^c = \Pi^{(L)}\{s^c : s \in S\}$. By hypothesis, there exists $s \in S$ such that $a = s^c$. But $a = s^c \geq s$. Also $a \leq s$, hence $a = s$. Thus $a \in \mathcal{M}(L)$.

7. THE CASES THAT ^{c} IS MINIMAL ($L^c = L$) AND ^{k} IS MINIMAL ($L^k = L$) FOR $L \in \mathcal{D}_r$.

Recall (1.5) that if $L \in \mathcal{D}_r$, then $L^c = L \Longleftrightarrow L^u = L$ and dually. We will see that in the finite case the conditions $L^c = L$ and $L^k = L$ are equivalent for $L \in \mathcal{D}_r$ but that is not the case for L infinite. Within a different context Bandelt [6] has characterized the finite distributive lattices L for which $L^u = L_v = L$. Our approach will be different and cover both the finite and infinite case. We first provide the following lattice-theoretic characterization.

THEOREM 7.1. [6]. Let $L \in \mathcal{D}_r$. The following are equivalent: (i) $L^c = L$; (ii) $\mathcal{J}(L) \supseteq \mathcal{M}(L)$; (iii) $\mathcal{J}(L)$ is meet dense in L, and dually.

<u>Remark</u>. The equivalence of (i) and (iii) was proven in [6].

Proof of Theorem 7.1. (i) $\Rightarrow$ (ii) If $a \in \mathfrak{M}(L)$ then by (1.5) there exists $b \in L$ such that $b^u = a$. But $b = \Pi S$ for some $S \subseteq \mathfrak{M}(L)$ and thus $a = b^u = \Pi\{s^u : s \in S\}$. By hypothesis there exists $s \in S$ such that $a = s^u \in \mathfrak{J}(L)$. (ii) $\Rightarrow$ (iii) If $a \in L$, then $a = \Pi S$ for some $S \subseteq \mathfrak{M}(L) \subseteq \mathfrak{J}(L)$. (iii) $\Rightarrow$ (i) We must show that ^{u} is onto. Let $a \in L$, then $a = \Pi S$ for some $S \subseteq \mathfrak{J}(L)$. Let $b = \Pi\{s_v : s \in S\}$, then $b^u = a$.

Remark. It follows immediately from Theorem 7.1 that if L is finite, $L \in \mathfrak{D}_r$, then the conditions $L^c = L$ and $L^k = L$ are equivalent [6]. If L is infinite then this need not be true. Indeed, let $L = \omega \oplus \{\sim\}$ where $\mathfrak{J}(L) \subset \mathfrak{M}(L)$.

Recall (section 4) that the categories $\mathfrak{P}$ and $\mathfrak{D}_r$ are dually equivalent. In the remaining part of this section, and in the next section we will characterize those partially ordered sets P which satisfy both conditions that $L^c = L$ and $L^k = L$, where $L = \mathcal{S}(P)$. The more general case that only $L^c = L$ (or only $L^k = L$) will not be treated in detail but the technique which will be presented can be used with slight modifications for those cases as well.

We start with defining on a partially ordered set P two partial unary operations which coincide with the operations ^{u} and ^{v} if P is a complete lattice. Therefore we will use these symbols to denote the partial operations as well.

$$\text{(7.2)}\qquad \begin{aligned} &\text{For } a \in P, \text{ let } a^u = \text{glb}\{s : s \nleq a,\ s \in P\} \text{ if it exists} \\ &\qquad\qquad\qquad a_v = \text{lub}\{s : s \ngeq a,\ s \in P\} \text{ if it exists.} \end{aligned}$$

Note that ^{u} and $_v$ preserve order. If $P, Q \in \mathfrak{P}$ and $Q \subseteq P$ (where the partial order on Q is the restriction of the partial order on P, to Q), we will also write for $a \in Q$, a_{v_P} and a_{v_Q}, provided they exist, in order to distinguish between the operation $_v$ on P and Q respectively, and dually for ^{u}. Note $a_{v_Q} \leq a_{v_P}$ and dually. We will say that Q is closed under $_v$ if whenever a_{v_P} exists for $a \in P$, then $a_{v_P} \in Q$ and thus $a_{v_Q} = a_{v_P}$, and dually. We now introduce the following terminology:

(7.3) A $_v$-set is a partially ordered set P for which a_v exists and $a_v \neq a$ for each $a \in P$. ^{u}-sets are defined dually.

(7.4) A $^u{}_{,v}$-set is a partially ordered set which is both a ^{u}-set and a $_v$-set. The void set will also be considered to be a $^u{}_{,v}$-set.

We first establish some properties of ^{u}-sets and $_v$-sets which, in most cases, we will only formulate for $_v$-sets.

LEMMA 7.5. If P is a $_v$-set then $_v : P \longrightarrow P$ is an order embedding, and dually.

<u>Proof</u>. Immediate.

LEMMA 7.6. Let P be a finite partially ordered set. The following are equivalent: (i) P is a $_v$-set; (ii) P is a ^{u}-set. Moreover, if P is a finite $^u{}_{,v}$-set then ^{u} and $_v$ are automorphisms which are the inverse of another.

<u>Proof</u>. If P is a $_v$-set, then by Lemma 7.5, $_v$ is an order embedding. But P is finite, thus $_v$ is an automorphism. It is easy to show that P is a ^{u}-set and that ^{u} is the inverse of $_v$.

LEMMA 7.7. If P is a $_v$-set then the following are equivalent: (i) $_v$ is onto; (ii) P is a ^{u}-set. Moreover if (i) and (ii) are satisfied, then ^{u} is the inverse of $_v$.

<u>Proof</u>. It follows from Lemma 7.5 that if P is a $_v$-set and $_v$ is onto, then $_v$ is an automorphism. It is again easy to show that P is a ^{u}-set and that ^{u} is the inverse of $_v$.

LEMMA 7.8. Suppose P is a ${}^{u}{}_{,v}$-set. Then (i) $(a_v)^u = (a^u)_v = a$ for $a \in P$; (ii) ^{u} and ${}_{v}$ are automorphisms which are the inverse of another.

<u>Proof</u>. (i) $a \nleq a_v \Rightarrow (a_v)^u \leq a$. It is easy to see that $(a_v)^u \geq a$, hence $(a_v)^u = a$. (ii) By Lemma 7.5, ^{u} and ${}_{v}$ are order embeddings which are, by (i), onto and thus automorphisms which are, again by (i), the inverse of another.

The following theorem reveals the relation which exists between complete rings of sets L for which $L^c = L$ and ${}_{v}$-sets.

THEOREM 7.9. Let $L \in \mathscr{D}_r$. The following are equivalent: (i) $L^c = L$; (ii) $\mathscr{J}(L)$ is closed under ${}_{v}$; (iii) $\mathscr{J}(L)$ is a ${}_{v}$-set.

<u>Proof</u>. (i) $\Rightarrow$ (ii) If $a \in \mathscr{J}(L)$ then, by Lemma 2.5, $a_{v_L} \in \mathfrak{M}(L)$ and thus by Theorem 7.1, $a_{v_L} \in \mathscr{J}(L)$. (ii) $\Rightarrow$ (iii) Immediate, since for $a \in \mathscr{J}(L)$, $a_{v_L} \geq a$; (iii) $\Rightarrow$ (i) By Theorem 7.1, it suffices to show that $\mathscr{J}(L) \supseteq \mathfrak{M}(L)$. Let $a \in \mathfrak{M}(L)$ and let $b = a^{u_L}$. Then by Lemma 2.5, $b \in \mathscr{J}(L)$. By hypothesis, $b_{v_{\mathscr{J}(L)}}$ exists and is $\ngeq b$. On the other hand, by Lemma 2.8, $b_{v_L} = \Sigma^L\{s : s \ngeq b,\ s \in \mathscr{J}(L)\}$. But $s \in \mathscr{J}(L)$, $s \ngeq b \Rightarrow s \leq b_{v_{\mathscr{J}(L)}} \Rightarrow b_{v_L} \leq b_{v_{\mathscr{J}(L)}}$. Also, $b_{v_{\mathscr{J}(L)}} \leq b_{v_L}$ and thus $b_{v_L} = b_{v_{\mathscr{J}(L)}}$. Therefore, $b_{v_L} \in \mathscr{J}(L)$. But $b_{v_L} = a$ and we conclude that $a \in \mathscr{J}(L)$.

The following theorem establishes the relationship between ${}^{u}{}_{,v}$-sets and the class of complete rings of sets L for which $L^c = L^k = L$ (^{c} and ^{k} both minimal).

THEOREM 7.10. Let $L \in \mathscr{D}_r$. The following are equivalent: (i) $L^c = L^k = L$; (ii) $\mathscr{J}(L)$ is a ${}^{u}{}_{,v}$-set; (iii) $\mathfrak{M}(L)$ is a ${}^{u}{}_{,v}$-set; (iv) $\mathscr{J}(L) = \mathfrak{M}(L)$.

Proof. Recall that $_v : \mathcal{J}(L) \longrightarrow \mathcal{M}(L)$ and $^u : \mathcal{M}(L) \longrightarrow \mathcal{J}(L)$ are isomorphisms. Thus $\mathcal{J}(L)$ is a $_v$-set if and only if $\mathcal{M}(L)$ is a $_v$-set. Apply Theorems 7.1 and 7.9.

8. $^u{,}_v$-SETS.

We will use in this section the notion of ordinal sum of partially ordered sets in the following sense. If $\{P_i : i \in I\}$ is a set of disjoint partially ordered sets and P is a partially ordered set, then P is an ordinal sum of $\{P_i : i \in I\}$ if (i) $P = \cup\{P_i : i \in I\}$; (ii) I admits a lineal order such that for $a, b \in P$, $a \leq b$ in P if and only if either $a, b \in P_i$ for some $i \in I$ and $a \leq b$ in P_i, or $a \in P_i$, $b \in P_j$, $i \neq j$ and $i < j$. If P is a $^u{,}_v$-set and $Q \subseteq P$ then we call Q a $^u{,}_v$-subset of P if Q is closed under ^{u} and $_v$. Obviously, if Q is a $^u{,}_v$-subset of P then Q itself is a $^u{,}_v$-set. Also, if P is a $^u{,}_v$-set and $Q \subseteq P$ then the intersection of all $^u{,}_v$-subsets of P containing Q, is a $^u{,}_v$-subset of P which will be denoted by $[Q]$ and we will say that $[Q]$ is the $^u{,}_v$-subset of P generated by Q. We will write $[a, b, \ldots]$ instead of $[\{a, b, \ldots\}]$ for $a, b, \ldots \in P$. It will be useful to introduce the following notation.

Let P be a $^u{,}_v$-set and let N denote the set of all integers. For $a \in P$, $n \in N$, a_n is defined by

$$(8.1) \qquad a_o = a, \ a_n = (a_{n-1})_v \ \text{ for } \ n \geq 1, \ a_n = (a_{n+1})^u \ \text{ for } \ n \leq -1.$$

It is immediate from Lemma 7.8 and from the properties of the maps ^{u} and $_v$ that for $a, b \in P$, $n, m \in N$, we have $(a_n)_m = a_{n+m}$; $a_{n+m} \geq a_n$; $a_n = b_n \Longleftrightarrow a_{n+m} = b_{n+m}$; $a_n < b_n \Longleftrightarrow a_{n+m} < b_{n+m}$.

We now introduce the notion of cyclic $^u{,}_v$-sets. These sets will turn out to be the "building stones" of $^u{,}_v$-sets. Let P be a $^u{,}_v$-

set. P is cyclic if $P = [a]$ for some $a \in P$. Note that in that case $P = \{a_n : n \in N\}$. Also by (8.2), P is generated by each of its elements. If P is a ${}^{u}{}_{,v}$-set then a subset Q of P is called a cyclic subset of P if $Q = [a]$ for some $a \in Q$. Finally, note that if P is a ${}^{u}{}_{,v}$-set then P is the union of disjoint cyclic ${}^{u}{}_{,v}$-sets. On the other hand it is obvious that an ordinal sum of ${}^{u}{}_{,v}$-sets is again a ${}^{u}{}_{,v}$-set. Therefore the question arises whether each ${}^{u}{}_{,v}$-set is an ordinal sum of cyclic ${}^{u}{}_{,v}$-sets. We will see that this not the case. We will therefore, characterize the class of all ${}^{u}{}_{,v}$-sets which are ordinal sums of cyclic ${}^{u}{}_{,v}$-sets and we will exhibit a class of ${}^{u}{}_{,v}$-sets which are not an ordinal sum of cyclic ${}^{u}{}_{,v}$-sets. Our first main result will be that there are exactly three non-isomorphic cyclic ${}^{u}{}_{,v}$-sets.

LEMMA 8.3. Let P be a ${}^{u}{}_{,v}$-set and let $a \in P$. Then one of the following holds: (i) $a_1 < a_o$; (ii) $a_o = a_2$; (iii) $a_o < a_2$.

Proof. If (i) does not hold, then $a_1 \nless a_o$. But $a_1 \ngtr a_o$, thus $a_1 \nless a_o \Rightarrow a_1 \ngtr a_o \Rightarrow a_o \leq a_2$. On the other hand if (ii) or (iii) holds then (i) does not hold. Indeed if $a_o \leq a_2$ and $a_1 < a_o$ then $a_1 \leq a_2$. Contradiction.

Remark. It is obvious that if (i) holds for $a \in P$, then (i) holds for each $b \in [a]$. Similarly, for (ii) and (iii).

THEOREM 8.4. Suppose P is a cyclic ${}^{u}{}_{,v}$-set, $P = [a]$. (i) If $a_1 < a_o$ then P is the chain N of all integers (Figure 1). (ii) If $a_o = a_2$, then P consists of 2 elements which are unordered (Figure 2). (iii) If $a_o < a_2$, then P is the partially ordered set of Figure 3.

Proof. We will repeatedly, without further reference, use the properties of ^{u} and $_v$ (cf. 8.2). (i) $a_1 < a_o \Rightarrow a_{n+1} < a_n$ for all $n \in N$, hence $P = N$. (ii) We have $a_1 \nleq a_o$ but also $a_o \nleq a_1$. It is easy to see that $a_o = a_2$ implies $a_n = a_o$ or a_1 for all $n \in N$. (iii) We show that for $n \in N$, $a_{n+1} \nleq a_n$, $a_n < a_{n+2}$, $a_n < a_{n+3}$ and thus $a_n < a_m$ for $m \geq n+2$. It suffices to show this for $n = 0$ (because of (8.2)). If $a_o < a_2$ then $a_1 \nleq a_o$. Indeed $a_1 \leq a_o \Rightarrow a_1 < a_o \Rightarrow a_1 < a_2$. Contradiction. Next, note that by hypothesis, $a_o < a_2$. Finally, to show $a_o < a_3$. But $a_o < a_2 \Rightarrow a_2 \nleq a_o \Rightarrow a_o \leq 3$. Suppose $a_o = a_3$. But then $a_1 \nleq a_o \Rightarrow a_1 \nleq a_3 \Rightarrow a_o \nless a_2$. Contradiction. It now follows easily that P is the partially ordered set of Figure 3.

Recall that for $P \in \mathcal{P}$, $\mathcal{J}(\mathcal{S}(P)) \cong P$. Thus we have the following result.

COROLLARY 8.5. There are three non-isomorphic complete rings of sets L for which $\mathcal{J}(L)$ is a cyclic $^u{}_{,v}$-set: (i) $L = \{-\sim\} \oplus \{\sim\}$ (Figure 4); (ii) L = the 4-elements Boolean algebra (Figure 5); (iii) L is the lattice of Figure 6.

Remark. Note that Figure 6 contains the free Heyting algebra on one generator.

We now turn our attention to $^u{}_{,v}$-sets which are not cyclic and we first establish some results.

LEMMA 8.6. Let P be a $^u{}_{,v}$-set, $P = [a,b]$, $b \notin [a]$. Then P is an ordinal sum of $[a]$ and $[b] \Leftrightarrow b > a_n$ for all $n \in N$ or $b < a_n$ for all $n \in N$.

Proof. $\Rightarrow$ Trivial. $\Leftarrow$ Suppose $b > a_n$ for all $n \in N$. Let $m_o, n_o \in N$. Then $b > a_{m_o - n_o}$ and thus $b_{m_o} > a_{n_o}$.

LEMMA 8.7. Let P be a $^{u}{}_{,v}$-set, $a, b \in P$, $b \notin [a]$. Then each of the conditions (i) and (ii) of Lemma 8.3 implies that $[a,b]$ is an ordinal sum of $[a]$ and $[b]$.

Proof. First, we assume that (i) holds. Note that $b \notin [a]$ implies $b \neq a_n$ for all $n \in N$. Also note that $[a]$ is the partially ordered set of Figure 1. By Lemma 8.6, it suffices to show that either $b < a_n$ or $b > a_n$ for all $n \in N$. Suppose there exists $n_o \in N$ such that $b \not> a_{n_o}$. But $b \not> a_{n_o} \Longrightarrow b \not\geq a_{n_o} \Longrightarrow b < a_{n_o+1} < a_{n_o} \Longrightarrow b < a_n$ for all $n \leq n_o + 1$.

Suppose there exists an $n \in N$ such that $b \not< a_n$, then there is a smallest n, say n_1 such that $b \not< a_n$ and where $n_1 > n_o+1$. But $b \not< a_{n_1} \Longrightarrow b \not\leq a_{n_1} \Longrightarrow a_{n_1-1} < b \Longrightarrow a_{n_1} < b$. But also $b \not< a_{n_1-1} \Longrightarrow b \not\leq a_{n_1-1} \Longrightarrow b < a_{n_1}$. Contradiction. It follows that $b < a_n$ for all $n \in N$. Next, assume that (ii) holds. By Lemma 8.3, $[a] = \{a_o, a_1\}$, $a_o = a_2$, $a_1 \not\leq a_o$, $a_o \not\leq a_1$. Note $a_{-1} = (a_2)_{-1} = a_1$. By Lemma 8.6 it suffices to show that either (1) $b < a_o$ and $b < a_1$ or (2) $b > a_o$ and $b > a_1$. Suppose (1) does not hold, then either $b \not< a_o$ or $b \not< a_1$. But $b \not< a_o \Longrightarrow b \not\leq a_o \Longrightarrow a_o \leq b_1 \Longrightarrow a_o < b_1$ $\Longrightarrow a_{-1} < b \Longrightarrow a_1 < b \Longrightarrow b \not\leq a_1 \Longrightarrow a_1 < b_1 \Longrightarrow a_o < b$. Thus (2) holds. Again, $b \not< a_1 \Longrightarrow a_1 < b_1 \Longrightarrow a_o < b \Longrightarrow a_o \not\geq b \Longrightarrow a_2 \not\geq b \Longrightarrow a_1 \not\geq b_{-1}$ $\Longrightarrow a_1 < b$. Thus, again (2) holds.

Theorem 8.4 and Lemma 8.7 now yield the following partial characterization of $^{u}{}_{,v}$-sets.

COROLLARY 8.8. Let P be a $^{u}{}_{,v}$-set. Suppose $a_o \not< a_2$ for each $a \in P$. Then P is an ordinal sum of copies of Figures 1 and 2.

The previous results also enable us to characterize the finite $^{u}{}_{,v}$-sets and thus the finite distributive lattices L, for which $L^c = L^k = L$. Indeed, if P is a finite $^{u}{}_{,v}$-set and $a \in P$ then it follows from Theorem 8.4 that $[a]$ consists of two unordered elements.

Thus P is an ordinal sum of a finite set of copies of Figure 2.

COROLLARY 8.9. (cf. [6])(i) Let P be a finite ${}^{u}{}_{,v}$-set, $P \neq \emptyset$. Then P is an ordinal sum of a finite set of copies of Figure 2; (ii) Let L be a finite distributive lattice, $L \neq \{1\}$ for which $L^c = L^k = L$ (or equivalently, $\mathcal{J}(L) = \mathfrak{m}(L)$). Then L is the lattice of Figur 7 (finite "sum" of 4-elements Boolean algebras).

If P is a ${}^{u}{}_{,v}$-set and $P = [a,b].[b] \notin [a]$. $a_o < a_2$. $b_o < b_2$. then P need not be an ordinal sum of [a] and [b]. The following establishes what P is in that case.

THEOREM 8.10. Let P be a ${}^{u}{}_{,v}$-set, $P = [a,b]$, $b \notin [a]$, $a_o < a_2$, $b_o < b_2$ and suppose P is not an ordinal sum of [a] and [b]. Then P is the partially ordered set of Figure 8.

Proof. Let $N_1 = \{n : n \in N,\ b < b_n\}$. By Lemma 8.6, there exists $n \in N$ such that $b \nparallel a_n$ and thus $b < a_{n+1}$. Thus $N_1 \neq \emptyset$. Also by Lemma 8.6, $N_1 \neq N$. We show there exists a smallest n for which $b < a_n$. Suppose not. Let $n_1 \in N$. Inspection of Figure 3 shows that $a_{n_1-3} < a_{n_1}$. But then there exists $n_2 \in N$ such that $n_2 < n_1 - 3$ and $b < a_{n_2}$. Again by Figure 3, $a_{n_2} < a_{n_1}$ and it follows that $b < a_{n_1}$. Hence $b < a_n$ for all $n \in N$. Contradiction. There is no loss of generality if we assume that the smallest $n \in N$ such that $b < a_n$ is $n = 2$. Thus $b < a_2$ but $b \nless a_1$. Also $a_1 \nless b$ since $a_1 < b \Rightarrow a_1 < b = b_o < a_2$. Contradiction. Again, $b = b_o < a_2 \Rightarrow a_2 \nparallel b_o \Rightarrow b_o < a_3$. Thus $a_o < b_o < a_2$, $b_o < a_3$ and it follows that $a_1 < b_1 < a_3$. Also $a_o < b_o \Rightarrow a_o \nparallel b_o \Rightarrow a_o < b_1 \Rightarrow a_{-1} < b_o$. Inspection of Figure 3 and repeated application of the inequalities establised in part (iii) of the proof of Theorem 8.4 yield that P is of the type of Figure 8.

Remark. The complete ring of sets corresponding to Figure 8 is depicted in Figure 9.

We will now summarize the results obtained so far. First, we introduce some notation. Let P be a ${}^{u},{}_{v}$-set and define $P_1 = \{a : a_1 < a_0,\ a \in P\}$, $P_2 = \{a : a_0 = a_2,\ a \in P\}$ and $P_3 = \{a : a_0 < a_2,\ a \in P\}$. Note $P = P_1 \cup P_2 \cup P_3$, $P_i \cap P_j = \emptyset$ for $i \neq j$. The previous results can now be summarized as follows.

COROLLARY 8.11. (i) $P_1 \cup P_2$ is an ordinal sum of Figures 1 and 2; (ii) if $a \in P_3$ then $P_1 \cup P_2 \cup [a]$ is an ordinal sum of copies of Figures 1 and 2 and a copy of Figure 3; (iii) if $a, b \in P_3$, $[a] \neq [b]$, then $[a,b]$ is either an ordinal sum of two copies of Figure 3 or $[a,b]$ is Figure 8.

We now first focus our attention on P_3. We introduce a binary relation $\sim$ on P_3 defined by $a \sim b \iff [a] = [b]$ or $[a,b]$ is not an ordinal sum of $[a]$ and $[b]$. It is easy to see that $\sim$ is an equivalence relation. Now let for $a \in P_3$, $P_a = \{b : b \sim a,\ b \in P_3\}$, then $P_3 = \cup\{P_a : a \in P_3\}$.

LEMMA 8.12. Suppose $P_a \neq P_b$ for $a, b \in P_3$. Then $P_a \cup P_b$ is an ordinal sum of P_a and P_b.

Proof. Since $P_a \neq P_b$, $[a,b]$ is an ordinal sum of $[a]$ and $[b]$ and we may therefore assume $b > a_n$ for all $n \in N$. Suppose $a' \in P_a$, $a' \notin [a]$. Then by Lemma 8.10, $[a,a']$ is a copy of Figure 8. It follows that $b > a'_n$ for all $n \in N$. Thus $b > x$ for all $x \in P_a$. Again, let $b' \in P_b$, $b' \notin [b]$, then $[b,b']$ is a copy of Figure 8. But $b > x$ for all $x \in P_a$, so $b' > x$ for all $x \in P_a$. It follows that for $x \in P_a$, $y \in P_b$, $y > x$ and thus $P_a \cup P_b$ is an ordinal sum of P_a and P_b.

We now return to the general situation.

LEMMA 8.13. Let P be a ${}^{u}{}_{,v}$-set. Let $a \in P_1 \cup P_2$, $b \in P_3$, then $[\{a\} \cup P_b]$ is an ordinal sum of $[a]$ and P_b.

<u>Proof</u>. By Lemma 8.7, $[a,b]$ is an ordinal sum of $[a]$ and $[b]$. So we may assume $a < b_n$ for all $n \in N$. Let $c \in P_b$, $[c] \neq [b]$. Again, $[a,c]$ is an ordinal sum of $[a]$ and $[c]$. But if $c < a_n$ for all $n \in N$, then $c < b_n$ for all $n \in N$ and not $c \sim b$. Contradiction. Thus $c > a_n$ for all $n \in N$. It follows that $[\{a\} \cup P_b]$ is an ordinal sum of $[a]$ and P_b.

COROLLARY 8.14. If P is a ${}^{u}{}_{,v}$-set, then P is an ordinal sum of copies of Figures 1 and 2 and of the partially ordered sets P_a, $a \in P_3$.

9. SPECIAL CASES OF ${}^{u}{}_{,v}$-SETS.

In this section we will, in addition to the ${}^{u}{}_{,v}$-set of Figure 8, provide 3 more examples of ${}^{u}{}_{,v}$-sets P for which $P = P_3$ and which consists of only one equivalence class under the equivalence relation $\sim$. Thus $a_0 < a_2$ for each $a \in P$ and for $a,b \in P$, $[a] \neq [b]$, $[a,b]$ is not an ordinal sum of $[a]$ and $[b]$ (thus $[a,b]$ is the ${}^{u}{}_{,v}$-set of Figure 8). Figures 10 and 12 are two examples of this kind and each are generated by 3 elements a, b and c. The corresponding rings of sets are depicted in Figures 11 and 13. Note that a sublattice of Figure 13 is the free distributive lattice on 3 generators. Finally, as a last example on this category, let P be a set which is the union of two disjoint copies R and $\overline{R}$ of the real numbers. Corresponding elements of R and $\overline{R}$ (thus elements representing the same real number) will be denoted by x and $\overline{x}$ respectively. We extend the linear orderings of R and $\overline{R}$ to a partial ordering on P by: for $x \in R$, $y \in \overline{R}$, $x < \overline{y} \iff x < y$ and $x > \overline{y} \iff x - 1 > y$. It is easy to see that P is a ${}^{u}{}_{,v}$-set of the above type and that P is generated by the half open unit interval. Note that for $x \in R$ and $n \in N$, $x_{2n} = \overline{x}_{2n-1} = x+n$ and $x_{2n+1} = \overline{x+n} = x_{2n}$. The corresponding complete ring of sets L which is depicted in Figure 15, can be described as follows. Let, for each $x \in R$, I_x denote the unit interval in which every element σ is split into two elements σ' and σ, $\sigma' < \sigma$, and let for each $x \in R$, $L_x = I_x \times \underline{2}(\underline{2} = \{0,1 : 0 < 1\}$. Let $L = \bigcup\{L_x : x \in R\}$. If $a \in L$, then a is an element of L_x for some $x \in R$. As an element of L_x, $a = (\sigma',0)$ or $(\sigma,0)$ or $(\sigma',1)$ or $(\sigma,1)$. Therefore the elements of L can be denoted by $(x,\sigma',0)$ or $(x,\sigma,0)$ or $(x,\sigma',1)$ or $(x,\sigma,1)$ for $x \in R$ and $0 \leq \sigma \leq 1$. We endow L with a partial ordering which is an extension of the ordering of each L_x. Define for $x,y \in R$, $x \neq y$, $0 \leq \sigma \leq 1$, $0 \leq \tau \leq 1$, $(x,\sigma',1) < (y,\tau',0)$ $\Longleftrightarrow (x,\sigma',1) < (y,\tau,0) \Longleftrightarrow (x,\sigma,1) < (y,\tau,0) \Longleftrightarrow x < y$, $\sigma \leq \tau + y - x$, and $(x,\sigma,1) < (y,\tau',0) \Longleftrightarrow x < y$, $\sigma < \tau + y - x$.

J. Berman has provided a method (communication to the author) to construct a class of ${}^{u}{}_{,v}$-sets of the type considered in this section. This class contains the examples of Figures 8, 10, 12, and 14. Let Q be a partially ordered set and let for each integer $n \in N$, Q_n be an isomorphic copy of Q. If $a \in Q$, then we will denote the element of Q_n corresponding to the isomorphism by a_n. Let $P = \bigcup\{Q_n : n \in N\}$. We extend the partial ordering of each Q_n to a partial ordering of P as follows. For $a,b \in Q$, $a_n < b_{n+1} \Longleftrightarrow a_n \geq b_n$ for $n \in N$ and $a_n < b_{n+j}$ for $n,j \in N$, $j \geq 2$. It is easy to check that under this ordering P is a ${}^{u}{}_{,v}$-set and that for $a \in Q$, $n \in N$, $a_{n+1} = (a_n)_v$ and also that P is of the type considered in this section. In fact, for $a \in Q$, $n_o \in N$, $\left[a_{n_o}\right] = \{a_n : n \in N\}$ and for $a,b \in Q$, $a \neq b$, $[a_n,b_m]$ is Figure 8 for $n,m \in N$. Finally, one easily checks that Figure 8 is obtained by taking $Q = \{a,b : a < b\}$: Figure 10 is obtained by taking $Q = \{a,b,c;\ a,b,c$ totally unordered$\}$, Figure 14 is obtained by taking for Q the half open unit interval. The question whether all $^{u}{}_{,v}$-sets of this type can be obtained in this way remains unanswered.

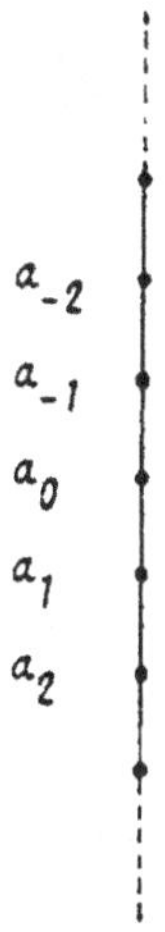

Figure 1

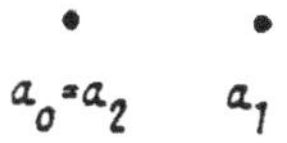

Figure 2

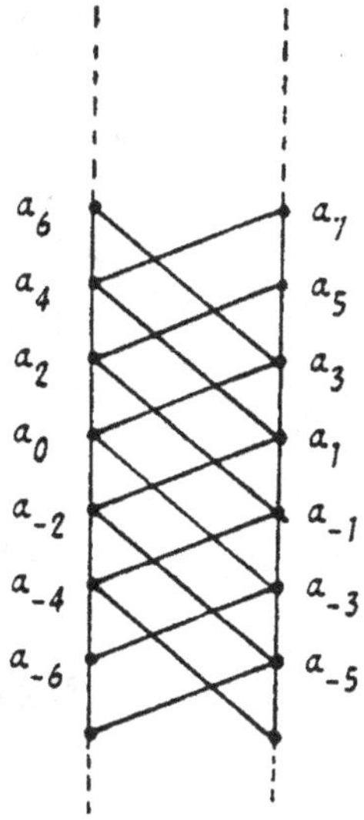

Figure 3

Figure 4

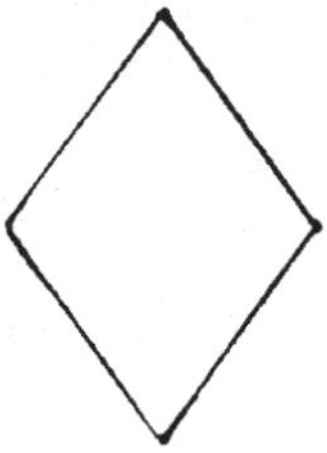

Figure 5

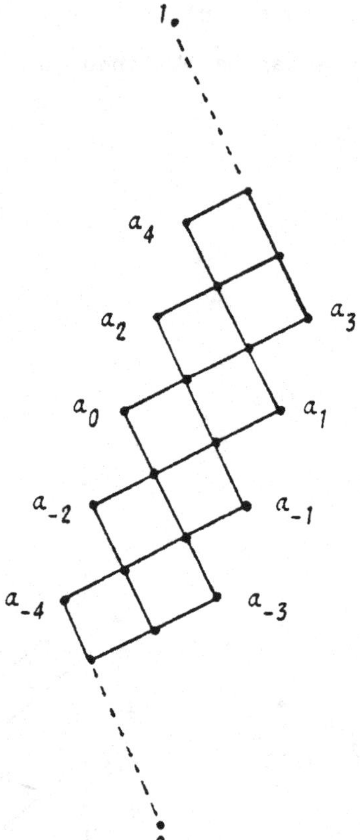

Figure 6

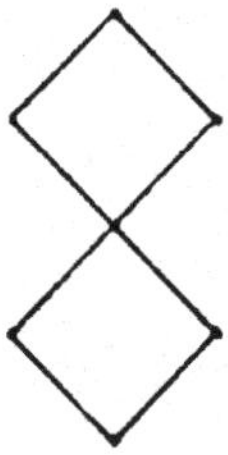

Figure 7

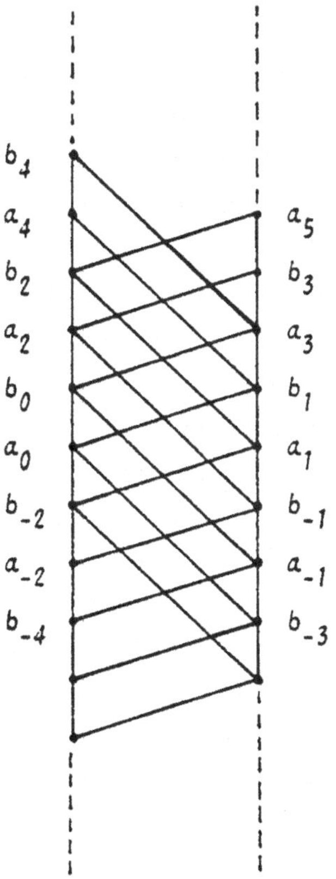

Figure 8

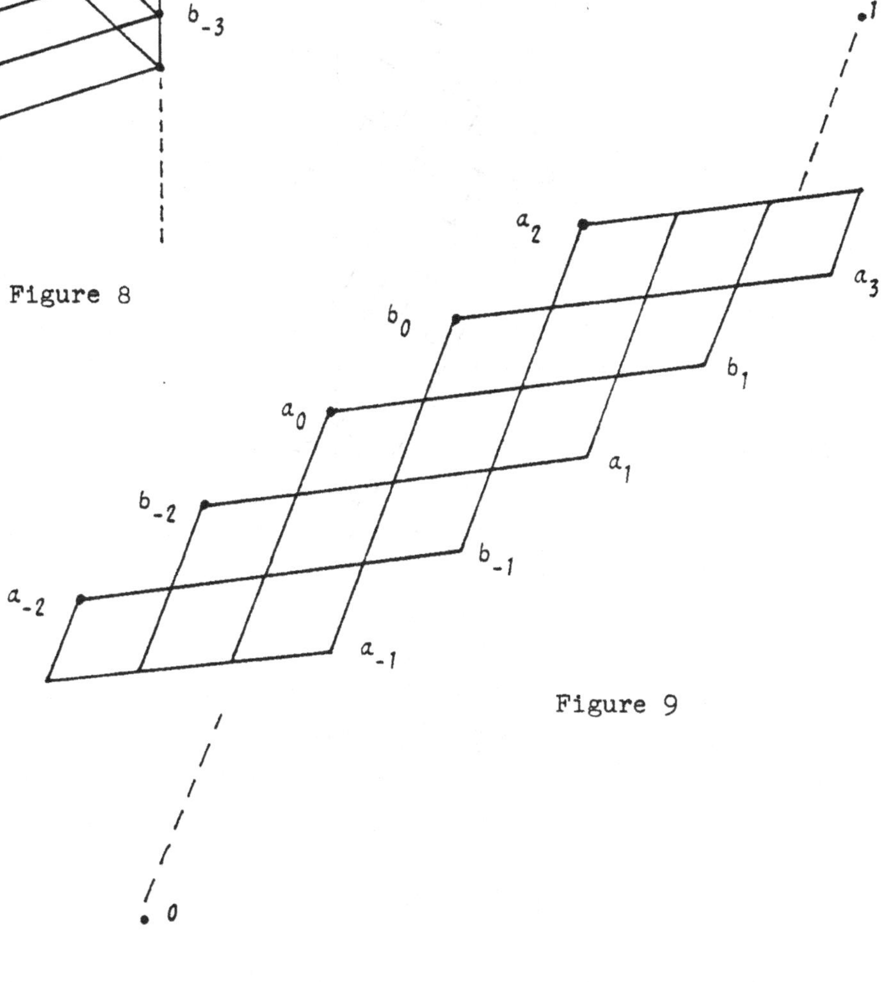

Figure 9

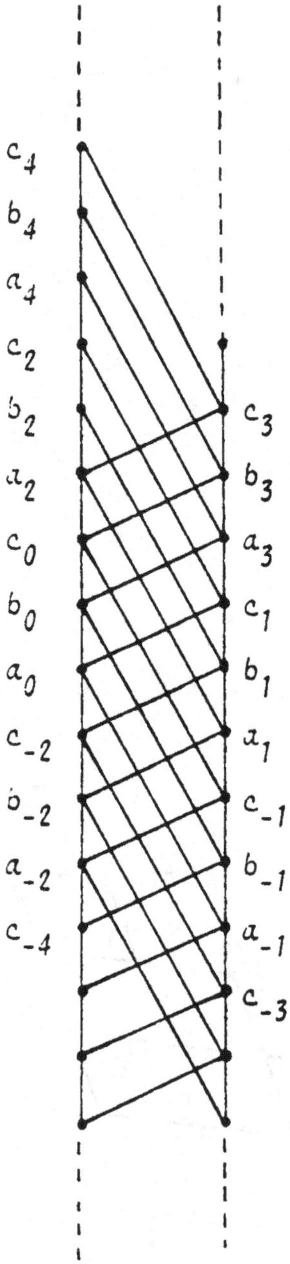

Figure 10

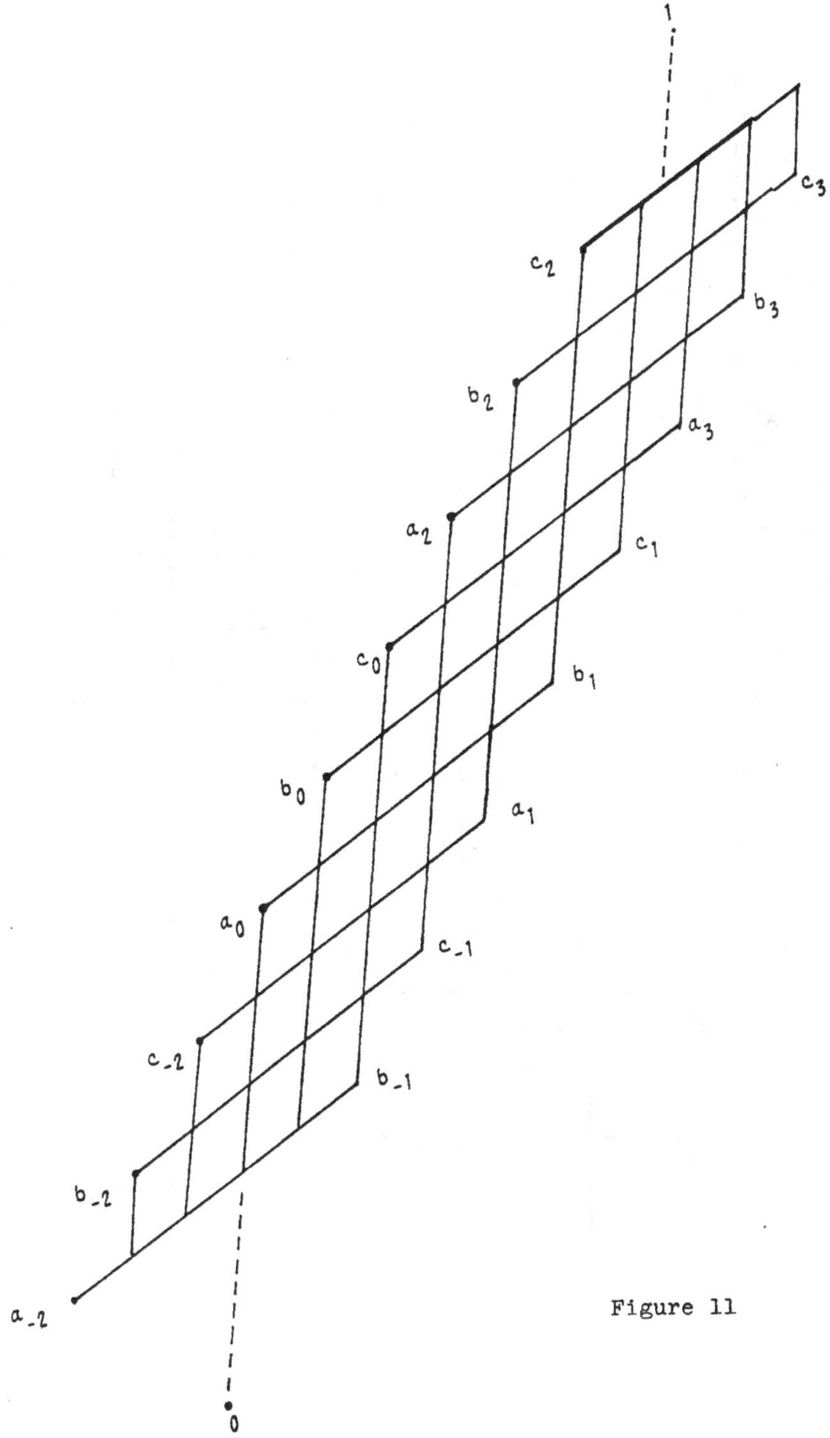

Figure 11

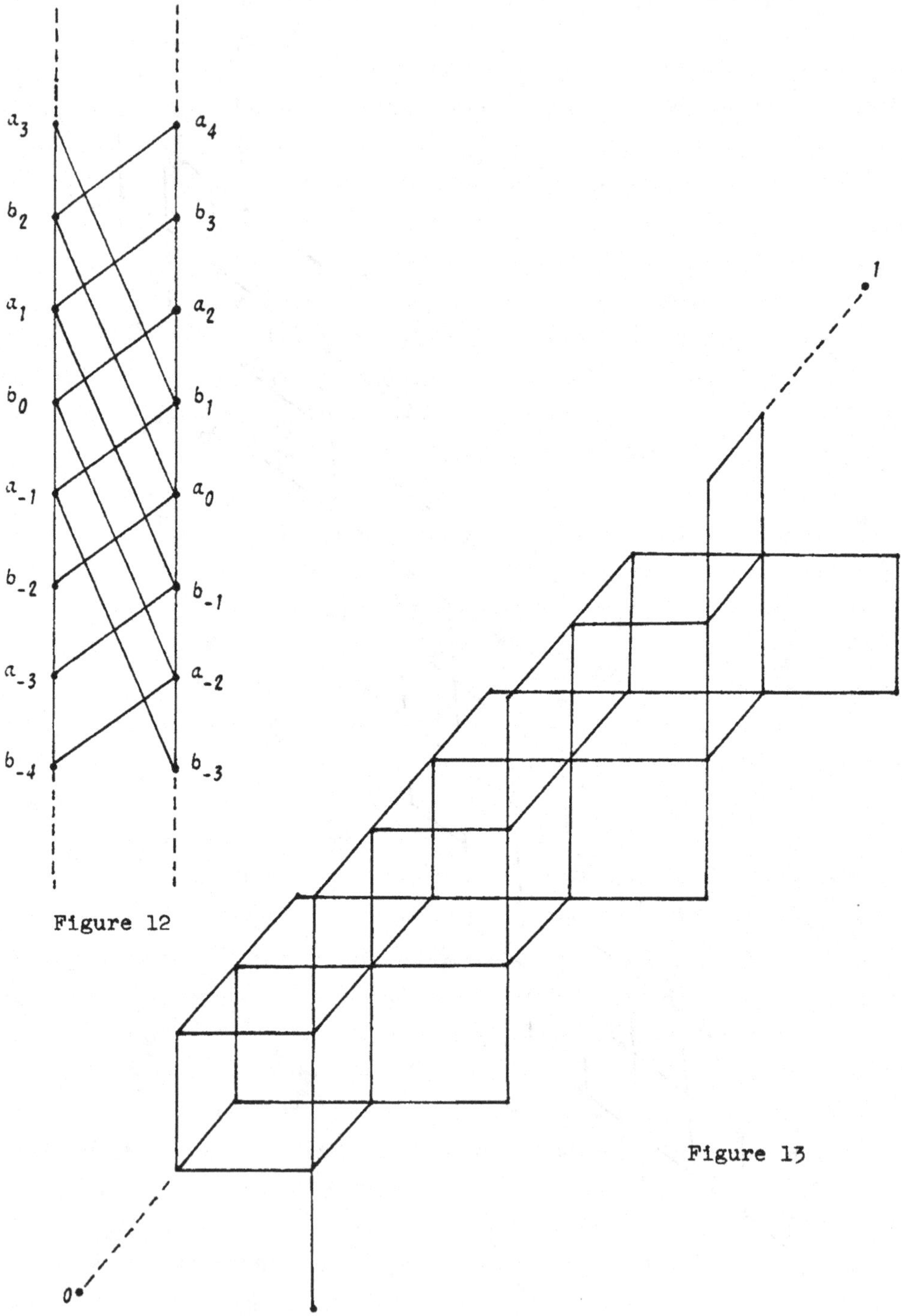

Figure 12

Figure 13

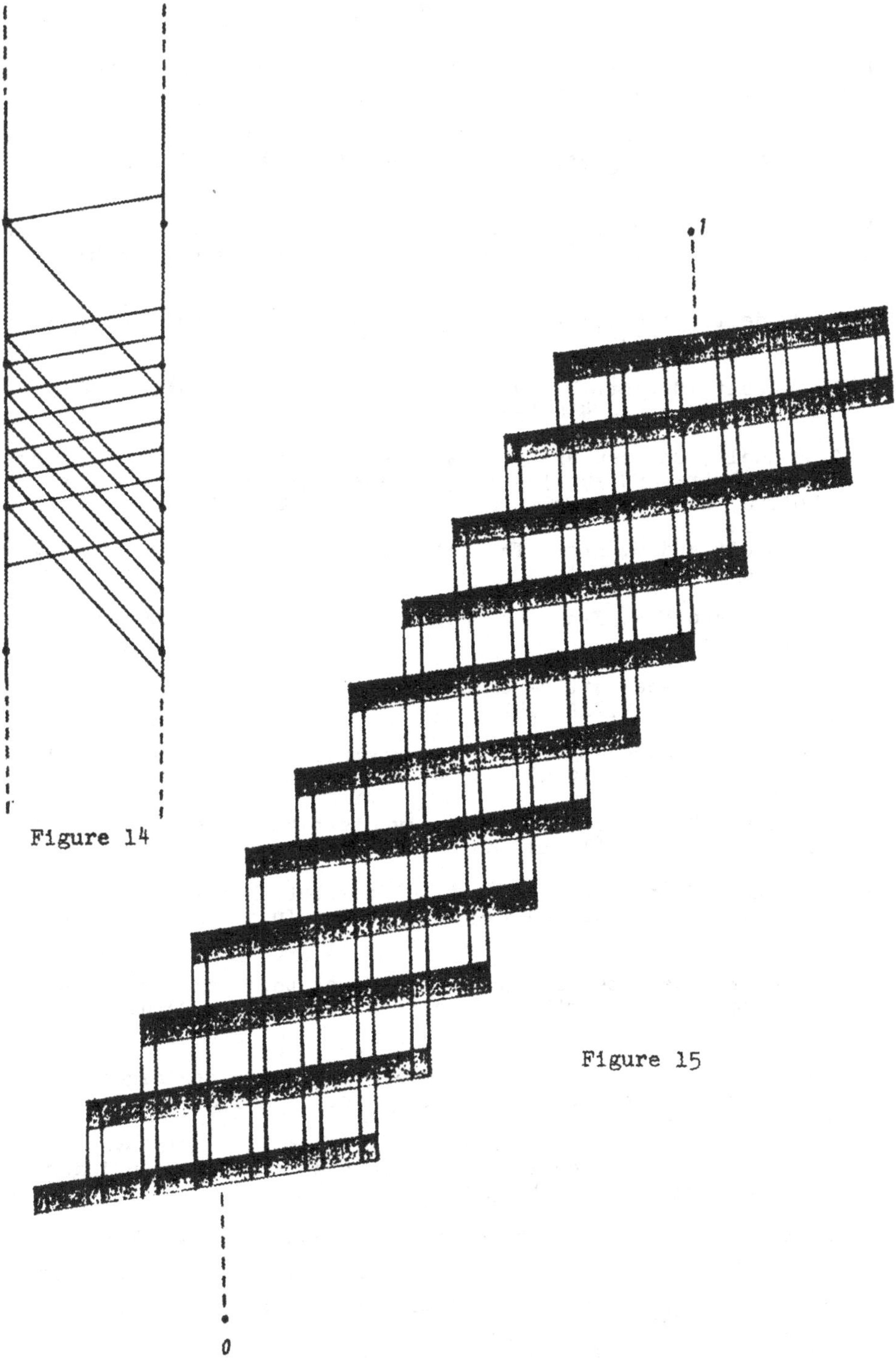

Figure 14

Figure 15

REFERENCES

[1] R. Balbes and Ph. Dwinger, Distributive lattices. University of Missouri Press, 1974.

[2] B. Banaschewski and G. Bruns, Categorical characterization of the MacNeil completion, Archiv der Math. XVIII (1967), Fasc. 4, 369-377.

[3] H.J. Bandelt, On complete distributivity and maximal d-intervals. Coll. Math. Soc., Janos Bolay: 14 Lattice Theory, Szeged, 1974, 29-43.

[4] H.J. Bandelt, Regularity and complete distributivity. Semigroup Forum 19 (1980), 123-126.

[5] H.J. Bandelt, M-distributive lattices. Arch. der Math., 39 (1982), Fasc. 5, 436-441.

[6] H.J. Bandelt, On regularity classes of binary relations. Universal Algebra and applications. Banach Center Publications, Vol. G, PWN- Pol. Sci. Publ., Warsaw, 1982, 329-333.

[7] H.J. Bandelt and M. Erne, The category of Z-continuous posets. J.Pure Appl. Algebra 30 (1983), no. 3, 219-226.

[8] T.S. Blyth and M.F. Janowitz, Residuation theory. Pergamon Press, Oxford, 1972.

[9] B.A. Davey, On the lattice of subvarieties. Houston J. of Math., Vol. 5, 2, 1979, 183-192.

[10] Ph. Dwinger, Classes of completely distributive complete lattices. Indag. Math. 41 (1979), no. 4, 411-423.

[11] Ph. Dwinger, Structure of completely distributive complete lattices Indag. Math. 43 (1981), no. 4, 361-373.

[12] Ph. Dwinger, Characterization of the complete homomorphic images of a complete distributive lattice I. Indag. Math. 44, (1982), no. 4, 403-414.

[13] Ph. Dwinger, Characterization of the complete homomorphic images of a complete distributive lattice II. Indag. Math. 45, (1983), no. 1, 43-49.

[14] L. Geisinger and W. Graves, The category of complete algebraic lattices. J. of Comb. Theory, 13, no. 3, 1972, 332-338.

[15] G. Gierz, K.H. Hofmann, K. Keimel, J.D. Lawson, M. Mislove, D.S. Scott, A compendium of continuous lattices. Springer Verlag, Berlin-Heidelberg-New York, 1980.

[16] G. Gratzer, General lattice theory, Birkhauser Verlag, Basel-Stuttgart, 1978.

[17] G. Markowsky, Free completely distributive lattices. Proc. Amer. Math. Soc. 74 (1979), 227-228.

[18] G.N. Raney, A subdirect-union representation for completely distributive complete lattices. Proc. Amer. Math. Soc., 97 (1953), 518-522.

[19] G.N. Raney, Tight Galois connections and complete distributivity Trans. Amer. Math. Soc., 97 (1960), 418-426.

University of Illinois at Chicago
325 UH (M/C 228)
Post Office Box 4348
Chicago, Illinois 60680

CONNECTED COMPONENTS OF THE COVERING RELATION IN FREE LATTICES

Ralph Freese
University of Hawaii
Honolulu, Hawaii 96822

In a free lattice we let $\sim$ denote the equivalence relation generated by the covering relation, i.e., $u \sim v$ if there is a sequence $v_0 = v, \ldots, v_n = u$, where each v_i covers or is covered by v_{i+1}. The connected components are just the blocks of this equivalence relation viewed as a subposet of the free lattice. We completely characterize these connected components. Although there are some rather complicated examples, they are still quite restricted. For example, neither the 3 element chain nor the 4 element partially ordered set that looks like a Y is a component.

If $X = \{x, y, z\}$, then the connected component of x in FL(X) is the following

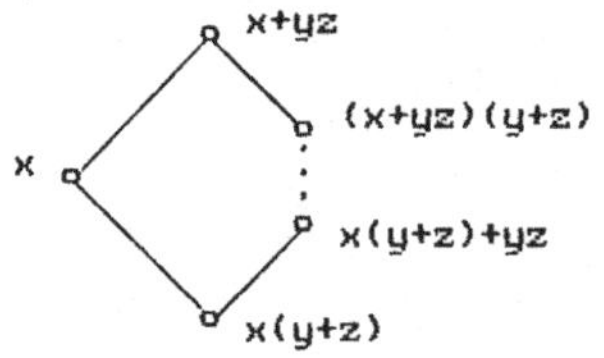

FIGURE 1.

All of the coverings indicated in this diagram are coverings in FL(X) except $(x+yz)(y+z)$ does not cover $x(y+z)+yz$. This is indicated with a dotted line. The unlabelled version of this particular figure plays a special role in the proofs and we will denote it by $\tilde{N}_5$. Also note that the order relation cannot be recovered from the covering relation alone. For this reason we think of the connected components as subposets of FL(X) with the covering relation of FL(X) indicated by solid lines.

This paper makes extensive use of the techniques and results of the author's joint paper with J. B. Nation [5]. That paper develops the theory of covers in free lattices and gives efficient algorithms for finding them. George McNulty suggested the problem of finding the connected components after hearing me lecture on [5]. As an outgrowth of that paper I had developed a LISP program which implements most of the algorithms of [5] using muLISP, a fast LISP for microcomputers. This program, which is capable of finding and drawing the connected component of any lattice word, was used in the present work.

Using [5], it is easy to find the connected components of 0 in $FL(X)$. If $|X| = 2$, $FL(X)$ is 2^2 and is its only component. If $|X| = 3$, the connected component of 0 is

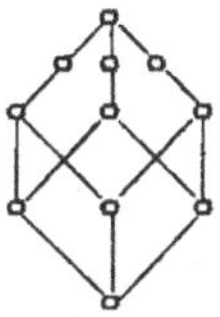

FIGURE 2.

When $|X| = n \geq 4$, *the connected component of* 0 *consists of* 0, *the* n *atoms* $\overline{x}_j = \prod_{i \neq j} x_i$, $j = 1,\ldots,n$, the $\binom{n}{2}$ pairwise joins of the atoms, and $\binom{n}{2}$ additional elements of the form $\prod_{k \neq i,j} (\overline{x}_i + \overline{x}_j + \overline{x}_k) \succ \overline{x}_i + \overline{x}_j$. See the diagrams in [5].

If k is a whole number, let $\bar{N}_5(k)$ be k copies of $\bar{N}_5$, with all the least elements and the elements labelled x Figure 1 above identified. For each k, $\bar{N}_5(k)$ is a connected component. For $k = 3$, this component is

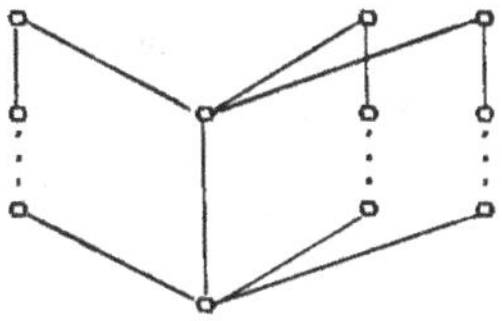

FIGURE 3.

In fact, this is the connected component of xyz in $FL(X)$, $|X| \geq 5$. (xyz is the element in the center, but not the least element, of the figure.) Of course, the dual of this configuration is also a connected component. However, these are the only connected components which contain $\bar{N}_5$. They must be glued exactly as in the diagram.

Let $C(m, k)$ be the poset consisting of m chains of length two, and k chains of length one with their greatest elements identified. Take $C(0, 0)$ to be the one element poset. $C(2, 1)$ is diagrammed below:

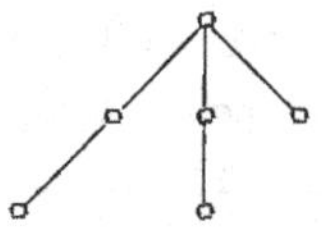

FIGURE 4.

With these preliminaries we can state our main theorem.

THEOREM *Let* C *be a connected component in a free lattice* FL(X). *Then one of the following or its dual holds.*

(a) $|X| = 2$ *and* C = FL(X),

(b) $|X| = 3$ *and* $\emptyset \in C$ *(and* C *is given in Figure 2)*,

(c) $|X| > 3$ *and* $\emptyset \in C$ *(and* C *is as descibed above)*,

(d) C *is* $\bar{N}_5(k)$ *for some* k,

(e) C *is* C(m,k) *for some* $m \geq \emptyset$ *and* $k \geq \emptyset$.

Moreover, all of the above occur as connected components except C(1,∅), *which is the three element chain, and* C(2,∅) *which is two three element chains with their top elements identified.*

1. A REVIEW OF COVERINGS IN FREE LATTICES. In [10] Whitman solved the word problem for free lattices. We assume the reader is familiar with his solution. Part of his solution is the condition (W) : if $v = v_1 \cdots v_n$ and $u = u_1 + \dots + u_m$ then $v \leq u$ in FL(X) if and only if there is an i with $v_i \leq u$ or there is a j with $v \leq u_j$. Note this implies that every element of FL(X) is either meet or join *irreducible*. Whitman also showed that for each $w \in FL(X)$ there is a term of *minimal* length representing w, unique up to commutivity and associativity. If $w = w_1 + \dots + w_k$ is this representation, where each w_i is not formally a join, and if $w = u_1 + \dots + u_m$, then for all $i \leq k$ there is a $j \leq m$, such that $w_i \leq u_j$. In this situation we say that $\{w_1,\dots,w_k\}$ *refines* $\{u_1,\dots,u_m\}$ and we write $\{w_1,\dots,w_k\} \ll \{u_1,\dots,u_m\}$. Also we call $\{w_1,\dots,w_k\}$ the *canonical joinands* of w. It follows from this that *if* $a \prec b$ *in* FL(X) *then there is a unique largest element* c *with* $c \geq a$ *but* $c \ngeq b$. In particular, if w is completely join irreducible, *i.e.,* w *is join irreducible and has a unique lower cover (always denoted)* w_*, then there is a unique largest element, denoted $\kappa(w)$, satisfying $\kappa(w) \geq w_*$ and $\kappa(w) \ngeq w$. If q is completely meet irreducible, we let $\kappa'(q)$ be the map dual to κ. Note κ and κ' are inverses of each other.

In [5] J. B. Nation and the author proved several theorems about covers in free lattices which we will need. If w has the form

$$w = \prod_i (\sum_j w_{ij}) \prod_k x_k$$

where the x_k 's are generators, then $J(w)$ is defined recursively to be $\{w\}$ if w is a meet of variables and $\{w\} \cup \bigcup_{i,j} J(w_{ij})$ otherwise. Theorem 4.3 of [5] characterizes completely join irreducible elements in terms of $J(w)$. In particular, it is shown that *if* w *is completely join irreducible then so is every element of* $J(w)$. Theorem 4.4 shows that if w has the form displayed above and w is completely join irreducible then for each i there is exactly one j with $w_{ij} \not\leq w$. Theorem 4.7 shows that if w is completely join irreducible then the canonical meetands of w_* are $\{\kappa(w)\} \cup \{w_i : w_i \not\geq \kappa(w)\}$.

An element a in a lattice is *totally atomic* if $a < b$ implies there is a c with $a \prec c \leq b$ and the dual condition holds. Lemma 8.1 of [5] shows that there is a strong connection between totally atomic elements and chains of covers in free lattices. If the middle element w of a chain of covers of length two is a meet and w_1 is the canonical meetand of w not above the top of the chain, then w_1 is totally atomic.

Theorem 10.2 gives a useful characterization of three element intervals in free lattices. If w is join irreducible and the middle element of a three element interval $w_* \prec w \prec u$ and $q = q_1 + \ldots + q_m$ $(q_1 \not\leq w)$ is the canonical meetand of w not above u, and $p_1, \ldots, p_k$ are the canonical joinands of u which lie below w, then

$$w = qu = q(\kappa'(q) + \Sigma\, p_i) \quad \text{canonically}$$

and

$$\sum_{i=2}^{m} q_i \leq \sum_{i=1}^{k} p_i < q.$$

2. PROOF OF THE THEOREM. Let C be a connected component. As pointed out above, if $0 \in C$ then it follows from Theorem 6.1 of [5] that either (a), (b), or (c) holds. Thus we assume that neither 0 nor 1 is in C. Moreover, using the algorithm of [5] it is not hard to show that if $|X| \geq 3$ then the connected component of $x \in X$ is $\bar{N}_5$ (see Figure 1.). Hence we also assume that x is not in C. By Theorem 9.2 of [5] the only chains of covers of length greater than two are contained in the components which contain either 0 or 1. Thus we may assume that C *has no chains of covers with more than* 3 *elements*. For the first part of the proof we assume that $\bar{N}_5$ is contained in C and prove that (d) holds in this case. We shall label the elements of the $\bar{N}_5$ as below.

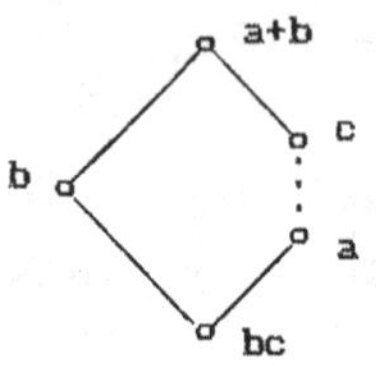

FIGURE 5

LEMMA 1. a *has no upper cover and no lower cover except* bc. *Similarly,* c *has no lower cover and no upper cover except* $a + b$.

Proof. Suppose $e \prec a$, $e \neq bc$. If $ebc \prec bc$, then we obtain a covering chain of length 3, which as pointed out above, cannot occur. So let $ebc < u < bc$. Then $u + e = a > bc$, violating (W). Thus a has no lower cover except bc. Similarly, c has no upper cover except $a + b$.

Now suppose that a is a meet. Then a is completely join irreducible with lower cover $a_* = bc$. Let a_1 be a canonical meetand of a not above c. Let d be the canonical meetand of bc not above b. Then $bd = bc$ and it follows that $c \leq d$. Now $bd = bc = a_* \leq a_1$. If $b \leq a_1$ then $c \leq a + b \leq a_1$, a contradiction. If $d \leq a_1$ then $c \leq d \leq a_1$. Thus neither b nor d is below a_1. In particular, a_1 is not a generator (as generators are meet prime). Let $a_1 = a_{11} + a_{12} + \ldots + a_{1k}$ canonically. By Theorem 4.4 of [5] we may assume that $a_{12} + \ldots + a_{1k} \leq a_*$. Apply (W) to

$$bd \leq a_1 = a_{11} + (a_{12} + \ldots + a_{1k}).$$

If $a_{11} \geq bd = a_*$, then $a_{11} \geq a_{12}$, a contradiction. So we must have $a_* \leq a_{12} + \ldots + a_{1k} \leq a_*$. Thus, since $a_* = bc$ is join irreducible, $k = 2$ and $a_{12} = a_* = bc$. But $a_{12} \in J(a)$ and a is completely join irreducible. Hence, a_{12} has a lower cover a_{12*}. But then $a_{12*} \prec bc \prec b \prec a + b$ is a covering chain of length three, a contradiction. Thus a *is not a meet*. In particular, *there is no element* $e \succ a$ *with* $e \nleq c$. By duality c *is not a join*.

By duality we may assume that b is a meet. (Recall that we are assuming that C contains no generators.) Hence, b is completely join irreducible with unique lower cover $b_* = bc$. Now suppose there an element e with $c \succ e$. Since c is a meet, we must have $e \geq a$. Let $u = a + b$. If there is an element v with $e < v < u$, $v \neq c$, then by the coverings $u = b + v = c + v$. So by Jonsson's semidistributive law, $u = v + bc = v$, a contradiction. Thus in this case $|u/e| = 3$.

Let q be the canonical meetand of c not above u. By Lemma 8.1 of [5] q is totally atomic. Note also that $q^* = q + u$, so $b + q = q^*$. But $b_* = bc \leq c \leq q$.

Hence, $\kappa'(q) = b$. If q were a generator then $b = \kappa'(q)$ would be an atom, contrary to our assumption on C.

Let p_1 be the canonical joinand of u with $p_1 \nleq b = \kappa'(q)$. Then $p_1 \leq a$, so $p_1 + b_* < u$. It follows that $u = p_1 + b$ canonically. Now by Theorem 10.2 of [5], $q = q_1 + \ldots + q_m$ canonically with $q_2, \ldots, q_m \leq p_1$. If for some i, $q_i < p_1$, then $q_i \leq p_{1*} = bp_1 \leq b_* = q\kappa'(q)$. However, it is proved in the proof of Lemma 8.2 that $q_i \leq q\kappa'(q)$ is impossible for totally atomic elements unless $\kappa'(q) = q^*$, which is clearly not the case. Thus $m = 2$ and $q_2 = p_1$.

Let b_1 be the canonical meetand of b not above u. Then b_1 is totally atomic by Lemma 8.1 of [5] and it is easy to see that $\kappa(p_1) = \kappa(q_2) = b_1$. Since q_2 is also totally atomic, Lemma 8.4 of [5] implies that $q_2 = y_0 \cdots y_m$, $y_i \in X$. It follows from the description of totally atomic elements in section 7 of [5] that $q_1 = y_0 \cdots y_{m-1} y_{m+1}$. As is shown in the proof of Lemma 8.4 of [5],

$$b = \kappa'(q) = \prod_{j=0}^{m-1} (\overline{y}_j + \overline{y}_m + \overline{y}_{m+1})$$

(recall $\overline{y}_j = \Pi\, X - \{y_j\}$). This implies that $b_1 = \overline{y}_j + \overline{y}_m + \overline{y}_{m+1}$, for some j. However, $\overline{y}_j + \overline{y}_m + \overline{y}_{m+1}$ is not totally atomic, a contradiction.

Now suppose a has an upper cover. Since a is a join the upper cover is a^* and $a^* \leq c$. As before a^*/bc is a three element chain. Let q be the canonical joinand of a not below bc. Let r be the canonical meetand of b not above $u = a + b$. Then $q_* = qbc \leq r$ and $rq = ruq = qb = qab = q_*$, and $r^* \geq u$. Hence, $r = \kappa(q)$. By Lemma 8.1 of [5] both q and $\kappa(q)$ are totally atomic. By Lemma 8.4 $q = y_0 \cdots y_m$ and $\kappa(q) = x_0 + \ldots + x_k$, where $X = \{y_0, \ldots, y_m, x_0, \ldots, x_k\}$. Using Theorem 4.4 of [5] we may assume that $x_1 + \ldots + x_k \leq b_* = bc$ and $a^* \leq y_1 \cdots y_m$. This implies that either $k = 0$ or $m = 0$. If $m = 0$, $q = y_0$ and $a \geq y_0 + x_1 + \ldots + x_k$. This implies a contains all but one of the generators and so a is in the connected component containing 1, a contradiction. If $k = 0$, then bc is contained in all but one of the generators, again giving a contradiction. This completes the proof of Lemma 1. □

LEMMA 2. *Let* C *contain a copy of* $\bar{N}_5$ *labelled as in Figure 5. If* $e \succ bc$, $e \neq b$, *then* $e + b \succ b$.

<u>Proof</u>. If $e = a$ then the lemma is obviously true. Otherwise, $eb = ab$, so by semidistributivity, $bc = ab = e(a + b)$. Since $u = a + b \succ b$, $b = u(e + b)$. Hence b is completely join irreducible and $b_* = bc$. Let b_{*1} be the canonical meetand of b_* not above e. Since $b_* \prec e$, and b_{*1} is the largest element above

b_* not above e, $b_{*1}(e + b) \prec e + b$. So it suffices to show that $b = b_{*1}(e + b)$. Suppose $b_{*1}(e + b) > b$ and let b_1 be a canonical meetand of b not containing $b_{*1}(e + b)$. Clearly, $b_1 \not\geq e$ so $b_1 e = b_*$ as $e \succ b_*$. Hence $b_1 \leq b_{*1}$. Since $e \leq \kappa(b)$, $b_1 \not\geq \kappa(b)$. Hence, by Theorem 4.7 of [5], $b_1 = b_{*1}$. However, this contradicts $b_1 \not\geq b_{*1}(e + b)$, proving the lemma. □

LEMMA 3. *Suppose that* C *contains* $\bar{N}_5$ *labelled as in Figure 5. Suppose* $e \succ bc$, $e \neq a$, $e \neq b$. *Then* C *contains a copy of* $\bar{N}_5(2)$ *labelled as below.*

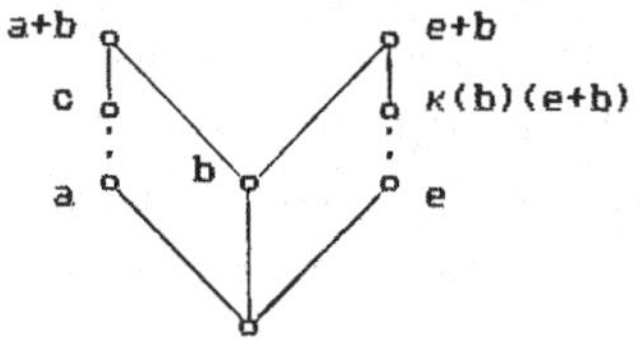

FIGURE 6

Proof. Using the above lemmas, this is routine. □

LEMMA 4. *Suppose that* C *contains* $\bar{N}_5$ *labelled as in Figure 5. Suppose* $d \succ b$ *and let* d_1 *be the canonical joinand of* d *not below* b. *Let* $e = d_1 + bc$. *Then* $e \succ bc$ *and* C *contains the configuration of Figure 6 with* $d = e + b$.

Proof. If the interval d/b_* was a three element chain, then d would be a canonical meetand of b by Theorem 10.2 of [5]. Since $b_* \prec b \prec u$, Lemma 8.1 of [5] implies that d is totally atomic and hence has an upper cover. This implies C contains a covering chain of length three, a contradiction. If $d\kappa(b) = b_*$, then it is easy to see that d/b_* would be a three element chain. Since this is not the case, we obtain $d_1 \leq d\kappa(b)$. The rest is routine. □

LEMMA 5. *If* C *contains a copy of* $\bar{N}_5$ *then* $C = \bar{N}_5(k)$ *for some* k.

Proof. By duality we may assume that b is a meet. Since C contains no chains of length three, $u = a + b$ has no upper cover and bc has no lower cover. By the dual Lemma 2 (or 3) if u had a lower cover other than b or c then b would be a join, contrary to our assumption. By Lemma 1, the only possible new elements of C must come from covers of bc or of b. Moreover, a cover of bc determines a cover of b, and vice versa. By Lemma 3, these two elements are part of an $\bar{N}_5$ attached to the original $\bar{N}_5$ at b and bc. The lemma now follows easily. □

We will now assume that $\bar{N}_5$ is not contained in C and show (e) of the Theorem holds.

LEMMA 6. *Suppose that* $a \prec b$, $c \prec d$, $a < d$, $b \not\leq d$, *and* $c \not\geq a$ *in* FL(X). *Then either there is an* $e \succ d$ *or an* $f \prec a$.

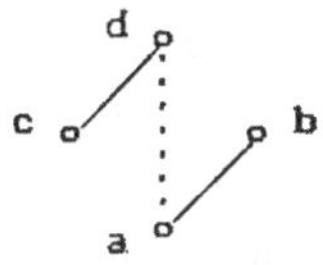

FIGURE 7

Proof. We have $d < d + b$ and $ac < a$. If neither of these is a cover, choose $d < d_1 < d + b$ and $a > a_1 > ac$. Then $d_1 b = a < d = c + a_1$, violating (W). □

LEMMA 7. *If* $a, b \in C$ *and the sequence of coverings witnessing this alternates at least twice (between up and down), then* $\bar{N}_5$ *is contained in* C.

Proof. The hypotheses guarantee that there is either a sequence of the form $a \prec b \succ c \succ d \prec e$, or a similar sequence without c. Assume the former. Since C has no covering chain of length three, Lemma 6 says that either $a \geq d$ or $b \geq e$. By duality, we may assume $a \geq d$. So we have

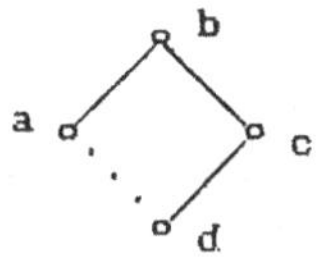

FIGURE 8

Let f be the smallest element below b but not below c. Then $d + f \succ d$, as usual. Also $d + f \leq a$. In fact, we must have $d + f < a$, since the four element covering Boolean algebra can only occur at the bottom or top of the free lattice (cf. Theorem 10.1 of [5]). Now $\{a, b, c, d, d + f\}$ forms an $\bar{N}_5$. With the aid of Lemma 6, the other case can be handled similarly. □

LEMMA 8. *Let* $a, b \in C$ *with* $a \leq b$. *If* $\bar{N}_5$ *is not contained in* C *then either* $a \prec b$ *or there is a* c *with* $a \prec c \prec b$.

Proof. By the previous lemma we must have one of the following two situations (with solid lines indicating coverings).

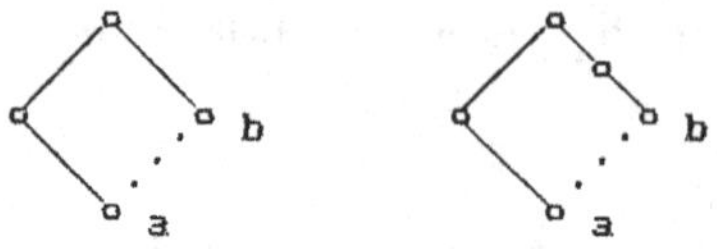

FIGURE 9

Note that in either case we obtain the first configuration and this easily gives $\bar{N}_5$. □

LEMMA 9. *If* C *does not contain* $\bar{N}_5$, *then* C *is a tree.*

Proof. If $a, b, c, d \in C$ with $d \geq a, b$, and $c \leq a, b$, then by the previous lemma, we would get a covering four element Boolean algebra. As pointed out above these only occur at the top and bottom of the free lattices, and are not connected components even there. □

Finally, to show that (e) of the main theorem holds all we need to do is show that the configuration below cannot occur.

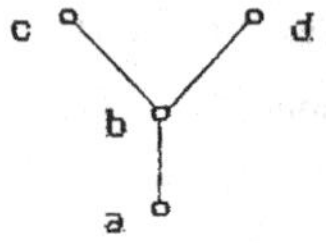

FIGURE 10

However, it easy to see that if $\bar{N}_5$ is not contained in C then c/a in the figure must be a three element chain. Now by Theorem 10.2 of [5], c is a canonical meetand of b. This implies that $c \prec c + d$, which gives a chain of length three in C, a contradiction.

LEMMA 10. $C(1,0)$, *the three element chain, is never a connected component.*

Proof. Let w be the middle element of a three element covering chain. We may assume that w is a meet so that w is completely join irreducible with lower cover w_*. Let u be the cover of w in the chain. As we have seen before, if u/w_* is not a three element chain, then $\bar{N}_5$ is contained in C. By Theorem 10.2 of [5], u is a canonical meetand of w. Hence, each canonical joinand of u is in $J(w)$. Since w has a lower cover, each joinand of u does. Now by Theorem 4.4 and Corollary 2.4 of [5], u has a lower cover other than w. Hence, $\{w_*, w, u\}$ is not a connected component. □

LEMMA 11. $C(2,\emptyset)$ *is not a connected component.*

Proof. Suppose we had the following connected component:

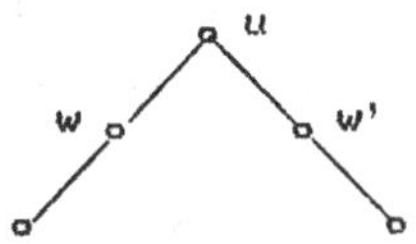

FIGURE 11

As usual, if the interval from u to the lower cover of w is not a three element chain, then $\bar{N}_5$ is contained in C. By the proof of the last lemma, if w were a join then the lower cover of w would have an upper cover different from w. Thus w and w' are both meets. By Theorem 10.2 of [5] there are totally atomic elements q and q' with $w = qu$ and $w' = q'u$ canonically. Moreover,

$$u = \kappa'(q) + \kappa'(q')$$

canonically. Indeed, by Theorem 10.2 both $\kappa'(q)$ and $\kappa'(q')$ are canonical joinands of u. Also by Theorem 10.2 each canonical joinand of u is completely join irreducible. So the number of canonical joinands of u is the number of lower covers, two.

So in the notation of Theorem 10.2 of [5], $k = 1$ and $p_1 = \kappa'(q')$. By that theorem $q_2,\ldots,q_m \le p_1 = \kappa'(q')$. This inequality is strict unless $m = 2$ and $q_2 = \kappa'(q')$. But q_2 is totally atomic. So by Lemma 8.4 of [5], this can only happen *if* q' *is a join of generators.*

Suppose $q' = y_1 + y_2 + \ldots + y_m$, $m > 1$, and $q_2 = p_1 = \kappa'(q') = x_1\ldots x_n$. By the nature of totally atomic elements, this forces $q_1 = y_1x_2\ldots x_n$, which implies $\kappa'(q) = \prod_{i=2}^{n} (\bar{x}_i + \bar{x}_1 + \bar{y}_1)$. By Theorem 10.2, for $i \ge 2$, $y_i = q_i' \le p_1' = \kappa'(q)$, which implies $y_2 \le \prod (\bar{x}_i + \bar{x}_1 + \bar{y}_1)$, a contradiction.

If $q' = y_1$ and $q_2 = p_1 = \kappa'(q')$, then q_2 is the meet of all the generators except y_1. Hence $q = q_1 + q_2$ is the join of two atoms. But $w \le q$, implies we are in the connected component containing $\emptyset$ which is certainly not $C(2,\emptyset)$.

Now the following conditions hold:

(1) $$q_i \le q' \text{ and } q_i' \le q \quad \text{for } i > 1,$$

(2) $$q_1 \nleq q' \text{ and } q_1' \nleq q$$

(3) $$\{q_1, q_2, \ldots, q_m\} \cap \{q'_1, q'_2, \ldots, q'_{m'}\} = \emptyset$$

Since $q_i < \kappa'(q')$, $q_i \le \kappa'(q')_* \le q'$ and thus (1) holds. Since q and q' are incomparable, (2) holds. To see (3), note by (2) that q_1 cannot be in the intersection. If q_i were, then $q_i \le \kappa'(q)$, which cannot happen with a totally atomic element by Lemma 8.2 of [5]. Now conditions (1) and (2) together with Lemma 7.6 of [5] imply that $p = qq'$ is totally atomic. Moreover, it is easy to use Whitman's test for canonical form, to see that the canonical form of p is qq'. However, the description of totally atomic elements given in §7 of [5] show that (3) cannot hold, proving the lemma. □

The only thing remaining in the proof of the Theorem is to show that the connected components described in (d) and (e) do occur. If w is the meet of k generators and $|X| \ge k+2$ then the connected component of w in FL(X) is $\bar{N}_5(k)$. To realize $C(m,\emptyset)$, with m at least 2, choose X with at least four elements and $|X| \ge m$. Let w be the join of m distinct atoms of FL(X). Then the connected component of w is $C(m,\emptyset)$. To realize $C(m,k)$, $m, k \ge 1$, let w be the join of m atoms of FL(X), $|X| \ge 4$, and let $Y = \{y_1, y_2, \ldots, y_k\}$ be disjoint from X. Then $u = w + y_1 + \ldots + y_k$ has $C(m,k)$ as its connected component in FL(X ∪ $\{y_1, \ldots, y_k\}$). For $C(\emptyset,k)$, $k \ge 3$, let w be the join of k atoms in FL(X) with $|X| \ge 4$. Let Y be disjoint from X and be nonempty. Then the connected component of w in FL(X ∪ Y) is $C(\emptyset,k)$. The two element chain, $C(\emptyset,1)$, is acheived by $(x + yz)(y + xz)$ in FL(3). Finally, $C(\emptyset,2)$ is acheived by

$$(x + yz)(y + xz) + (r + st)(s + rt)$$

in FL(6).

To see that the connected component C of $w = x_1 \ldots x_k$ in FL(X) is $\bar{N}_5$ when $|X| \ge k + 2$, simply note that since w is totally atomic it has a lower cover and exactly k upper covers. Since $|X| \ge k + 2$, neither 0 nor 1 is in C. Hence either (d) or (e) of the theorem applies, and it follows that $C = \bar{N}_5$. We leave the verification of the other facts as an exercise.

3. REMARKS. There are several open problems on connected components. For example, exactly where do they occur, i.e., what are the labelled connected components? Also what are the connected components of FL(X) for X fixed? Since the number of totally atomic elements of FL(X) is finite, there is a bound on k in condition (d) and a bound on m in (e).

REFERENCES

[1] P. Crawley and R. P. Dilworth, *Algebraic Theory of Lattices*, Prentice-Hall, Englewood Cliffs, NJ, 1973.

[2] A. Day, Splitting lattices generate all lattices, Algebra Universalis, **7**(1977), 163-169.

[3] R. Freese, Some order theoretic questions about free lattices and free modular lattices, Proc. of the Banff Symposium on Ordered Sets, D. Reidel Publishing Company, Holland, 1982.

[4] R. Freese and J. B. Nation, Projective lattices, Pacific J. Math., **75**(1978), 93-106.

[5] R. Freese and J. B. Nation, Covers in free lattices, Trans. Amer. Math. Soc., to appear.

[6] G. Gratzer, *General Lattice Theory*, Academic Press, New York, 1978.

[7] B. Jonsson, Varieties of lattices: Some open problems, Colloq. Math. Soc. Janos Bolyai, **29**(1982), Contributions to Universal Algebra (Esztergom), North Holland, 421-436.

[8] B. Jonsson and J. B. Nation, A report on sublattices of a free lattice, Colloq. Math. Soc. Janos Bolyai, **17**(1977), Contributions to Universal Algebra (Szeged), North Holland, 223-257.

[9] R. McKenzie, Equational bases and nonmodular lattice varieties, Trans. Amer. Math Soc., **174**(1972), 1-43.

[10] P. Whitman, Free lattices, Annals of Math., **42**(1941), 325-330.

[11] P. Whitman, Free lattices II, Annals of Math., **43**(1942), 104-115.

VARIETIES WITH LINEAR SUBALGEBRA GEOMETRIES

Bernhard Ganter and Thomas Ihringer
Technische Hochschule
6100 Darmstadt
Federal Republic of Germany

1. INTRODUCTION. One obtains the *subalgebra geometry* of an algebra by taking all subalgebras as subspaces of the geometry. This concept appeared under several different denotations in the literature; see e.g. Ganter, Werner [4] and [5], Osborn [10], Quackenbush [13], Stein [14] and Wille [17]. In this paper the properly 2-generated subalgebras will also be called *blocks* of the subalgebra geometry. A subalgebra geometry is called *linear* if two blocks never contain each other properly. The topic of this paper are *subalgebra linear varieties*, i.e. varieties in which each algebra has a linear subalgebra geometry. The main result will be a classification of all subalgebra linear varieties which satisfy the additional property that all blocks of all algebras in such a variety are finite (Theorem 3.8). The classification is given in terms of the 2-generated free algebras of these varieties. This algebra is either trivial, in the sense that it does not contain a properly binary term, or all its term functions are idempotent polynomial functions of a nearfield, or all its term functions are polynomial functions of a vector space. Thus one does not obtain any "new" examples of subalgebra linear varieties.

Such varieties with at most one algebraic constant have been investigated by Quackenbush [13] and Ganter, Werner [4]. Their interest was focused on the coordinatization of block designs: The subalgebra geometries in subalgebra linear varieties can be regarded as block designs (with the properly 1-generated subalgebras as points). This aspect will be treated in the last part of this paper (Theorem 3.9). Wille [17] and Müller [9] obtained results closely related to those given here. They examined subalgebra linear varieties with further restrictions (but without any finiteness assumptions).

Section 2 generalizes a result used e.g. in Wille [17], Pasini [12], and also in [7] and [8] for the examination of linear congruence class geometries. As a consequence of Theorem 2.2 one obtains that each unary admissible operation of a finite algebra is either constant or a permutation provided that this algebra has "sufficiently many" congruences with pairwise trivial intersection. The proof of the non-idempotent case of Theorem 3.8 depends to a great extent on the application of Theorem 2.2 and of a strong result of Pálfy [11]. However, the idempotent case was already settled in [4].

For the denotations of universal algebra the reader is referred to Burris, Sankappanavar [1] and Grätzer [6].

2. UNARY ADMISSIBLE OPERATIONS OF SEMINETS. Let A be a set and let $\Omega \subseteq Eq(A)$ be a set of equivalence relations of A. The pair (A,Ω) is called a *seminet* if $id_A \notin \Omega$ and if any two different equivalence relations $\Theta_1, \Theta_2 \in \Omega$ intersect trivially, i.e. if $\Theta_1 \cap \Theta_2 = id_A$. The equivalence classes of equivalence relations of Ω can be regarded as *lines*, and the elements of A as *points*. Equivalently, a seminet can be defined as a set of points with certain subsets called lines such that any two distinct lines intersect in at most one point, and with the additional property that the lines can be partitioned into parallel classes. The points $a,b \in A$ are called *collinear*, in symbols $a \sim b$, if $a \Theta b$ for some $\Theta \in \Omega$. A mapping $\delta: A \to A$ is called Ω-*admissible* if $a \Theta b$ with $\Theta \in \Omega$ always implies $\delta a \Theta \delta b$. For each subset $S \subseteq A$ of points let $E(S) := S \cup \{c \in A \mid \exists\, \Theta_1, \Theta_2 \in \Omega,\ \Theta_1 \neq \Theta_2,\ a,b \in S: c \in [a]\Theta_1 \cap [b]\Theta_2\}$. Let $S_o := S$ and, for each positive integer i, $S_i := E(S_{i-1})$. Define a closure operator C_Ω on A by $C_\Omega(S) := \bigcup_{i=o}^{\infty} S_i$. A subset T of A is called an Ω-*subspace* if $T = E(T)$. Obviously, $C_\Omega(S)$ is the smallest Ω-subspace containing S.

Let δ and γ be unary Ω-admissible operations on A with the property that their restrictions $\delta|_S$ and $\gamma|_S$ onto S are equal. Then one can conclude $\delta|_{C_\Omega(S)} = \gamma|_{C_\Omega(S)}$: Assume $\delta|_{S_{i-1}} = \gamma|_{S_{i-1}}$, and let $c \in S_i \setminus S_{i-1}$. Then there are distinct $\Theta_1, \Theta_2 \in \Omega$ and $a,b \in S_{i-1}$ with $c \in [a]\Theta_1 \cap [b]\Theta_2$. Hence $\{\delta c, \gamma c\} \subseteq [\delta a]\Theta_1 \cap [\delta b]\Theta_2$, and thus $\delta c = \gamma c$. This implies $\delta|_{S_i} = \gamma|_{S_i}$, and induction on i yields $\delta|_{C_\Omega(S)} = \gamma|_{C_\Omega(S)}$. Assume now all Ω-subspaces of (A,Ω) to be *trivial* (i.e. to contain either at most one or all points of A), and let S be a two-element set. Then $C_\Omega(S) = A$, and $\delta|_S = \gamma|_S$ implies already $\delta = \gamma$. This completes the proof of the following lemma, thus generalizing a result on p.38 of Wille [16].

LEMMA 2.1. *Let all Ω-subspaces of the seminet (A,Ω) be trivial. Then each unary Ω-admissible operation is uniquely determined by the images of any two distinct points. In particular, each such operation is either constant or injectice.*

Assume the seminet (A,Ω) now to be finite, i.e. A to be a finite set. Let $m := |A|$ and $g := \max_{a \in A, \Theta \in \Omega} |[a]\Theta|$, i.e. m is the number of *all* points and g the the maximum number of points contained in a line. For each point b denote the set of all points being *not* collinear with b by $N(b)$. Let $n := \max_{b \in A} |N(b)|$, i.e. n is the smallest number such that each point is noncollinear with at most n points.

THEOREM 2.2. *Let the finite seminet (A,Ω) satisfy*

(1) $\qquad 2n \leq m - 2g + 1$.

Then each unary Ω-admissible operation is either a constant mapping or a permutation. Hence the nonconstant such operations form a permutation group on A.

Proof. In order to apply Lemma 2.1 it is sufficient to prove that all Ω-subspaces are trivial. Assume on the contrary that T is a nontrivial Ω-subspace.

Choose $c \in A \setminus T$ and $\Theta \in \Omega$ with $h := |[c]\Theta \cap T| = \max_{y \in A\setminus T, \Psi \in \Omega} |[y]\Psi \cap T|$. Let $a,b \in T$, $a \neq b$. For a,b collinear let G be the line through a and b, and define $G := \{a,b\}$ if a and b are noncollinear. Furthermore, assume $a,b \in [c]\Theta$ if $h \geq 2$, i.e. in this case $G = [c]\Theta$. Then

$$T \supseteq ([c]\Theta \cap T) \cup E(a,b) = ([c]\Theta \cap T) \cup \{a,b\} \cup [A \setminus (N(a) \cup N(b) \cup G)],$$

and thus $t := |T| \geq m - 2n - g + \max\{2,h\} \overset{(1)}{\geq} g - 1 + \max\{2,h\}$. Hence $g - 1 \geq h$ implies

(2) $\qquad t > h, \quad t \geq 2h.$

From $T \subseteq N(c) \cup ([c]\Theta \cap T)$ one obtains

(3) $\qquad t \leq n + h.$

Now the number of elements of $W := \{(x,y) \in T \times (A\setminus T) \mid x \sim y\}$ will be counted in two different ways. Since $W \supseteq (T \times (A\setminus T)) \setminus \bigcup_{x \in T}(\{x\} \times N(x))$, one obtains $|W| \geq t(m-t) - tn$. On the other hand, $W = \bigcup_{y \in A\setminus T}\{(x,y) \mid x \in T, x \sim y\}$ implies $|W| \leq (m-t)h$. Therefore $t(m-t) - tn \leq (m-t)h$, i.e. $m - t \leq \frac{t}{t-h}n$ by (2). Hence $t \geq m - \frac{t}{t-h}n \overset{(1)}{\geq} 2n + 2g - 1 - \frac{t}{t-h}n$ $= 2g - 1 + \frac{t-2h}{t-h}n > 2h + \frac{t-2h}{t-h}$. Thus $t - 2h > \frac{t-2h}{t-h}n$, and (2) yields $1 > \frac{n}{t-h}$, contradicting (3). □

Let (A,F) be a finite algebra and $\Omega \subseteq \mathrm{Con}(A,F)$ a set of congruences of (A,F) such that (A,Ω) is a seminet with $2n \leq m - 2g + 1$. Then the above theorem implies all nonconstant unary admissible operations of (A,F) to be permutations. In this situation (and if, moreover, $|A| \geq 3$ and if (A,F) has an essentially binary term function) one can apply a result of Pálfy [11]: There is a vector space structure $(A,+,K)$ on A (with K the associated field) such that all term functions of (A,F) are of the form $(x_1,x_2,\dots,x_n) \mapsto a + \lambda_1 x_1 + \lambda_2 x_2 + \dots + \lambda_n x_n$, with $a \in A$, $\lambda_1,\lambda_2,\dots,\lambda_n \in K$. Notice that, for the special case of algebras with linear congruence class geometries, this result was proved in [8].

3. SUBALGEBRA LINEAR VARIETIES. For each positive integer k, the k-generated free algebra of a variety V will be denoted by $F_k(V)$ or F_k. In the rest of this paper $F_0(V)$ will denote the subalgebra of $F_1(V)$ consisting of all constants. Recall from Section 1 that the subalgebra geometry of an algebra (A,F) is called *linear* if two properly 2-generated subalgebras never contain each other properly. A variety is called *subalgebra linear* if each of its algebras has a linear subalgebra geometry. The following proposition can easily be proved, using the defining properties of subalgebra linearity.

PROPOSITION 3.1. *A variety V is subalgebra linear if and only if the 2-generated free algebra $F_2(V)$ has a linear subalgebra geometry.*

PROPOSITION 3.2. *Let* V *be a subalgebra linear variety with* $F_2(V)$ *finite. Then*

a) $F_1(V)$ *is freely generated by each* $t \notin F_0(V)$,

b) $F_2(V)$ *is freely generated by each pair* s,t *with* $s \notin \langle t\rangle$ *and* $t \notin F_0(V)$.

Proof. Let x and y be the free generators of F_2, and assume $t \in F_2 \setminus \langle y\rangle$. Let ϕ be the endomorphism of F_2 with $\phi x = t$ and $\phi y = y$. From $\phi F_2 = \langle t,y\rangle \supsetneq \langle y\rangle \cong F_1$ and the finiteness of F_2 one obtains that ϕF_2 cannot be 1-generated. Linearity therefore implies $\phi F_2 = F_2$. Hence the finiteness of F_2 proves ϕ to be an automorphism, and thus F_2 to be freely generated by t and y. Analogously, one can now replace the free generator y by any $s \notin \langle t\rangle$, thus proving part b) of the proposition. In order to settle part a), assume now $t \in \langle x\rangle \setminus \langle y\rangle$. The restriction of ϕ to $\langle x\rangle$ is then an injective endomorphism and thus also an automorphism of $\langle x\rangle \cong F_1$, i.e. $\langle x\rangle = \phi\langle x\rangle = \langle t\rangle$. □

By the above two propositions it is obvious that each algebra in a subalgebra linear variety V with finite $F_2(V)$ co-ordinatizes a *block design* which consists of the properly 2-generated subalgebras of this algebra as *blocks* and the properly 1-generated subalgebras as *points*, with set inclusion as incidence. In this way one obtains $(2,k)$-designs with k the number of properly 1-generated subalgebras of $F_2(V)$. In particular, *all* block designs co-ordinatized by V have the same block size k. In Theorem 3.9 it will be shown that either k or $k-1$ must be a prime power.

The following corollary is due to Evans, see e.g. [3].

COROLLARY 3.3. *Let* x *and* y *be the free generators of the free algebra* $F_2(V)$. *Then, under the assumptions of Proposition 3.2, each term* $p(x,y) \notin \langle x\rangle \cup \langle y\rangle$ *of* $F_2(V)$ *is a quasigroup term of* V. *As a consequence, if* $F_2(V) \neq \langle x\rangle \cup \langle y\rangle$ *then all congruences of algebras of* V *are uniform, i.e. all congruence classes of each such congruence have the same cardinality.*

Notice that the next two lemmas are valid without any restriction on the variety V. The proofs are an easy exercise.

LEMMA 3.4. *Let* U *be a subalgebra and* Θ *a congruence of the algebra* (A,F). *Then* $[U]\Theta := \bigcup_{u\in U}[u]\Theta$ *is a subalgebra of* (A,F).

LEMMA 3.5. *Let* V *be a variety, and* a,b *a pair of free generators of* $F_2(V)$. *Then each congruence class of* $\Theta(a,b)$ *intersects the subalgebra* $\langle a\rangle$ *in exactly one element.*

LEMMA 3.6. *Let the variety* V *satisfy*

(i) all congruences of algebras of V *are uniform,*

(ii) $F_2(V)$ *is freely generated by each pair* s,t *with* $s \notin \langle t\rangle$, $t \notin F_o(V)$.

Then any two different congruences of $F_2(V)$ *contained in* $\Omega := \{\Theta(a,b) \mid a \notin \langle b\rangle,\ b \notin F_o\}$ *intersect trivially, i.e.* $(F_2(V),\Omega)$ *forms a seminet.*

Proof. Let a,b and c,d satisfy $a \notin \langle b\rangle$, $b \notin F_o$ and $c \notin \langle d\rangle$, $d \notin F_o$, and let $\Psi := \Theta(a,b) \cap \Theta(c,d)$. By Lemma 3.4 $[\langle a\rangle]\Psi$ is a subalgebra of F_2. By property (ii) either $[\langle a\rangle]\Psi = \langle a\rangle$, or $[\langle a\rangle]\Psi = F_2$. In the first case congruence uniformity and Lemma 3.5 imply $\Psi = id_{F_2}$. In the second case Lemma 3.5 yields $\Theta(a,b) = \Theta(c,d)$. □

LEMMA 3.7. *Let the variety* V *satisfy the assumptions of Lemma 3.6, and let* a,b *be a pair of free generators of* $F_2(V)$.

If $F_o(V) = \emptyset$, *then each congruence class of* $\Theta(a,b)$ *intersects each subalgebra* $\langle c\rangle$ *with* $c \in F_2(V)$ *in exactly one element.*

If $F_o(V) \neq \emptyset$, *then there is an* $r \in F_2(V) \setminus F_o(V)$ *with* $[F_o(V)]\Theta(a,b) = \langle r\rangle$, *and each congruence class of* $\Theta(a,b)$ *intersects each subalgebra* $\langle c\rangle$ *with* $c \in F_2(V) \setminus \langle r\rangle$ *in exactly one element.*

Proof. At first, let $F_o = \emptyset$. Then $[\langle a\rangle]\Theta(a,b) = F_2$ by Lemma 3.5. Hence for each $c \notin \langle a\rangle$ there is an element $d \in \langle a\rangle$ with $c\ \Theta(a,b)\ d$. Lemma 3.6 then implies $\Theta(a,b) = \Theta(c,d)$, and by Lemma 3.5 $\Theta(c,d)$ has the claimed property.

Let now $F_o \neq \emptyset$. The Lemmas 3.4 and 3.5 and the congruence uniformity of V imply $F_o \subsetneq [F_o]\Theta(a,b) \subsetneq F_2$. Hence $[F_o]\Theta(a,b) = \langle r\rangle$ for some $r \in F_o$. The rest can be shown similarly as in the case $F_o = \emptyset$. □

THEOREM 3.8. *Let* V *be a subalgebra linear variety with* $F_2(V)$ *finite. Assume* $F_2(V) \neq \langle x\rangle \cup \langle y\rangle$ *with* x *and* y *the free generators of* $F_2(V)$.

If V *is an idempotent variety, then a nearfield structure* $(F_2(V),+,\cdot)$ *with subtraction* $-$ *can be defined on* $F_2(V)$ *such that the binary term functions of* $F_2(V)$ *are exactly the mappings of the form*

$$(x,y) \longmapsto x-(x-y)\lambda$$

with $\lambda \in F_2(V)$.

If V *is not idempotent, then a vector space structure* $(F_2(V),+,K)$ *over some field* K *with vector addition* $+$ *can be defined on* $F_2(V)$ *such that each term function* t *of* $F_2(V)$ *is a mapping of the form*

$$(x_1,x_2,\dots,x_n) \longmapsto a+\lambda_1x_1+\lambda_2x_2+\dots+\lambda_nx_n$$

with n *the arity of* t *and* $a \in F_2(V)$, $\lambda_1,\lambda_2,\dots,\lambda_n \in K$.

Proof. The idempotent case was already settled in [4], using Zassenhaus' results on the equivalence of sharply doubly transitive permutation groups with nearfields.

By Proposition 3.2 and Corollary 3.3 the variety V satisfies the assumptions of the last two lemmas. For V nonidempotent it will be shown that the seminet $(F_2(V),\Omega)$ of Lemma 3.6 satisfies the assumptions of Corollary 2.3. Let $f_i := |F_i(V)|$ for $i=0, 1,2$. At first, assume $f_0 = 0$ (and $f_1 > 1$, as V is nonidempotent). By Lemma 3.7 each congruence class of a congruence of Ω intersects each 1-generated subalgebra of F_2 in exactly one element. Hence g is the number of such subalgebras, $m = f_2 = f_1 g$ and $n = f_1 - 1$. Moreover, $f_1 > 1$ and $F_2 \neq \langle x\rangle \cup \langle y\rangle$ imply $3 \leq g \leq f_1$. Therefore $1 \leq (f_1-2)(g-2)$, and thus $2n \leq m - 2g + 1$. Assume now $f_0 \neq 0$. As above one obtains $m = f_2 = f_1 g$ and, similarly, $f_1 = f_0 g$. Let $d \in F_0$. By Lemma 3.7 there is an $r \in F_2 \setminus F_0$ such that d and r are collinear in the seminet (F_2,Ω). The automorphism group of F_2 operates transitively on $F_2 \setminus F_0$. Hence d is collinear with each element of $F_2 \setminus F_0$, and thus $n = f_0 - 1$. From $0 \leq (g-1)^2$ one obtains $2g-3 \leq g^2-2 \leq f_0(g^2-2) = f_1 g - 2f_0$, implying $2n \leq m - 2g + 1$. □

Theorem 3.8 (and its proof) yield the existence of a prime power q and of a nonnegative integer d such that

either $\quad f_0 = 0, \quad f_1 = q^d, \quad f_2 = q^{d+1},$

or $\quad f_0 = q^d, \quad f_1 = q^{d+1}, \quad f_2 = q^{d+2}.$

(Conversely, for each such q and d a subalgebra linear variety with the above parameters can be constructed.) An easy calculation shows that $F_2(V)$ contains either q or $q+1$ properly 1-generated subalgebras. Therefore one obtains

THEOREM 3.9. *Let* V *be a subalgebra linear variety with* $F_2(V)$ *finite. Then there is a positive integer* k *such that each algebra of* V *co-ordinatizes a* $(2,k)$*-design. Moreover, either* k *or* $k-1$ *is a prime power.*

There are results of Quackenbush [13] and of Ganter, Werner [4] showing that for each k as in the above theorem there exists a variety co-ordinatizing *all* $(2,k)$-designs.

Notice that Theorem 3.9 is also valid for $F_2(V) = \langle x\rangle \cup \langle y\rangle$. In this case one has $k = 2$. An example of a subalgebra linear variety with $F_2(V) = \langle x\rangle \cup \langle y\rangle$ and with a properly ternary term is given in Świerczkowski [15] (see also [5]).

REMARK. The referee told us that E.W. Kiss has recently found a proof of Theorem 3.8 basing on the application of Pálfy's theorem on the 1-generated free algebra $F_1(V)$.

REFERENCES

1. S. Burris, H.P. Sankappanavar, *A course in universal algebra*. Springer, New York Heidelberg Berlin, 1981.
2. P. Dembowski, *Finite geometries*. Springer, Berlin Heidelberg New York, 1968.
3. T. Evans, B. Ganter, Varieties with modular subalgebra lattices. *Bull. Austral. Math. Soc.* 28 (1983), 247-254.
4. B. Ganter, H. Werner, Equational classes of Steiner systems. *Algebra Universalis* 5 (1975), 125-140.
5. B. Ganter, H. Werner, Co-ordinatizing Steiner systems. *Annals of Discrete Mathematics* 7 (1980), 3-24.
6. G. Grätzer, *Universal algebra*. 2nd edition, Springer, New York Heidelberg Berlin, 1979.
7. Th. Ihringer, On groupoids having a linear congruence class geometry. *Math. Z.* 180 (1982), 394-411.
8. Th. Ihringer, On finite algebras having a linear congruence class geometry. *Algebra Universalis* 19 (1984), 1-10.
9. H.P. Müller, *Unteralgebrenräume mit Austauschaxiom*. Diplomarbeit, Darmstadt, 1979.
10. J.M. Osborn, Vector loops. *Illinois J. Math.* 5 (1961), 565-584.
11. P.P. Pálfy, Unary polynomials in algebras, I. *Algebra Universalis* 18 (1984), 262-273.
12. A. Pasini, On the finite transitive incidence algebras. *Boll. Un. Mat. Ital.* (5) 17-B (1980), 373-389.
13. R. Quackenbush, Near vector spaces over GF(q) and (v,q+1,1)-BIBD's. *Linear Algebra and Appl.* 10 (1975), 259-266.
14. S.K. Stein, Homogeneous quasigroups. *Pacific J. Math.* 14 (1964), 1091-1102.
15. S. Świerczkowski, Algebras independently generated by every n elements. *Fund. Math* 49 (1960), 93-104.
16. R. Wille, *Kongruenzklassengeometrien*. Lecture Notes in Mathematics 113, Springer, Berlin Heidelberg New York, 1970.
17. R. Wille, *Allgemeine geometrische Algebra*. Manuscript, Darmstadt, 1977.

GENERALIZED COMMUTATIVITY

Octavio C. García
Instituto de Matemáticas
Universidad Nacional Autónoma de México
04510 México, D. F., México

Walter Taylor
University of Colorado
Boulder, Colorado 80309-0426, USA

In this paper we will consider laws of the form

$$L(\sigma) \qquad f(x_1, \cdots, x_m) \approx f(x_{\sigma(1)}, \cdots, x_{\sigma(m)})$$

for σ a permutation of $\{1, \cdots, m\}$. Such laws (with f interpreted as the ordinary sum of integers $x_1 + \cdots + x_m$) are well known in everyday life: a column of numbers may be added in any order. [Notice, however, that we are not assuming any of the further laws satsified by this example, such as generalized associativities, etc.]

We are actually interested in sets of such laws for σ in a group G of permutations of $\{1, \cdots, m\}$. That is, we consider a subgroup G of the full permutation group S_m, and define Σ_G^m to be the set of all laws $L(\sigma)$ for $\sigma \in G$. Recall from our Memoir [2] that we say that one equational theory Σ (with, say, one m-ary primitive operation f) is ***interpretable*** in another equational theory Γ iff there exists a Γ-term $\alpha(x_1, \cdots, x_m)$ which obeys all the laws of Σ. (I.e., if we form $\Sigma(\alpha)$ from Σ by replacing every occurrence of f in Σ by α, then $\Sigma(\alpha)$ can be deduced from Γ.) In this paper we will investigate the relationship

$$G \le H$$

(with $G \subseteq S_m$ and $H \subseteq S_n$), which will mean simply that the theory Σ_G^m is interpretable in the theory Σ_H^n. As was originally proved in 1974 by W. D. Neumann [5], the interpretability relation defines a lattice ordering on the class of all equational theories (after the identification of mutually interpretable theories). Thus we may view the relation $G \le H$ as measuring the relative strengths of the theories Σ_G^m and Σ_H^n.

The groups we will consider include

S_m the ***symmetric*** group

A_m the ***alternating*** group

$\mathbb{Z}_m$ the ***cyclic*** group generated by the single cycle $(1,2,\cdots,m)$

D_m the ***dihedral*** group, which is generated by $\mathbb{Z}_m$ and the "reflection" $(1,m)(2,m-1)(3,m-2)\cdots$.

We always regard groups as subgroups of S_m for some specific m, for the properties here depend very much both on m and on how G is embedded into S_m ; they are not invariant under isomorphism.

Let us illustrate the definition of $\leq$ by exhibiting one rather elementary way in which the definition can be satisfied, and then another way which is somewhat less obvious. If g is a km-ary operation obeying $g(x_1, \cdots ,x_{km}) \approx g(x_2, \cdots ,x_{km},x_1)$, then we may define

$$f(x_1, \cdots ,x_m) \; := \; g(x_1,\cdots,x_1, \;\cdots\; ,x_m,\cdots,x_m) ,$$

with each x_i occurring k times. One easily checks that f must satisfy $f(x_1, \cdots ,x_m) \approx f(x_2, \cdots ,x_m,x_1)$; this law of course tells us that $\mathbb{Z}_m \leq \mathbb{Z}_{km}$. The same interpretation also yields $A_m \leq A_{km}$, $S_m \leq S_{km}$ and even $S_m \leq A_{km}$ for even numbers k. We will call an interpretation of this type ***trivial***; for a typical example of non-trivial interpretation, we look at $S_4 \leq S_6$. If g is a ***fully symmetric*** 6-ary operation (i.e., it obeys $L(\sigma)$ for each $\sigma \in S_6$), then, as the reader may easily check, the term

$$g[\, g(x,x,x,y,y,y),\; g(x,x,x,z,z,z),\; g(x,x,x,w,w,w),$$
$$g(y,y,y,z,z,z),\; g(y,y,y,w,w,w),\; g(z,z,z,w,w,w)\,]$$

is fully symmetric in the four variables x, y, z, w.. All of the positive relationships $G \leq H$ of this paper are deduced from the interpretations which are either trivial or very similar to the 4-6 type just described, or formed by combining these two basic kinds of interpretations, either by transitivity (Proposition 1.2) or by an additive combination (Theorem 1.6). For a full account, see §1.

As is often the case for interpretability theory, to establish a negative relation $G \nleq H$, requires an analysis of the word problem for Σ^n_H (to see that the laws $\Sigma^m_G(\sigma)$ cannot be deduced from Σ^n_H). In the case at hand, this word problem is easily analyzed (see Lemma 2.1); as a corollary of this analysis, we will obtain our main tool for proving $G \nleq H$, namely

COROLLARY 2.4. *If* $G \leq S_n$, *and no point of* $\{1, \cdots ,m\}$ *is fixed by every permutation in* G, *then there exists a homomorphism* $\mu : G \longrightarrow S_n$ *with no common fixed point, i.e. such that*

$$\bigcap_{g \in G} \{ x : 1 \le x \le n \text{ and } \mu(g) \text{ fixes } x \} = \emptyset.$$

(See Theorem 2.2 for a more complicated statement which holds with a general H replacing S_n.)

Thus a major portion of this investigation turned out to revolve about the existence of homomorphisms $\mu : G \longrightarrow H$ with no common fixed points; §3 and §4 are devoted to this topic. In fact our main negative result is the fact (Corollary 4.5) that for n not a multiple of m, and $12 \le m \le n < \frac{1}{2}m(m-1)$, every homomorphism $f : A_m \longrightarrow S_n$ has a common fixed point (thus making $A_m \le S_n$ impossible). Incidentally, we can see from these results of §4 that, e.g. if $12 \le m < n < 2m$, the only homomorphisms $f : A_m \longrightarrow S_n$ are the utterly trivial ones formed by having $f(\alpha)$ act on some m-element subset of $\{1, \cdots ,n\}$ exactly as α acts on $\{1, \cdots ,m\}$.

An important tool in our proof has turned out to be an analysis, due to Bertrand and Serret more than 100 years ago, of the possible indices $[A_m:G]$ of relatively large subgroups of the alternating group A_m. See the proof of Lemma 4.3 below.

We finish this introduction with a brief summary of the results: the numbers in parentheses after each result indicate where it may be found in the body of the paper.

1. *If* $G \subsetneq H$, *then* $G \le H$ (1.1).

2. *If* G *is simple and* $G \le S_n$, *then* $|G| \le n!$ (2.4).

3. $(m \ge 5)$ *If* $A_m \le S_n$ *then* $m \le n$. [Hence the same conclusion holds for $A_m \le A_n$ and $S_m \le S_n$.] (Corollary of the previous results; see 3.1.)

4. $\mathbb{Z}_m \le \mathbb{Z}_n$ *iff each prime factor of* m *also divides* n (3.4).

5. $(m \ge 3)$ *If* $A_m \le S_n$, *then every prime factor of* m *also divides* n. (3.5)

6. *For each* m *there exists* N *such that if* $n \ge N$ *and all prime factors of* m *also divide* n, *then* $S_m \le S_n$ *and* $A_m \le A_n$ (1.11). [In particular, if $4|m$ and $n = \frac{1}{2}m(m-1)$, then $S_m \le S_n$ (1.8).]

7. On the other hand, *if* $12 \le m \le n < \frac{1}{2}m(m-1)$, *then* $A_m \le S_n$ *holds if and only if* m *divides* n (4.5).

8. $A_4 < A_6$ (remarks above and 3.5).

9. *With the exception noted in 8, if* $3 \le m < n < 2m$, *then* $A_m \nleq A_n$ *and* $S_m \nleq S_n$ (4.6).

10. *In* 9, *unless* $m = 9$ *and* $n = 15$, *we have the stronger conclusion that* $A_m \nleq S_n$ (4.6).

11. $D_m = \mathbb{Z}_2 \vee \mathbb{Z}_m$ *in our lattice* (1.12).

We remark that one motivation for the study of the theories Σ_G^m was our continuing investigation (see [2]) of whether 0 is infinitely meet irreducible in the Neumann lattice mentioned above (it is known to be finitely meet irreducible). This investigation necessarily involves a search for theories which are non-trivial but still very weak from the point of view of this lattice, a good example being the theories Σ_G^m. Nevertheless the theories Σ_G^m have not turned out to be particularly helpful for solving our problem about meet irreducibility.

Part of this research was done while the first author was visiting McMaster University. He gratefully acknowledges both the support of the Instituto de Matemáticas of the Universidad Nacional Autónoma de México, which made that sabbatical possible, and the further support from the Natural Sciences and Engineering Research Council of Canada operating grant of his host, B. Banaschewski. The second author gratefully acknowledges support from the National Science Foundation of the USA under grant number MCS-8301065. For their valuable help during various stages of development of this paper, we thank Joel Brenner, Peter Elliott, George Bergman and David Goldschmidt. And we especially thank Jan Mycielski for his continuing interest and helpful suggestions throughout the period of this research.

1. The known cases of $\leq$

The symmetric group S_n consists of all permutations of $\{1, \cdots n\}$. By a ***permutation group*** we mean a pair (G,n) with G a subgroup of S_n. Generally we write only G, and say, "G is a group," but it should be borne in mind that each G is a group of permutations of $\{1, \cdots ,n\}$ for one specific value of n. (Thus. e.g., $(S_5,5)$ and $(S_5,6)$ are distinct objects of this theory.)

In our introduction, we stated the main definitions which are required to understand the notion of $G \leq H$ for permutation groups G and H; for a uniform notation, we will usually have G a subgroup of S_m and H a subgroup of S_n. Taking f to stand for the primitive n-ary operation of the theory Σ_H^n, what we must do in order to prove that $G \leq H$ is to find a term α in f and the variables $x_1, \cdots ,x_m$ which obeys all the laws of Σ_G^m, i.e., all the laws $L(\sigma)$ for $\sigma \in G$. (That is to say, these laws must be deducible from Σ_H^n.) In fact, in all cases where we directly prove $G \leq H$, the term α has a particularly simple form, namely

$$f[\ f(u_{11},u_{12},\cdots),\ f(u_{21},u_{22},\cdots),\ \cdots\ ]$$

where each u_{ij} is one of the variables $x_1, \cdots ,x_m$. For our first proposition, α has an even simpler form, namely $\alpha = f(x_1, \cdots ,x_m)$.

PROPOSITION 1.1 *If* $m = n$ *and* $G \subsetneq H$, *then* $G \leq H$. ∎

PROPOSITION 1.2 *If* $G \leq H$ *and* $H \leq K$, *then* $G \leq K$. ∎

We now focus our attention on $G = \mathbb{Z}_m$ and $H = \mathbb{Z}_n$ (cyclic groups — recall from the introduction that for us these have definitions as specific subgroups of S_m and S_n).

LEMMA 1.3 *If* $m|n$, *then* $\mathbb{Z}_m \leq \mathbb{Z}_n$.

Proof. Define

$$\alpha := f(x_1, \cdots, x_1, \cdots, x_m, \cdots, x_m),$$

with each x_i appearing equally often. ∎

The converse is false, as will be evident from the next lemma.

LEMMA 1.4 $\mathbb{Z}_{n^2} \leq \mathbb{Z}_n$.

Proof. Define

$$\alpha := f[\,f(x_1, x_{n+1}, x_{2n+1}, \cdots),\ f(x_2, x_{n+2}, x_{2n+2}, \cdots),\ \cdots\,]. \ ∎$$

THEOREM 1.5 *If every prime divisor of* m *also divides* n, *then* $\mathbb{Z}_m \leq \mathbb{Z}_n$.

Proof. The hypothesis implies that $m|n^{2^k}$ for some k, and thus the Theorem follows from Proposition 1.2 and the two lemmas. ∎

The converse of Theorem 1.5 holds, as we shall see in Corollary 3.4 below; thus this simple prime-divisor criterion characterizes the relation $\mathbb{Z}_m \leq \mathbb{Z}_n$. For the full symmetric groups the situation is more complicated: the prime-divisor criterion remains necessary (see §3), but it is sufficient only asymptotically in n. That is, as we shall see in Theorem 1.11 below, if n is sufficiently large relative to m, then the prime divisor criterion yields $S_m \leq S_n$. Complicated as the situation may be, we nevertheless have the following additivity property. (The corresponding assertion for $\mathbb{Z}_m$'s follows trivially from 1.5 and 3.4.)

THEOREM 1.6 *If* $S_m \leq S_n$ *and* $S_m \leq S_r$, *then* $S_m \leq S_{n+r}$.

Proof. We first give an informal version of our proof, but since it is so important for our work, we will follow this with a more formal proof. Let us call an m-ary operation f ***fully symmetric*** iff it obeys all the commutativity laws $L(\sigma)$ of the Introduction, for all permutations σ of $\{1, \cdots, m\}$. Our hypothesis is that we have fully symmetric terms $\alpha(x_1, \cdots, x_m)$ and $\beta(x_1, \cdots, x_m)$, with α constructed from a fully symmetric n-ary operation f and β constructed from a fully symmetric r-ary operation g. We will now use α and β to construct a fully symmetric $\gamma(x_1, \cdots, x_m)$ from a fully symmetric $(n+r)$-ary operation h.

We first observe that to specify γ (or any uniform term), we merely have to state which variable is to appear at the end of each of its $(n+r)^{2k-1}$ branches. Let us pick a branch b and define corresponding branches b_α and b_β in the terms α and β. As we follow this branch from the root to its leaf, we will go through $2k-1$ branchings. Simultaneously, let us define branches of α and β, as follows. If we take branching i in the γ-tree, with $1 \leq i \leq n$, then we will take the ith branching in the α-tree. On the other hand, if we take branching $(n+i)$ in the γ-tree, with $1 \leq i \leq r$, then we will take the ith branching in the β-tree. Thus we have [partially] defined the branches b_α and b_β. In going $2k-1$ steps, exactly one of b_α and b_β must define a complete branch, i.e must reach one of the variables $x_1, \cdots, x_m$. It is this variable which we attach in γ to the branch b.

We have now defined the term γ. Its full symmetry is almost obvious, and so we leave the remaining details of this informal argument to the reader.

Looking now at a more formal version of the above argument, we let A and B be disjoint sets of n and r elements, respectively. We will let $W_j(A)$ be the subset of the free monoid consisting of 'words' of length j (and thus $W_0(A)$ contains only the 'empty word' $\Box$). The full free monoid is of course the disjoint union $W(A) = W_0(A) \cup W_1(A) \cup W_2(A) \cup \cdots$. Now a uniform term of depth k in one n-ary function symbol and in m variables corresponds exactly to a function $\alpha : W_k \longrightarrow m$. The fact that this term is fully symmetric translates, in the case at hand, to the following rather lengthy fact about α, which will occupy all of the next paragraph. [This translation looks obvious, but in fact makes implicit use of our solution to the word problem (Lemma 2.1 below).]

Full symmetry. For each permutation σ of m, each $j < k$, and each word $w \in W_j(A)$, there exists a permutation $\pi_A(\sigma,w)$ of A such that $\sigma \circ \alpha = \alpha \circ \overline{\sigma}$, where $\overline{\sigma}_A$ is defined as follows. We will recursively define $\overline{\sigma}_A$ as a permutation of $W_j(A)$, for $0 \leq j \leq k$. Naturally, $\overline{\sigma}_A(\Box) = \Box$, and for the recursive step, we let $\overline{\sigma}_A(a_1 \cdots a_{j+1}) = \overline{\sigma}_A(a_1 \cdots a_j)b_{j+1}$, where the last juxtaposition denotes concatenation (product in the free monoid), and where $b_{j+1} = \pi_A(\sigma, a_1 \cdots a_j)(a_{j+1})$.

Now likewise for B there is a mapping $\beta : W_k(B) \longrightarrow m$ and a definition of permutations $\pi_B(\sigma, b_1 \cdots b_j)$ such that $\sigma \circ \beta = \beta \circ \overline{\sigma}_B$ for all σ. Our objective is to define $\gamma: W_{2k-1} \longrightarrow m$ and permutations $\pi_{A\cup B}(\sigma, c_1 \cdots c_j)$ (for $j \leq 2k-1$ and $c_i \in A \cup B$), such that if $\sigma_{A\cup B}$ is defined in like manner, then we will have $\sigma \circ \gamma = \gamma \circ \overline{\sigma}_{A\cup B}$.

Defining γ is not very difficult; notice first that a $(2k-1)$-letter word w

in letters from $A \cup B$ must have the majority of its letters either in A or in B. If the A-letters are in the majority, and they are say $a_1, \cdots, a_{k+r}$ (interspersed with various B-letters), then we define $\gamma(w)$ to be $\alpha(a_1 \cdots a_k)$. And, of course, there is a corresponding definition involving β if the B-letters have the majority.

The definitions of $\pi_{A\cup B}(\sigma, c_1 \cdots c_s)(c_{s+1})$ depends on whether c_{s+1} is in A or in B. If $c_{s+1} \in A$, then we discard all elements of B from $c_1 \cdots c_s$, thereby forming a [possibly empty] word $a_1 \cdots a_j$ in A-letters only. We then define $\pi_{A\cup B}(\sigma, c_1 \cdots c_s)$ to be $\pi_A(\sigma, a_1 \cdots a_{k-1})(c_{s+1})$ if $j < k$, and to be $\pi_A(\sigma, a_1 \cdots a_{k-1})(c_{s+1})$ otherwise. There is a similar definition, involving π_B, for $c_{s+1} \in B$.

Now we prove the desired equality that $\sigma(\gamma(w)) = \gamma(\bar{\sigma}_{A\cup B}(w))$ for each $w \in W_{2k-1}(A\cup B)$. Without loss of generality, we assume that the A-letters are in the majority in w, and then we let w_0 denote the word which is formed from w by deleting all B-letters and enough A-letters from the right hand end of w to leave a word in $W_k(A)$. By definition we have $\gamma(w) = \alpha(w_0)$, and so

$$\sigma(\gamma(w)) = \sigma(\alpha(w_0)) = \alpha(\bar{\sigma}_A(w_0)).$$

Now in order to prove the desired equality, it will be enough for us to prove that $\gamma(\bar{\sigma}_{A\cup B}(w)) = \alpha(\bar{\sigma}_A(w_0))$. It follows immediately from the definitions involved that the A-letters have the majority in $\bar{\sigma}_{A\cup B}(w)$, and so, for the last equality, it remains only to show that the first k A-letters appearing in $\bar{\sigma}_{A\cup B}(w)$ form precisely the word $\bar{\sigma}_A(w_0)$. This last fact is obvious from the definition of $\bar{\sigma}_{A\cup B}$. ∎

COROLLARY 1.7 *If* $m|n$, *then* $S_m \le S_n$. ∎

Of course, if m has no multiple prime factors, then Corollary 1.7 is adequate for sufficiency of the prime-divisor condition which we discussed prior to Theorem 1.6. The next lemma is our major tool for discussing the more complex case of multiple prime factors. In fact, its proof is a simple extension of the discussion of $m = 4$ and $n = 6$ which we gave in the introduction.

LEMMA 1.8 *If* p *is prime,* $p^2|m$ *and* $n = \binom{m}{p}$, *then* $S_m \le S_n$.

<u>Proof</u>. Writing m as p^2r and writing out the well known expression for $\binom{m}{p}$, we easily see that $p|n$; let us write $n = pk$. Let $Q_1, \cdots, Q_n$ be a list of the $n = \binom{m}{p}$ p-element subsets of $\{x_1, \cdots, x_m\}$. Now to interpret $\Sigma^m_{S_m}$ in

$\Sigma^n_{S_n}$, we take

$$\alpha \quad = \quad f[\ f(u_{11},u_{12},\cdots),\ f(u_{21},u_{22},\cdots),\ \cdots\ ]$$

where, for each i, the variables $u_{i1}, \cdots, u_{in}$ consist of the p variables in Q_i, each taken to occur k times. The full symmetry of α is now readily apparent. ∎

LEMMA 1.9 *If* p *is prime,* r, n *are positive integers and* p *does not divide* r*, then* p^n *does not divide* $\begin{bmatrix} p^n r \\ p \end{bmatrix}$. ∎

The following lemma is rather well known, and hence we omit its proof. It is stated, for example, in [5].

LEMMA 1.10 *If* J *is any subset of* $\omega = \{0, 1, 2, \cdots\}$ *which is closed under addition, then* J *is ultimately an ideal of* $(\mathbb{Z},+)$. (By this we mean that for each such J there exists $K, s \in \omega$ such that, for all $n \geq K$, we have $n \in J \longleftrightarrow s|n$.) ∎

THEOREM 1.11 *For every* $m \geq 2$, *there exists* N *such that if* $n > N$ *and if every prime factor of* m *also divides* n*, then* $S_m \leq S_n$.

Proof. Fix m and define

$$J \quad = \quad \{ n : S_m \leq S_n \} .$$

By Theorem 1.6, J is a subset of ω closed under addition, and therefore, by Lemma 1.10, there exist N and k such that for $n \geq N$, $n \in J$ iff $k|n$. By Corollary 1.7, $m^s \in J$ for every s, and clearly s may be chosen so that $m^s \geq N$. Therefore $k|m^s$ for some s, which means that every prime factor of k is also a factor of m. To finish the proof, it will be enough to establish that k has no repeated prime factors.

By way of contradiction, let us suppose that p divides k exactly e times, with $e \geq 2$. This readily implies that there exists $n \in J$ such that $n \geq N$ and p divides n exactly e times. Thus, by Lemma 1.8, $S_n \leq S_t$, where $t = \begin{bmatrix} n \\ p \end{bmatrix}$, and certainly $S_m \leq S_n$ by the definition of J. By transitivity (Proposition 1.2), we have $S_m \leq S_t$, i.e., $t \in J$. Clearly $t \geq N$, and so $k|t$, but this is a contradiction, since p divides $t = \begin{bmatrix} n \\ p \end{bmatrix}$ at most $e-1$ times, by Lemma 1.9. ∎

Remark. It will be apparent from Corollary 3.5 below that the k which appeared in this proof is precisely the product of all prime divisors of m, i.e., the largest square-free divisor of m.

It would of course be interesting to know something about the smallest value of

N which is required to make Theorem 1.11 true. One can easily check that (for m with few prime factors) the value of N which is implicit in the proof grows exponentially with m, but we do not know how far this might be from the best possible value. On the other hand, the main result of §4 will show that (for $m \geq 12$) one must have $N \geq \frac{1}{2} m(m-1)$.

Nevertheless, the ideas of the last proof allow us to obtain some representative values of N. For example, for $m = 8$, we successively obtain the following elements of J:

$$\begin{bmatrix}8\\2\end{bmatrix} = 28 = 3\cdot 8 + 4 \in J$$

$$\begin{bmatrix}28\\2\end{bmatrix} = 47\cdot 8 + 2 \in J$$

$$\begin{bmatrix}8\\2\end{bmatrix} + \begin{bmatrix}28\\2\end{bmatrix} = 50\cdot 8 + 6 \in J.$$

Now by adding multiples of 8 (using the additivity of J, it is easy to see that J contains every even number $\geq 50\cdot 8 = 400$. We now demonstrate some similar calculations for $m = 36$:

$$\begin{bmatrix}36\\2\end{bmatrix} = 17\cdot 36 + 8$$

$$\begin{bmatrix}36\\3\end{bmatrix} = 198\cdot 36 + 12$$

$$\begin{bmatrix}36\\2\end{bmatrix} + \begin{bmatrix}36\\3\end{bmatrix} = 215\cdot 36 + 30$$

$$2\cdot\begin{bmatrix}36\\3\end{bmatrix} = 396\cdot 36 + 24$$

$$\begin{bmatrix}36\\2\end{bmatrix} + 2\cdot\begin{bmatrix}36\\3\end{bmatrix} = 414\cdot 36 + 6.$$

Thus it follows of course that we could use $N = 314\cdot 36 = 19{,}904$, but we wonder if this number is not extravagantly large.

There is another way to discover results of the form $G \leq H$, which we have only begun to explore. In this approach, one begins with an H-symmetric n-ary operation f, i.e., an operation f satisfying the laws $L(\tau)$ for all $\tau \in H$. One then simply picks a likely f-term $\alpha(x_1, \cdots, x_m)$ and defines G to be the subgroup of S_m consisting of all those symmetries of α which can be deduced from the laws $L(\tau)$. Of course, in order to obtain a result of the desired sort, one must be able to give an independent description of G; we will illustrate with one example where we have found this to be possible.

Our example is slightly more complex than we just now indicated, since we will work with the ***join*** V of the two theories $\Sigma^2_{\mathbb{Z}_2}$ and $\Sigma^m_{\mathbb{Z}_m}$ in our lattice. For a detailed discussion of joins, we must refer the reader to our Memoir [2]; suffice it to say that in the join theory $\Sigma^2_{\mathbb{Z}_2} \vee \Sigma^m_{\mathbb{Z}_m}$ we have both a $\Sigma^2_{\mathbb{Z}_2}$-operation f and a

$\Sigma^m_{\mathbb{Z}_m}$-operation g. No laws are assumed for f or for g other than the laws which we already have (namely $\Sigma^2_{\mathbb{Z}_m} \cup \Sigma^m_{\mathbb{Z}_m}$). Well, given such an f and g, let us define the m-ary term

$$\alpha(x_1, \cdots, x_m) = f(\ g(x_1, \cdots x_m),\ g(x_m, \cdots, x_1)\ ).$$

We claim that this α is D_m-symmetric, where D_m is the dihedral group (which was defined in the introduction). For the $\mathbb{Z}_m$-symmetry of g obviously gives $\mathbb{Z}_m$-symmetry to α, and the $\mathbb{Z}_2$-symmetry of f obviously gives σ-symmetry to α, where σ is the other generator of D_m, namely $\sigma(1) = m$, $\sigma(2) = m-1$, etc. These observations of course establish that $\Sigma^m_{D_m} \leq \Sigma^2_{\mathbb{Z}_2} \vee \Sigma^m_{\mathbb{Z}_m}$. In fact we have the following Theorem.

THEOREM 1.12 $\Sigma^m_{D_m} \leq \Sigma^2_{\mathbb{Z}_2} \vee \Sigma^m_{\mathbb{Z}_m}$, *with equality holding if* m *is even.*

<u>Proof</u>. We have already proved the inequality $\leq$. For the opposite inequality we need only prove that $\Sigma^m_{\mathbb{Z}_m} \leq \Sigma^m_{D_m}$ and that $\Sigma^2_{\mathbb{Z}_2} \leq \Sigma^m_{D_m}$. The first of these inequalities is evident from Proposition 1.1, and the second from Lemma 1.3. ∎

In fact, if m is even, then $\Sigma_2 \leq \mathbb{Z}_m$, and so the above join is simply $\Sigma^m_{\mathbb{Z}_m}$, and so we have

COROLLARY 1.13 *If* m *is even, then* $\Sigma^m_{D_m} = \Sigma^m_{\mathbb{Z}_m}$ (as elements of our lattice). ∎

2. The main method for $\not\leq$

To prove $G \not\leq H$, we need to establish that no Σ^n_H-term can define an m-operation with the full G-symmetry. And so we need an analysis of what sorts of laws can be deduced from Σ^n_H; we are able to give a complete analysis, since we have a complete (albeit elementary) solution of the ***word problem*** for this variety.

For $H \subseteq S_n$ and for t, w any two terms in a single n-ary operation symbol F and variables $x_0, x_1, x_2, \cdots$, we recursively define $t \sim_H w$ to mean that either $t = w$ (i.e., t and w are formally identical) or the following conditions hold for some F-terms u_i, v_i and for some $\sigma \in H$:

$$t = F(u_1, \cdots, u_n)$$

$$w = F(v_1, \cdots, v_n)$$

$$u_i \sim v_{\sigma(i)} \quad (1 \le i \le n).$$

LEMMA 2.1 $\Sigma_H^n \vdash t = w$ *iff* $t \sim_H w$.

<u>Proof</u>. Let us define

$$R_H = \{ (t,w) : t \sim_H w \}.$$

One easily checks, by direct inspection, and by induction, that

$$\Sigma_H^n \subseteq R_H \subseteq \{ (t,w) : \Sigma_H^n \vdash t = w \}.$$

To complete the proof, we need to show that the second inclusion here is really an equality of sets. Since $\{ (t,w) : \Sigma_H^n \vdash t \approx w \}$ is the ***smallest equational theory*** containing Σ_H^n, it will be enough to show that R_H is itself an equational theory. In other words, we need to see that R_H is closed under G. Birkhoff's five rules of equational deduction. See §14 of S. Burris and H. P. Sankappanavar [1] for these five rules; as they explain there, it is equivalent to prove that R_H is a ***fully invariant equivalence relation*** on the Σ_H^n-free algebra generated by $x_0, x_1, \cdots$.

The proof that R_H is an equivalence relation is an easy induction on the formation of $\sim_H$ (we omit the details); and the fact that $\sim_H$ is a congruence relation is immediate from the definition. Thus, it remains for us to show that R_H is fully invariant, or substitutive, i.e., that if $t(x_1, \cdots ,x_r) \sim_H w(x_1, \cdots ,x_r)$, then $t(s_1, \ldots ,s_r) \sim_H w(s_1, \cdots ,s_r)$ for any terms $s_1, \cdots , s_r$. Our proof is by induction on the formation of $\sim_H$. Certainly, if $t = w$, then the conclusion is immediate. Alternatively, we have F-terms u_i, v_i, and a permutation $\sigma \in H$ such that

$$t(x_1, \cdots ,x_r) = F[\, u_1(x_1,\cdots,x_r), \cdots ,u_n(x_1,\cdots,x_r) \,]$$

$$w(x_1, \cdots ,x_r) = F[\, v_1(x_1,\cdots,x_r), \cdots ,v_n(x_1,\cdots,x_r) \,]$$

$$u_i(x_1, \cdots x_r) \sim_H v_{\sigma(i)}(x_1, \cdots ,x_r) \qquad (1 \le i \le n).$$

By induction on complexity, we have

$$u_i(s_1, \cdots ,s_r) \sim_H v_{\sigma(i)}(s_1, \cdots ,s_r).$$

It then follows that

$$\begin{aligned} t(s_1, \cdots ,s_r) &= F[\, u_1(s_1,\cdots,s_r), \cdots ,u_n(s_1,\cdots,s_r) \,] \\ &\sim_H F[\, v_1(s_1,\cdots,s_r), \cdots ,v_n(s_1,\cdots,s_r) \,] \\ &= w(s_1, \cdots ,s_r). \ \blacksquare \end{aligned}$$

We are now almost ready to state and prove our major result which can yield $G \not\subseteq H$. Given an equivalence relation ρ on $\{1, \cdots ,n\}$ and $H \subseteq S_n$, we define

$$H^{\rho} = \{ \tau \in H : \text{ if } 1 \le i,j \le n \text{ and } (i,j) \in \rho, \text{ then } (\tau(i),\tau(j)) \in \rho \}$$

$$H_0^{\rho} = \{ \tau \in H : (i,\tau(i)) \in \rho \text{ for all } i \}.$$

The reader may easily check that H_0^{ρ} *is a normal subgroup of* H^{ρ}.

If $H_0 \subseteq H_1 \subseteq H \subseteq S_n$, with H_0 normal in H_1, and $\theta : G \longrightarrow H_1/H_0$ is a homomorphism, then a ***common fixed point of*** θ ***modulo*** H_0 is an element i of $\{1, \cdots, n\}$ which satisfies the following condition: for all $\sigma \in G$ there exists $\lambda \in H_1$ such that $\theta(\sigma) = \lambda H_0$ and $\lambda(i) = i$. If H_0 consists of the identity element, then we say that i is a ***common fixed point*** of the homomorphism $\theta : G \longrightarrow H_1$. Finally, if the identity map $G \longrightarrow G$ has a common fixed point, we say that the group G ***has a common fixed point***. The reader may easily check that an arbitrary permuation group G has a common fixed point if and only if [the mutual interpretability class of] Σ_G^m is the least element of the Neumann lattice mentioned in the Introduction.

THEOREM 2.2 *If* $G \le H$ *and* G *has no common fixed point, then there exists an equivalence relation* ρ *on* $\{1, \cdots, n\}$ *and a homomorphism* $\theta : G \longrightarrow H^{\rho}/H_0^{\rho}$ *with no common fixed points modulo* H_0^{ρ}.

<u>Proof</u>. Since $G \le H$, there exists a term $t(x_1, \cdots, x_m)$ in the single n-ary operation F, such that for each $\sigma \in G$ the identity

$$t(x_{\sigma(1)}, \cdots, x_{\sigma(m)}) \approx t(x_1, \cdots, x_m)$$

is a consequence of Σ_H^n; let us in fact take t to be of minimal complexity such that these identities are satisfied. Since Σ_G^m is non-trivial in the Neumann lattice, this t cannot consist of a single variable. By Lemma 2.1,

$$t(x_{\sigma(1)}, \cdots, x_{\sigma(m)}) \sim_H t(x_1, \cdots, x_m)$$

for each $\sigma \in G$, and this is easily seen to be equivalent to the assertion that, for each $\sigma \in G$,

$$\bar{\sigma}(t) \sim_H t,$$

where $\bar{\sigma}$ is the endomorphism of the Σ_H^n-free algebra on $\{x_1, \cdots, x_m\}$ which extends the map $x_i \longrightarrow x_{\sigma(i)}$ $(1 \le i \le m)$.

Since t does not consist of a single variable, we must have $t = F(u_1, \cdots, u_n)$ for some F-terms $u_i = u_i(x_1, \cdots, x_m)$ $(1 \le i \le m)$. Applying the recursive definition of $\bar{\sigma}$ to this expression for t, we now have

$$F(u_1, \cdots, u_n) = t \sim_H \bar{\sigma}(t)$$
$$= F(\bar{\sigma}(u_1), \cdots, \bar{\sigma}(u_n)),$$

and so, according to the recursive definition of $\sim_H$, for each $\sigma \in G$ there exists $\lambda \in H$ such that

$$(*) \qquad \bar{\sigma}(u_i) \sim_H u_{\lambda(i)} \qquad (1 \le i \le n).$$

Next, we will show that $\lambda \in H^{\rho}$, where the equivalence relation ρ is defined as the restriction to $\{u_1, \cdots, u_n\}$ of the equivalence relation $\sim_H$. (More precisely, for $1 \le i \le n$ we define $(i,j) \in \rho \longleftrightarrow u_i \sim_H u_j$.) Therefore, to get $\lambda \in H^{\rho}$, we need to show that if $u_i \sim_H u_j$, then $u_{\lambda(i)} \sim_H u_{\lambda(j)}$. By Lemma 2.1, $\sim_H$ is fully invariant, and so we have the calculation

$$u_{\lambda(i)} \sim_H \bar{\sigma}(u_i) \sim_H \bar{\sigma}(u_j) \sim_H u_{\lambda(j)}$$

which establishes that $\lambda \in H^{\rho}$. We leave it to the reader to check the obvious fact that λ is uniquely determined modulo the the normal subgroup H_0^{ρ}, and hence that $\sigma \longrightarrow \lambda H_0^{\rho}$ defines a function $\theta : G \longrightarrow H^{\rho}/H_0^{\rho}$. We also omit the easy proof that θ is a homomorphism.

To finish the proof, we assume, by way of contradiction, that i is a common fixed point of θ modulo H_0^{ρ}. This means that for any $\sigma \in G$ we can find $\lambda \in H^{\rho}$ such that (*) holds and such that $(i,\lambda(i)) \in \rho$. By definition of ρ, $u_{\lambda(i)} \sim_H u_i$, and so by (*) we have

$$\bar{\sigma}(u_i) \sim_H u_i$$

for all $\sigma \in G$ (and for this one value of i). In other words, Σ_H^n implies the identities

$$u_i(x_{\sigma(1)}, \cdots, x_{\sigma(m)}) \approx u_i(x_1, \cdots, x_m) \qquad (\sigma \in G)$$

with u_i of lesser complexity than t. These identities contradict the fact that t was taken, at the start of the proof, to be of minimal complexity for the validity of such identities, and this contradiction establishes our claim that θ has no common fixed point modulo H_0^{ρ}. ∎

A significant simplification is possible in the statement of the last theorem if the group extension

$$H_0^{\rho} \xrightarrow{\ i\ } H^{\rho} \xrightarrow{\ p\ } H^{\rho}/H_0^{\rho}$$

splits. [Here i is the inclusion mapping, and p is the natural map of H^{ρ} onto its quotient.] This means that there exists a homomorphism $\mu : H^{\rho}/H_0^{\rho} \longrightarrow H^{\rho}$ such that $p \circ \mu$ is the identity function. For more information on the notion of split extensions, see, e.g., pages 460–463 of [4]. Certainly such a splitting does not exist for arbitrary H and ρ, but in Corollary 2.4 below we will see that a splitting does exist for $H = S_n$.

COROLLARY 2.3 *If* $G \leq H$, *G has no common fixed point, and the above extension is split (with ρ as in Theorem 2.2), then there exists a homomorphism $\theta : G \longrightarrow H$ with no common fixed point.*

<u>Proof</u>. By Theorem 2.2, we have an equivalence relation ρ on $\{1, \cdots, n\}$ and a homomorphism $\theta : G \longrightarrow H^{\rho}/H^{\rho}_{0}$ with no common fixed point modulo H^{ρ}_{0}. Let us take μ to be as in the above definition of split extension. We will prove that the composite map $\mu \circ \theta$ (which may be viewed as a homomorphism into H) has no common fixed point. If i were a common fixed point of $\mu \circ \theta$, then i would be a fixed point of $\mu(\theta(\sigma))$ for each $\sigma \in G$, i.e., for each $\sigma \in G$ we would have $[\mu(\theta(\sigma))](i) = i$.

We now define λ to be the permutation $\mu(\theta(\sigma))$. At the end of the previous paragraph we saw that $\lambda(i) = i$, and from the defining property of a splitting we have $\theta(\sigma) = p(\mu(\theta(\sigma))) = p(\lambda)$, which of course is the coset λH^{ρ}_{0}. We have thus established the properties of λ appearing in the definition of 'common fixed point' prior to Theorem 2.2. Thus our assumption that i is a common fixed point of $\mu \circ \theta$ implies that i is also a common fixed point of θ modulo H^{ρ}_{0}, which contradicts what we already know about θ. This contradiction establishes our claim that $\mu \circ \theta$ has no common fixed point. ∎

COROLLARY 2.4 *If* $G \leq S_n$ *and G has no common fixed point, then there exists a homomorphism $G \longrightarrow S_n$ with no common fixed point.*

<u>Proof</u>. This will follow immediately from the previous Corollary, as soon as we establish that the extension

$$(S_n)^{\rho}_{0} \xrightarrow{\ i\ } (S_n)^{\rho} \xrightarrow{\ p\ } (S_n)^{\rho}/(S_n)^{\rho}_{0}$$

splits. To do this, we first exhibit a group $H \subsetneq (S_m)^{\rho}$ such that

$$\{1\} = H \cap (S_n)^{\rho}_{0}$$
$$(S_n)^{\rho} = H \cdot (S_n)^{\rho}_{0}$$

(where the latter equation means that each permutation $\alpha \in (S_n)^{\rho}$ is a product $\beta \cdot \gamma$ with $\beta \in H$ and $\gamma \in (S_n)^{\rho}_{0}$). It is easily checked that such an H is easily provided by

$$H_n^{\rho} = \{ \tau \in S_n : \tau \text{ maps each } \rho\text{-block monotonely onto another } \rho\text{-block} \}.$$

Now the theory of semidirect products [4, *loc. cit.*] immediately yields the splitting homomorphism λ. {For an $(S_n)_0^{\rho}$-coset $\alpha\cdot(S_n)_0^{\rho}$, one easily checks that $H \cap [\alpha\cdot(S_n)_0^{\rho}]$ is a singleton; we define $\lambda(\alpha)$ to be its unique member.} ∎

We remark that this Corollary (and also the previous Corollary and Theorem) depend only on the isomorphism class of G, and not on how G is embedded in S_m, except, of course, for the important requirement that G should have no common fixed point. Thus, to this extent, we may relax the warning, which we stated in the introduction, that we are not considering isomorphism-invariant properties.

3. Elementary applications of the main method

In this section we first present two Theorems (3.1 and 3.2) which follow directly from Corollary 2.4, and then the remainder of the section will consist of various Corollaries of Theorem 3.2. (A third, and more difficult, Theorem based on 2.4 will come later, in §4.) We have stated Theorem 3.1 only in the form which we need; we leave to the reader the obvious generalization to arbitrary simple groups.

THEOREM 3.1 $(m \geq 5)$ *If* $A_m \leq S_n$, *then* $m \leq n$.

Proof. Immediate from Corollary 2.4, the simplicity of A_m, and the obvious fact that A_m has no common fixed point. ∎

THEOREM 3.2 *If* $G \leq S_m$, G *has no common fixed point, and* G *is a direct sum of cyclic groups of prime orders* $p_1, p_2, \cdots, p_k$ *then*

$$n = \sum_{i=1}^{k} y_i p_i$$

for some non-negative integers y_i.

Proof. By Corollary 2.4 there exists a homomorphism $\mu : G \longrightarrow S_n$ with no common fixed point. In other words, $H = \mu[G]$ acts on $\{1, \cdots, n\}$ with no singleton orbits. Now it is well known and easy to check that the order of the orbit of a under the action of a group H is equal to the index $[H:H_a]$, where H_a is the ***stabilizer*** or ***isotropy*** subgroup

$$H_a = \{ \tau \in H : \tau(a) = a \}.$$

Therefore the power of every orbit is a proper divisor of the order of G. The conclusion of the Theorem is now evident. ∎

Our first Corollary is based on the special case of Theorem 3.2 in which G is is of prime order, i.e., $k = 1$, and the primes $p_2, p_3 \cdots$ do not occur. For this result, and various results which follow, the reader is asked to recall that the groups $\mathbb{Z}_m$, A_m and D_m are not defined abstractly, but as specific subgroups of S_m (details in the introduction). Moreover, $\subseteq$ refers specifically to the inclusion relation (e.g., for subsets of S_m), while $\leq$ retains the techical meaning which we have reserved for it in this paper.

COROLLARY 3.3 *If* $\mathbb{Z}_m \subseteq G \leq H \subseteq S_n$, *then every prime divisor of* m *also divides* n.

Proof. If p is any prime divisor of m, then, by Lemma 1.3, we have $\mathbb{Z}_p \leq \mathbb{Z}_m$. Therefore

$$\mathbb{Z}_p \leq \mathbb{Z}_m \leq G \leq H \leq S_n,$$

and so we may apply Theorem 5.2 (with $k = 1$ and $p_1 = p$) to the relation $\mathbb{Z}_p \leq S_n$. The conclusion $p|n$ is now immediate. ∎

COROLLARY 3.4 *For any integers* $m, n \geq 2$, *the following conditions are equivalent*:

(i) $\mathbb{Z}_m \leq \mathbb{Z}_n$;
(ii) $\mathbb{Z}_m \leq S_n$;
(iii) *every prime divisor of* m *also divides* n.

Proof. (i) $\longrightarrow$ (ii) *a fortiori*, and (ii) $\longrightarrow$ (iii) by Corollary 3.3. Finally, we notice that (iii) $\longrightarrow$ (i) was already proved in Theorem 1.5. ∎

Obviously, in the last Corollary, one may replace $\mathbb{Z}_n$ in (i), or S_n in (ii), by any group H with $\mathbb{Z}_n \subseteq H \subseteq S_n$. We leave the details to the reader.

COROLLARY 3.5 $(m \geq 3)$ *If* $A_m \leq S_n$, *then every prime divisor of* m *also divides* n.

Proof. If m is odd, this is immediate from Corollary 3.3. In any case, if p is an ***odd*** prime dividing m, say $m = ps$, then in A_m there exists a permutation α of order p with no fixed points (a product of s p-cycles). Applying Theorem 3.2 to G = the group generated by α, we immediately obtain $p|n$.

It remains to be seen that if m is even (say $m = 2s$), then n is also even. If s itself is even, then A_m contains a permutation α of order 2 with no fixed points, and we may conclude the proof as above. On the other hand, if $m =$

$2s$ with s odd, then $m \geq 6$, and so we may define

$$\alpha = (12)(34)$$
$$\beta = (34)(56)\cdots((m-1)m) .$$

Now α and β generate a subgroup G of A_m with no common fixed points and such that $G \cong \mathbb{Z}_2 \times \mathbb{Z}_2$. Therefore n is even, by Theorem 3.2. ∎

[Here and in the next section, our convention for multiplying permutations is that $(\alpha\beta)(i) = \alpha(\beta(i))$; thus, for instance, if $\alpha = (123)(45)$ and $\beta = (35)$, then $\alpha\beta = (12345)$.]

4. Embeddings $A_m \longrightarrow S_n$

For $m \geq 3$, it is clear that A_m has no common fixed points, and so our main principle, Corollary 2.4, tells us that if $A_m \leq S_n$, then there exists a homomorphism μ: $A_m \longrightarrow S_n$ with no common fixed point. Incidentally, the converse is false: $A_6 \nleq S_{15}$ (by Corollary 3.5), but such a μ: $A_6 \longrightarrow S_{15}$ does exist, in fact we even have μ: $S_6 \longrightarrow S_{15}$. (Each permutation of $\{1, \cdots ,6\}$ yields a permutation of the 15 two-element subsets of $\{1, \cdots ,6\}$.)

The main theorem of this section will allow us to apply this version of Corollary 2.4, since it states that such a homomorphism μ does not exist (for many values of m and n).

Of course, we already saw, in Theorem 3.1, that $A_m \leq S_n$ is impossible for $n < m$. But for $n \geq m$, the relation $A_m \leq S_n$ seems very complicated (i.e., the corresponding collection of ordered pairs (m,n) has no simple description known to us). Of course $A_m \leq S_m$ and $A_m \leq S_{km}$ for every positive integer k, as described in the introduction; likewise we already saw $A_4 \leq S_6$. The main result of this paper is that $A_m \nleq S_n$ in all remaining cases with $m < n < 2m$, and, more generally, if $12 \leq m \leq n \leq \frac{1}{2}m(m-1)$, then $A_m \leq S_n$ iff $m|n$.

To fix notation (and in slight disagreement with the notation laid out in the Introduction), for the next three lemmas we will say that:

A_m is the alternating group on $\{1, \cdots ,m\}$,
A_{m-1} is the alternating group on $\{2, \cdots ,m\}$ and
A_{m-2} is the alternating group on $\{3, \cdots ,m\}$.

LEMMA 4.1 A_{m-1} *is a maximal subgroup of* A_m. ∎

LEMMA 4.2 *If* $A_{m-2} \subseteq G \subseteq A_m$, *then* $G \cong A_{m-2}$, S_{m-2}, A_{m-1} *or* A_m.

Proof. If $G \neq A_{m-2}$, then G contains a permutation $\alpha \notin A_{m-2}$. Thus α has one or more cycles involving 1 or 2 or both. Since G is a group containing A_{m-2}, these cycles may be shortened to contain at most two members of $\{3,$

$\cdots, m\}$. If the cycle containing 1 is of even length (and hence is an odd permutation), then its length can be made odd, as we illustrate by the following example:

(13)(56) could be multiplied by (34)(56):

$$(13)(56)\cdot(34)(56) = (134).$$

Therefore G must contain one of the following four sorts of permutations:

(a) (12)(34)

(b) (134)(256)

(c) (234)

(d) (123).

Now clearly in case (a), if $\alpha = (12)(34)$, then $A_{m-2} \cup \{\alpha\}$ generates all permutations

$$A_{m-2} \cup (12)\cdot\left[S_{m-2} - A_{m-2} \right],$$

and these form a subgroup isomorphic to S_{m-2}. Either this group is all of G, in which case we're done, or G contains a permutation of type (b), (c) or (d). Therefore we need consider only cases (b), (c) and (d).

Case (b) yields

$$\left[(134)(256)\right]\cdot(478)\cdot\left[(134)(256)\right]^{-1} = (178),$$

and case (d) yields

$$(123)\cdot(345)\cdot(123)^{-1} = (145).$$

Therefore all three cases lead to a case like (c). But for case (c) we simply observe that $A_{m-2} \cup \{(234)\}$ generates A_{m-1}. Therefore this lemma now follows from Lemma 4.1. ∎

LEMMA 4.3 (Serret, Bertrand, *et al.*, prior to 1870.) *For* $m \geq 12$, *let* G *be any subgroup of* A_m. *If the index* $[A_m:G]$ *is* $\leq \frac{1}{2}m(m-1)$, *then* $[A_m:G]$ *is equal to* 1, m *or* $\frac{1}{2}m(m-1)$.

Proof. If $G \supseteq A_{m-2}$, then the result follows from Lemma 4.2. In fact, this line of reasoning applies to $G \supseteq A$ for A equal to the alternating group on any $(m-2)$-element subset of $\{1, \cdots, m\}$. Therefore we may assume that G contains no such alternating group. Therefore, according to an ancient result ('Théorème') found on page 68 of C. Jordan [3], we have $[A_m:G] \geq \frac{1}{2}m(m-1)$. ∎

THEOREM 4.4 *If* $12 \leq m \leq n < \frac{1}{2}m(m-1)$ *and* n *is not a multiple of* m, *then every homomorphism* $f\colon A_m \longrightarrow S_n$ *has a common fixed point.*

Proof. As we mentioned above, in the proof of 3.2), for each $i \in \{1, \cdots, n\}$, the number of points in the orbit of i under the action of A_m on $\{1, \cdots, n\}$ is the index $[A_m:G_i]$ of a certain subgroup $G_i \subseteq A_m$. Thus by Lemma 4.3, each of these orbits contains 1, m, or at least $\frac{1}{2}m(m-1)$ points. The last alternative is impossible since $n < \frac{1}{2}m(m-1)$, and so each orbit has either m points or 1 point. Since n is not a multiple of m, there clearly must be at least one singleton orbit, which is the same thing as a common fixed point of f. ∎

For some low values of m and n there does exist a homomorphism $A_m \longrightarrow S_n$ (even $S_m \longrightarrow S_n$) with no common fixed point. For instance if $m = 5$ and $n = 6$, there is a famous example, which we first learned from Barry Wolk and which occurs already on page 67 of Jordan [3]. The symmetric group on 5 letters has exactly 6 subgroups of order 5, and S_5 acts on these (by inner automorphism) without a common fixed point. And also S_6 can be made to act on S_{10} by considering the fact that there are exactly 10 partitions of $\{1, \cdots, 6\}$ into two 3-element sets. Nevertheless we have $A_5 \not\leq S_6$ and $A_6 \not\leq A_{10}$, both by Corollary 3.5. We do not in fact know the exact range of validity of the assertion in 4.4 for values of m below 12, but it will turn out that we have other ways of examining $A_m \leq S_n$ for low values of m (see 4.6 below).

COROLLARY 4.5 *If* $12 \leq m \leq n < \frac{1}{2}m(m-1)$, *then* $A_m \leq S_n$ *if and only if* n *is a multiple of* m. ∎

Clearly the upper bound $\frac{1}{2}m(m-1)$ is the best possible, for, as we saw in the Introduction, if m is a multiple of 4 then $A_m \leq S_n$ for $n = \binom{m}{2} = \frac{1}{2}m(m-1)$. Nevertheless we can improve somewhat the lower bound 12, at least for the range $m \leq n \leq 2m$.

THEOREM 4.6 *If* $4 < m < n < 2m$, *then* $A_m \not\leq A_n$ *and* $S_m \not\leq S_n$. *Moreover* $A_m \not\leq S_n$, *except possibly for the case* $m = 9$, $n = 15$.

Proof. For $m \geq 12$, the Theorem follows from Corollary 4.5, and for m squarefree it follows from Corollary 3.5. The only remaining m values above 4 are $m = 8$ and $m = 9$. By Corollary 3.5, therefore, the Theorem needs to be established only for the following values of (m,n): $(8,10)$, $(8,12)$, $(8,14)$, $(9,12)$ and $(9,15)$. We will prove in detail only the case $A_9 \not\leq A_{15}$, as it is typical of the proof in all the other cases.

By Corollary 2.4, it is enough to prove that any embedding of A_9 into A_{15}

has a common fixed point. It is convenient to let $\overline{\alpha}$ denote the image under this embedding of an arbitrary element α of A_9. Thus our objective is to find some $i \in \{1, \cdots, 15\}$ such that $\overline{\alpha}(i) = i$ for all $\overline{\alpha}$.

In A_9, let $\alpha = (12345)$, $\beta = (67)(89)$ and $\gamma = (678)$. Of course $\overline{\alpha}$ is a product of 5-cycles, and $\overline{\gamma}$ is a product of 3-cycles. We first claim at least one of $\overline{\alpha}$ and $\overline{\gamma}$ consists of a single cycle. For a proof by contradiction, we will assume that $\overline{\alpha}$ has at least two 5-cycles and $\overline{\gamma}$ has at least two 3-cycles. In other words, $|\operatorname{Supp} \overline{\alpha}| \geq 10$ and $|\operatorname{Supp} \overline{\gamma}| \geq 6$. Since the combined supports have at most 15 elements, we must have $\operatorname{Supp} \overline{\alpha} \cap \operatorname{Supp} \overline{\gamma} \neq \emptyset$. Now the reader may check that if μ and λ are any commuting permutations of orders 3 and 5 (or any two distinct primes), and if $x \in \operatorname{Supp} \mu \cap \operatorname{Supp} \lambda$, then the $\{\mu,\lambda\}$-orbit of x contains 15 (generally pq) points, on which μ acts as five 3-cycles and λ acts as three 5-cycles. Thus we see that, here, $\overline{\alpha}$ has three 5-cycles: w.l.o.g.

$$\overline{\alpha} = (1\ 2\ 3\ 4\ 5)\cdot(6\ 7\ 8\ 9\ 10)\cdot(11\ 12\ 13\ 14\ 15) .$$

Now $\overline{\beta}$ is a permutation of order 2, and, w.l.o.g., $1 \in \operatorname{Supp} \overline{\beta}$. Again by Lemmas 1.3 and 1.5, the $\overline{\alpha}\cdot\overline{\beta}$-orbit of 1 contains 10 elements and $\overline{\beta}$ acts on this orbit as a product of five disjoint 2-cycles. Since $\overline{\beta} \in A_{15}$, i.e. since $\overline{\beta}$ is even, $\overline{\beta}$ must have at least one more point in its support. This means that in fact there is a point, not in the $\overline{\beta}$-orbit of 1, which is moved by both $\overline{\alpha}$ and $\overline{\beta}$ This must lead to a second non-trivial $\overline{\alpha}\cdot\overline{\beta}$-orbit, also of 10 points. As a result, we obtain 20 distinct elements of $\{1, \cdots, 15\}$, a contradiction which establishes our claim that at least one of $\overline{\alpha}$ and $\overline{\gamma}$ consists of a single cycle.

We now claim that $\overline{\gamma}$ consists of a single 3-cycle. We know that either $\overline{\alpha}$ or $\overline{\gamma}$ is a single cycle; and if this is true of $\overline{\gamma}$, then our claim is established. We may therefore assume that $\overline{\alpha}$ consists of a single 5-cycle; from this we will deduce that in fact $\overline{\gamma}$ is a single 3-cycle.

By inner automorphism we see that in fact if γ is any 5-cycle, then $\overline{\gamma}$ must also consist of a single 5-cycle. Let $\delta = (1234567)$. From the equation

$$(1\ 2\ 3\ 7\ 4)\cdot(3\ 7\ 4\ 5\ 6) = (1\ 2\ 3\ 4\ 5\ 6\ 7)$$

we see that the support of $\overline{\delta}$ can contain at most 10 elements, and hence $\overline{\delta}$ must consist of a single 7-cycle. Again by inner automorphism, we see that if δ is any

7-cycle, then $\overline{\delta}$ must also consist of a single 7-cycle.

Continuing our proof that $\overline{\gamma}$ is a single 3-cycle for $\gamma = (678)$, we take $\sigma = (23456)$ and observe that $\sigma \cdot \gamma = (2345678)$, which is a 7-cycle. Therefore we have $\overline{\sigma}$ a 5-cycle and $\overline{\sigma} \cdot \overline{\gamma}$ a 7-cycle, and we may as well suppose that $\mathrm{Supp}\, \overline{\sigma} = \{1, \cdots, 5\}$. Similar reasoning shows that in fact $\overline{\sigma}^i \overline{\gamma}^j$ is a 7-cycle for $i = 1, \cdots, 4$ and $j = 1, 2$. Since 3 does not divide the order of $\overline{\sigma} \cdot \overline{\gamma}$, it is evident that each 3-cycle occurring in $\overline{\gamma}$ must contain one point of $\{1, \cdots, 5\}$. Let us prove that no such 3-cycle contains exactly two elements a, b of $\{1, \cdots, 5\}$. Now $\overline{\sigma}^i(a) = b$ for some i. Therefore, after renaming elements in $\{1, \cdots, 15\}$, we have $\overline{\sigma}^i = (12345)$ and $\overline{\gamma}^j$ contains (126). Therefore $\overline{\sigma}^i \cdot \overline{\gamma}^j$ is equal to

$$(1\ 2\ 3\ 4\ 5) \cdot (1\ 2\ 6) = (1\ 3\ 4\ 5) \cdot (2\ 6)$$

times some 3-cycles which do not involve 2 or 6. Therefore $\overline{\sigma}^i \cdot \overline{\gamma}^j$ contains a 2-cycle and hence has even order, in contradiction to the fact that it is known to have order 7. Therefore we know that none of the 3-cycles of $\overline{\gamma}$ contains exactly two points of $\{1, \cdots, 5\}$. Finally let us prove that none of the 3-cycles in $\overline{\gamma}$ has all its points in $\{1, \cdots, 5\}$. In this case some power $\overline{\sigma}^i$ will separate the two points of $\{1, \cdots, 5\}$ which are not in that $\overline{\gamma}$-cycle: w.l.o.g. $\overline{\sigma}^i = (12345)$ with $\overline{\gamma}$ containing the cycle (135) or (153). Therefore for some j, $\overline{\sigma}^i \cdot \overline{\gamma}^j$ contains

$$(1\ 2\ 3\ 4\ 5) \cdot (1\ 5\ 3) = (2\ 3) \cdot (4\ 5)$$

times some 3-cycles which each contain exactly one point of $\{2,4\}$. These extra 3-cycles can turn either $(2\ 3)$ or $(4\ 5)$ into a 4-cycle, but they retain the even order of $\overline{\sigma}^i \cdot \overline{\gamma}^j$, which, as before, is a contradiction.

Therefore we see that each 3-cycle of $\overline{\gamma}$ contains exactly one point in the orbit $\{1, \cdots, 5\}$ of $\overline{\sigma}$. Now from the fact that $\overline{\sigma} \cdot \overline{\gamma}$ is a 7-cycle it is immediate that there is only one 3-cycle to $\overline{\gamma}$. Thus, by inner automrphism, we have established that if γ is any 3-cycle in A_9, then $\overline{\gamma}$ is a single 3-cycle of A_{15}.

For the remainder of the proof, we define $\alpha = (123)$, $\beta = (3456789)$ and $\gamma = (123456789)$ in A_9. In A_{15} we have

$$\overline{\gamma} = \overline{\alpha} \cdot \overline{\beta},$$

and so $\mathrm{Supp}\, \overline{\gamma}$ is a subset of the 10-element set $\mathrm{Supp}\, \overline{\alpha} \cup \mathrm{Supp}\, \overline{\beta} = T$. If we now define

$$S := \{ \delta \in A_9 \ : \ \mathrm{Supp}\,\overline{\delta} \subseteq T \},$$

we see that S is a subgroup of A_9 containing $\alpha = (123)$ and $\gamma = (3456789)$. It is well known and easy to check that these two cycles generate S_9, and hence $S = A_9$. Therefore there is a 10-element subset supporting $\overline{\alpha}$ for every $\alpha \in A_9$, and therefore at least five common fixed points. ∎

We have now completed the proof of the main results of this paper. To summarize, we have a quadratic function $Q(m)$ (from Corollary 4.5) and an exponential function $E(m)$ (from §3) such that (1) in the range

$$m \leq n \leq Q(m)$$

the relation $A_m \leq S_n$ is characterized simply by the divisibility relation $m|n$, and (2) in the range

$$E(m) \leq n$$

$A_m \leq S_n$ holds if and only if each prime divisor of m also divides n. For the large range

$$Q(m) \leq n \leq E(m),$$

the relation $A_m \leq S_n$ remains mysterious.

References

1. S. Burris and H. P. Sankappanavar, ***A course in Universal Algebra***, Springer-Verlag, Berlin, 1980.

2. O. C. García and W. Taylor, ***The Lattice of Interpretability Types of Varieties***, AMS Memoirs, vol. 50, # 305, 1984, v + 125 pages.

3. C. Jordan, ***Traité des Substitutions et des Equations Algébriques***, Gauthier-Villars, Paris, 1870.

4. S. MacLane and G. Birkhoff, ***Algebra***, Macmillan, New York, 1968.

5. R. McKenzie, Negative solution of the decision problem for sentences true in every subalgebra of ⟨N,+⟩, J. Symbolic Logic 36 (1972), 607-609.

6. W. D. Neumann, On Mal'cev conditions, J. Austral. Math. Soc. 17 (1974), 376-384.

THE WORD AND ISOMORPHISM PROBLEMS IN UNIVERSAL ALGEBRA

A. M. W. Glass[1]
Bowling Green State University
Bowling Green, Ohio 43403

In this article I provide a short survey of recent developments in the study of word and isomorphism problems in algebra and their interrelationship. References are provided for those who wish to master the details which are omitted here.

Recall that a *finitely presented algebra* $\mathfrak{A} = \langle x_1,\ldots,x_m;\ p_i(\underset{\sim}{x}) = q_i(\underset{\sim}{x})\rangle_{1 \leq i \leq n}$ in a variety $\mathfrak{V}$ is the quotient of F the free in $\mathfrak{V}$ algebra on $x_1,\ldots,x_m$ by the congruence θ generated by the set of pairs $\{\langle p_i(\underset{\sim}{x}),q_i(\underset{\sim}{x})\rangle:\ 1 \leq i \leq n\}$. Such an algebra $\mathfrak{A}$ is said to have *soluble word problem* if there is a (recursive) algorithm which determines for any $p(\underset{\sim}{x}),\ q(\underset{\sim}{x}) \in F$ whether or not $\langle p(\underset{\sim}{x}),q(\underset{\sim}{x})\rangle \in \theta$; intuitively, if $p(\underset{\sim}{x}) = q(\underset{\sim}{x})$ in $\mathfrak{A}$. If every finitely presented algebra in $\mathfrak{V}$ has soluble word problem, then $\mathfrak{V}$ is said to have *soluble word problem;* if one algorithm works for all finitely presented algebras in $\mathfrak{V}$, I will say that $\mathfrak{V}$ has *uniformly soluble word problem.* If $\mathfrak{V}$ is recursively axiomatised, then there is a recursive listing $\{\mathfrak{A}_i:\ i \in \omega\}$ of all finitely presented algebras in the variety ($\omega = \{0,1,2,3,\ldots\}$). $\mathfrak{V}$ is said to have *soluble isomorphism problem* if there is an algorithm which determines whether or not two finitely presented algebras in $\mathfrak{V}$ are isomorphic; more formally, if $\{\langle i,j\rangle:\ \mathfrak{A}_i \cong \mathfrak{A}_j\}$ is a recursive set.

In most studied cases of recursively axiomatised varieties of algebras, the word and isomorphism problems have the same solution. For loops and abelian groups, for example, both are soluble [7,8; 20, Section 3.3]; for semigroups, groups and lattice-ordered groups, for example, both are insoluble [25,21,24,5,1,26,13,11]. Indeed, there is an underlying theorem here (due to Adian [1] and Rabin [26] for groups, but easily extendable to recursively axiomatised varieties of algebras).

I assume without further mention that $\mathfrak{V}$ is recursively axiomatised and

(i) there is a finitely presented algebra $\mathfrak{A}_0 \in \mathfrak{V}$ such that the free product (in $\mathfrak{V}$) of $\mathfrak{A}_0$ and any algebra $\mathcal{L} \in \mathfrak{V}$ is isomorphic to $\mathcal{L}$, and

(ii) for any $\mathfrak{A},\mathcal{L} \in \mathfrak{V}$, $\mathfrak{A}$ and $\mathcal{L}$ are naturally embedded in their free product in $\mathfrak{V}$.

Let $\mathfrak{A} = \langle x_1,\ldots,x_m;\ p_i(\underset{\sim}{x}) = q_i(\underset{\sim}{x})\rangle_{1 \leq i \leq n}$ and $p(\underset{\sim}{x}),\ q(\underset{\sim}{x}) \in F$, the free in $\mathfrak{V}$

---

[1]The author wishes to thank the organising committee for its invitation to speak at the meeting, and especially S. D. Comer for making the meeting both stimulating and enjoyable. I wish to thank The Citadel, N.S.F. and Bowling Green State University Faculty Research Committee for funding.

algebra on $x_1,\ldots,x_m$. I call $\mathfrak{A}(p,q) = \langle x_1,\ldots,x_m,x_{m+1},\ldots,x_r;\ p_i(x_1,\ldots,x_m) = q_i(x_1,\ldots,x_m), p'_j(x_1,\ldots,x_r) = q'_j(x_1,\ldots,x_r)\rangle_{1\le i\le n, 1\le j\le s}$ a *(p,q)-messuage*[2] of $\mathfrak{A}$ if

(a) $\mathfrak{A}(p,q) \cong \mathfrak{A}_0$ if $p(x_1,\ldots,x_m) = q(x_1,\ldots,x_m)$ in $\mathfrak{A}$, and

(b) $\mathfrak{A}$ can be embedded in $\mathfrak{A}(p,q)$ if $p(x_1,\ldots,x_m) \neq q(x_1,\ldots,x_m)$ in $\mathfrak{A}$.

Of course, p and q will usually play a role in some of the defining relations p'_j, q'_j $(1 \le j \le s)$.

A recursively axiomatised variety $\mathfrak{V}$ is said to satisfy the *Messuage Lemma* if: There is an effective uniform procedure which constructs messuages for all finitely presented algebras $\mathfrak{A} \in \mathfrak{V}$ and p,q in the free in $\mathfrak{V}$ algebra on the generators of $\mathfrak{A}$ [11].

The following theorem is easy to prove [1], [26] and [11]:

THEOREM 1. *Let* $\mathfrak{V}$ *be a recursively axiomatised variety containing a finitely presented algebra with insoluble word problem. If the Messuage Lemma holds for* $\mathfrak{V}$, *there is no algorithm to determine for an arbitrary finitely presented algebra in* $\mathfrak{V}$ *whether or not it is isomorphic to* $\mathfrak{A}_0$.

Indeed, a property $\mathfrak{P}$ of finitely presented algebras in $\mathfrak{V}$ is said to be *Markov* [21] if

(i) for all finitely presented $\mathfrak{A},\mathfrak{L} \in \mathfrak{V}$, $\mathfrak{A} \cong \mathfrak{L}$ and $\mathfrak{L}$ satisfies $\mathfrak{P}$ implies that $\mathfrak{A}$ satisfies $\mathfrak{P}$,

(ii) there is a finitely presented $\mathfrak{A}_1 \in \mathfrak{V}$ which enjoys $\mathfrak{P}$, and

(iii) there is a finitely presented $\mathfrak{A}_2 \in \mathfrak{V}$ which cannot be embedded in any finitely presented algebra in $\mathfrak{V}$ enjoying $\mathfrak{P}$.

Any hereditary property is Markov, as is being isomorphic to $\mathfrak{A}_0$ if $\mathfrak{V}$ is non-trivial. The proof of Theorem 1 actually establishes ([1], [26] and [11]):

THEOREM 2. *Under the hypotheses of Theorem 1, no Markov property is decidable; i.e., for any Markov property* $\mathfrak{P}$, *there is no algorithm which determines for an arbitrary finitely presented algebra in* $\mathfrak{V}$ *whether or not it satisfies* $\mathfrak{P}$.

The delicate part is to determine whether, in a particular variety, the Messuage Lemma holds. For groups see [19, §IV.4] and for lattice-ordered groups see [11]. Free modular lattices on at least four generators have insoluble word problem [9] and [16]. If the variety of all modular lattices satisfies the Messuage Lemma, all Markov properties (in particular, the isomorphism problem) will be undecidable for modular lattices. Similarly, the Messuage Lemma for lattice-ordered semigroups is worth investigating in view of the insolubility of the word problem [28].

The mere existence of Messuages can prove useful in surprising ways.

THEOREM 3. *If every countable algebra in* $\mathfrak{V}$ *can be embedded in a finitely*

[2] Messuage was "originally the portion of land intended to be occupied, or actually occupied, as a site for a dwelling house and its appurtenances." [Oxford English Dictionary]

generated algebra in $\mathfrak{V}$ *and every finitely generated member of* $\mathfrak{V}$ *has messuages in* $\mathfrak{V}$, *then every countable algebra in* $\mathfrak{V}$ *can be embedded in a finitely generated simple member of* $\mathfrak{V}$.

The technique is due to C. F. Miller III and can be found in [19, Theorem IV.3.5] for groups and [10] for lattice-ordered groups. It is sufficiently general that the reader will have no difficulty in adapting it to prove the above Theorem.

I next consider the naturally arising question: Is it possible for the word problem for a recursively axiomatised variety $\mathfrak{V}$ to be soluble but yet the isomorphism problem for $\mathfrak{V}$ is insoluble?

If the restriction that $\mathfrak{V}$ be a variety is removed, the answer is yes, since C. F. Miller III [23, p. 77] shows that there are recursive classes of finitely presented groups with uniformly soluble word problem but insoluble isomorphism problem. Also, we might expect the answer to be yes if the solution of the word problem for finitely presented algebras in $\mathfrak{V}$ is not uniform. However, in the finitely axiomatised variety $\mathfrak{V}_0$ of abelian lattice-ordered groups the universal theory is decidable [18]. This follows from the facts that any two non-trivial abelian linearly ordered groups satisfy the same universal sentences [15], every abelian lattice-ordered group is a subdirect product of linearly ordered abelian groups [3] and the Feferman-Vaught Theorem [6] (for this modification of Khisamiev's proof, see [12]). Hence the word problem for $\mathfrak{V}_0$ is uniformly soluble: $w(\underset{\sim}{x}) = 0$ in $\langle x_1,\dots,x_m;\ w_1(\underset{\sim}{x}) = 0,\dots,w_n(\underset{\sim}{x}) = 0\rangle$ if and only if the universal sentence $\forall x_1,\dots,x_m[w_1(\underset{\sim}{x}) = 0\ \&\ \dots\ \&\ w_n(\underset{\sim}{x}) = 0 \rightarrow w(\underset{\sim}{x}) = 0]$ holds in all abelian lattice-ordered groups, and Khisamiev's result allows me to determine whether or not this is the case.

Now the free abelian lattice-ordered group F_m on $x_1,\dots,x_m$ can be regarded as the sublattice subgroup of the additive group of continuous functions from $\mathbb{R}^m$ into $\mathbb{R}$ generated by the m projections ($\mathbb{R}$, the reals). For each $w(\underset{\sim}{x}) \in F$, $Z(w) = \{\underset{\sim}{\alpha} \in \mathbb{R}^m\colon\ w(\underset{\sim}{x})(\underset{\sim}{\alpha}) = 0\}$ is a rational closed polyhedral cone with vertex $\underset{\sim}{0}$ and any rational closed polyhedral cone with vertex the origin is equal to $Z(w)$ for some $w(\underset{\sim}{x}) \in F$. Furthermore, $\langle x_1,\dots,x_m;\ w(\underset{\sim}{x}) = 0\rangle \cong \langle x_1,\dots,x_m;\ u(\underset{\sim}{x}) = 0\rangle$ if and only if $Z(w)$ is piecewise homogeneous linear homeomorphic to $Z(u)$ [2]; from $w(\underset{\sim}{x})$ we can effectively obtain a finite presentation for $Z(w)$ and conversely [14]. But Markov [22] has proved that there is no algorithm to determine whether or not two arbitrary finitely presented polyhedra of dimension 4 are piecewise linearly homeomorphic. This immediately translates into:

THEOREM 4. [14] *The finitely axiomatised variety of abelian lattice-ordered groups has uniformly soluble word problem but there is no effective procedure to determine whether or not two arbitrary* 10 *generator* 1 *relator abelian lattice-ordered groups are isomorphic. However, there is an algorithm to determine whether or not an arbitrary finitely presented abelian lattice-ordered group has only one element; i.e., is isomorphic to* $\mathfrak{A}_0 = \langle x_1;\ x_1 = 0\rangle$.

Constrast the conclusion--especially the "however" versus the isomorphism problem--with Theorems 1 and 2.

The variety of abelian lattice-ordered groups is contained in every non-trivial variety of lattice-ordered groups [29]. Since $\{\langle x_1,\ldots,x_{10};\ w(\underset{\sim}{x}) = 1 \text{ and } x_ix_j = x_jx_i\rangle_{1\leq i,j\leq 10}:\ w(\underset{\sim}{x}) \in F\}$ is a recursive listing of all 10 generator 1 relator abelian lattice-ordered groups in any non-trivial variety $\mathfrak{V}$ of lattice-ordered groups (where F is the free in $\mathfrak{V}$ algebra on $x_1,\ldots,x_{10}$), I obtain:

COROLLARY 5. [14] *Any non-trivial recursively axiomatised variety of lattice-ordered groups has insoluble isomorphism problem.*

In particular, the variety of nilpotent class n lattice-ordered groups has insoluble isomorphism problem for each $2 \leq n \in \omega$. Since Kopytov [17] has shown that each such variety has uniformly soluble word problem, it would be interesting to determine which of these varieties have soluble conjugacy problem (a variety $\mathfrak{V}$ in a language which includes that of groups is said to have *soluble conjugacy problem* if for each finitely presented $\mathfrak{A} \in \mathfrak{V}$ there is an algorithm to determine for any $p(\underset{\sim}{x})$, $q(\underset{\sim}{x})$ belonging to F the free in $\mathfrak{V}$ algebra on the generators of $\mathfrak{A}$ whether or not there is $w(\underset{\sim}{x}) \in F$ such that $w(\underset{\sim}{x})^{-1}p(\underset{\sim}{x})w(\underset{\sim}{x}) = q(\underset{\sim}{x})$ in $\mathfrak{A}$.) Note that any recursively axiomatised variety of nilpotent groups has soluble conjugacy problem [4]

Finally, the isomorphism problem for commutative rings should be considered since this variety, like abelian lattice-ordered groups, has uniformly soluble word problem [27].

REFERENCES

1. S. I. Adian, On algorithmic problems in effectively complete classes of groups, *Dokl. Akad. Nauk. SSSR*, 123(1958), 13-16 (in Russian).

2. W. M. Beynon, Applications of duality in the theory of finitely generated lattice-ordered abelian groups, *Canad. J. Math.*, 24(1977), 243-254.

3. A. Bigard, K. Keimel and S. Wolfenstein, *Groupes et anneaux réticulés*, Lecture Notes in Math. 608 (Springer, Berlin 1977).

4. N. Blackburn, Conjugacy in nilpotent groups, *Proc. Amer. Math. Soc.*, 16(1965), 143-148.

5. W. W. Boone, The word problem, *Ann. of Math.*, 70(1959), 207-265.

6. C. C. Chang and H. J. Keisler, *Model Theory* (N. Holland, Amsterdam 1973).

7. T. Evans, The word problem for abstract algebras, *J. London Math. Soc.*, 26(1951), 64-71.

8. T. Evans, The isomorphism problem for some classes of multiplicative systems, *Trans. Amer. Math. Soc.*, 109(1963), 303-312.

9. R. Freese, Free modular lattices, *Trans. Amer. Math. Soc.*, 261(1980), 81-91.

10. A. M. W. Glass, Countable lattice-ordered groups, *Math. Proc. Cambridge Phil. Soc.*, 94(1983), 29-33.

11. A. M. W. Glass, The isomorphism problem and undecidable properties for finitely presented lattice-ordered groups, *Orders: description and roles*, Ann. of Discrete Math. (ed. M. Pouzet and D. Richard), (N. Holland, Amsterdam 1984), pp. 157-170.

12. A. M. W. Glass, The universal theory of lattice-ordered abelian groups, (unpublished notes).

13. A. M. W. Glass and Y. Gurevich, The word problem for lattice-ordered groups, *Trans. Amer. Math. Soc.*, 280(1983), 127-138.

14. A. M. W. Glass and J. J. Madden, The word problem versus the isomorphism problem, *J. London Math. Soc.*, 30(1984), 53-61.

15. Y. Gurevich and A. I. Kokorin, Universal equivalence of ordered abelian groups, *Algebra i Logika*, 2(1963), 37-39 (in Russian).

16. C. Hermann, On the word problem for the modular lattice with four free generators, *Math. Ann.*, 265(1983), 513-527.

17. V. M. Kopytov, Lattice-ordered locally nilpotent groups, *Algebra and Logic*, (14)4, 1975, 249-251.

18. N. G. Khisamiev, Universal theory of lattice-ordered abelian groups, *Algebra i Logika*, 5(1966), 71-76 (in Russian).

19. R. C. Lyndon and P. E. Schupp, *Combinatorial group theory*, Ergebnisse der Math. und ihrer Grenzgebiete, 89 (Springer, Berlin 1977).

20. W. Magnus, A. Karrass and D. Solitar, *Combinatorial group theory* (Wiley, New York 1966).

21. A. A. Markov, On the impossibility of certain algorithms in the theory of associative systems, *Dokl. Akad. Nauk. SSSR*, 55(1947), 583-586.

22. A. A. Markov, Insolubility of the problem of homeomorphy, *Proc., International Congress of Math.*, 1958 (ed. J. A. Todd, FRS, University Press, Cambridge, 1960), pp. 300-306.

23. C. F. Miller III, *On group-theoretic decision problems and their classification*, Annals of Math. Studies 68 (University Press, Princeton, 1971).

24. P. S. Novikov, On the algorithmic unsolvability of the word problem in groups, *Trudy Mat. Inst. Steklov*, 44, 143, 1955 (in Russian).

25. E. L. Post, Recursive unsolvability of a problem of Thue, *J. Symbolic Logic*, 12(1947), 1-11.

26. M. O. Rabin, Recursive unsolvability of group theoretic problems, *Ann. of Math.*, 67(1958), 172-194.

27. H. Simmons, The solution of a decision problem of several classes of rings, *Pacific J. Math.*, 34(1970), 547-557.

28. A. Urquhart, The undecidability of entailment and relevant implication, *J. Symbolic Logic*, 49(1984), 1059-1073.

29. E. C. Weinberg, Free lattice-ordered abelian groups I & II, *Math. Ann.*, 151(1963), 187-199; ibid., 159(1965), 217-222.

LINEAR LATTICE PROOF THEORY: AN OVERVIEW

Mark Haiman
Massachusetts Institute of Technology
Cambridge, Massachusetts 02139

1. Introduction. This paper concerns lattices representable by commuting equivalence relations (Jónsson's *type I* lattices), which we propose to call *linear lattices* in order to evoke their archetypal examples, the (coordinate) projective geometries. The results discussed here are taken from the author's doctoral thesis [**12**] and will appear in full detail in a separate publication adapted from it [**13**]. Therefore we give here only a brief historical introduction to the subject, followed by statements and proof sketches of the main results, with applications and examples.

The inventors and early exponents of modern lattice theory—Boole, C. S. Peirce, and Schröder; and later Dedekind, Öre, Birkhoff, Von Neumann, Dilworth, and others—intended their invention to bring a new common algebraic language and technique primarily to two fundamental examples. One was set theory, where the lattice operations were intersection and union of sets or, from another point of view, conjunction and disjunction of logical propositions. The other was algebra, where the quotient structures of groups, rings, modules, and vector spaces provided naturally occurring lattices. In the vector space case particularly, lattice operations comprehended the elementary operations of projective geometry: intersecting two flats; and forming their linear hull (generalizing the drawing of a line through two points).

It was soon realized that the distributive law was a defining property for lattices of sets (although Peirce seems to have thought it followed from the lattice axioms). Word problems for these lattices could be solved; whether by using conjunctive and disjunctive normal forms, by truth tables, or, of special relevance for us, by using such "axiom systems" for propositional logic as those of Hilbert or Gentzen in which the conclusion of an implication (i.e., the right-hand side of a distributive lattice inequality) is derived from the hypothesis (left-hand side) using deduction rules from a suitable list. The Venn diagram, moreover, provided a visual aid making distributive lattice calculations easy and intuitive.

For the other classical example—of quotient lattices in algebra—the situation was and remains not nearly so simple and satisfying. One problem is that Dedekind's "modular law" incompletely axiomatizes these lattices; some such stronger axiom as Jónsson's Arguesian law [**18, 19**] is needed even to prove Desargues' theorem of projective geometry lattice-theoretically. Another problem is the difficulty of the available proofs of Von Neumann's coordinatization theorem for complemented modular lattices [**1, 7, 15, 23, 29**]—the modular analog of Stone's Boolean algebra representation theorem. A final problem is that modular calculations are notoriously difficult, and in general impossible, in view of Freese's [**9**] result (later improved by Herrman [**14**]) that the free modular lattice word problem is unsolvable.

Since all the classical algebraic entities are Mal'cev algebras [**27**], their congruence lattices are linear (i.e., consist of commuting equivalence relations) by Mal'cev's theorem [**25, 11**]. Our guiding purpose in this paper and forthcoming ones is to advance the contention that linear lattices provide a better framework than modular lattices for the study of the classical quotient structures, and in particular of projective geometry; that in this framework many of the difficulties that plague modular lattices can be reduced or eliminated.

While the importance in universal algebra of linear *congruence* lattices is universally recognized, linear lattices *per se* have gone curiously unstudied since Jónsson [**20, 21, 22, 23**] characterized them by (finitary) Horn sentences, proved they are Arguesian, and fit them by degree of permutability into a neat taxonomy in which *type 0* = distributive lattices, *type I* = linear lattices, *type II* = modular lattices, and *type III* = all lattices.

In this paper we resurrect the study of linear lattices and continue it with two theorems which together form a linear lattice proof theory. Theorem 1 gives a linear analog of the Hilbert- or Gentzen-type deductive systems which solve the distributive lattice word problem. Theorem 2, a normal form theorem for the proofs given by Theorem 1, can be viewed as the linear analog of Gentzen's Hauptsatz in propositional logic. The "formulas" with which our proof theory deals are edge-labelled series-parallel graphs, which should be understood as linear analogs of Venn diagrams. These diagrams lend to linear lattice calculations an ease and intuitiveness not shared by modular ones. (interestingly, Czedli [**5**] introduced similar diagrams for a like purpose in general lattice calculations.)

Related to Theorems 1 and 2 are a number of observations, corollaries, and conjectures. The axiomatizability of linear lattices by finitary universal Horn sentences (originally due to Jónsson) is a corollary to Theorem 1. The proof theory of Theorem 1 also possesses a duality which explains why simple linear lattice theorems tend also to be dually valid (although most likely the dual of a linear lattice need not be linear). Theorem 2, while not solving the free linear lattice word problem, comes close and clearly isolates the difficulties. It also suggests a natural conjecture which if true would simultaneously solve the free linear lattice word problem and prove that the class of linear lattices is not self-dual and (hence) not equal to the variety of Arguesian lattices. Finally, to suggest the value of linear lattice theory as a tool in synthetic projective geometry, we produce a sequence of progressively (strictly?) stronger linear lattice identities expressing "higher-dimensional" generalizations of Desargues' theorem.

2. Graph Preliminaries. We shall deal only with undirected graphs (having, perhaps, loops and multiple edges) whose edges are labelled by *variables* from a fixed alphabet $\mathcal{A}$. The label on an edge e will be denoted $l(e)$. Usually, we distinguish two vertices, denoting them ε_0 and ε_1 (or ε_{G0}, ε_{G1} when the graph G is not clear from context). In discussing duality, it will be useful always to presume ε_0 and ε_1 connected by a special "marker" edge ε without a label.

Given a graph G with vertex set $V(G)$ and edge set $E(G)$, and an interpretation $\jmath\colon \mathcal{A} \to 2^{S\times S}$ of the variables by (symmetric) relations on a set S, there is a natural (not necessarily symmetric) relation $\jmath(G)$ defined on S by

$$x \sim y\,[\jmath(G)]$$

> if there is a map $f\colon V(G) \to S$ such that $f(\varepsilon_0) = x$, $f(\varepsilon_1) = y$, and whenever vertices $u, v \in V(G)$ are incident with an edge $e \in E(G)\setminus\{\varepsilon\}$, $f(u) \sim f(v)\,[\jmath(l(e))]$.

If G, H are graphs, the *parallel connection* $G \underline{\wedge} H$ is formed by identifying ε_{Gi} with ε_{Hi} ($i = 0, 1$). In the *series connection* $G \underline{\vee} H$, ε_{G1} and ε_{H0} are identified; ε_{G0} and ε_{H1} become ε_0 and ε_1 for $G \underline{\vee} H$. In either connection, the marker edges ε are removed from G and H and replaced by a new marker ε from ε_0 to ε_1. An *atomic series-parallel graph* $\underline{a}$ ($a \in \mathcal{A}$) has $V(\underline{a}) = \{\varepsilon_0, \varepsilon_1\}$

and two edges—ε and one with label a—connecting the two vertices. A *series-parallel graph* [**3**, **8**] is a graph built from atomic ones by series and parallel connections. Obviously $\jmath(\underline{a}) = \jmath(a)$, $\jmath(G\underline{\wedge}H) = \jmath(G)\cap\jmath(H)$, and $\jmath(G\underline{\vee}H) = \jmath(G)\circ\jmath(H)$. If $\jmath$ is an interpretation into a linear lattice, then $\jmath(G)\cap\jmath(H) = \jmath(G)\wedge\jmath(H)$ and $\jmath(G)\circ\jmath(H) = \jmath(G)\vee\jmath(H)$. Given a lattice polynomial P over the alphabet $\mathcal{A}$, underlining all symbols gives an expression defining its *associated series-parallel graph* $\underline{P}$. We have

LEMMA 1. *If P is a lattice polynomial over the alphabet $\mathcal{A}$ and $\jmath$ is an interpretation of $\mathcal{A}$ into a linear lattice, then $\jmath(\underline{P}) = \jmath(P)$, where $\jmath(P)$ denotes the evaluation of P under $\jmath$.*

3. Main Results.

THEOREM 1. *(Linear lattice proof theory). The (infinitary) universal Horn sentence*

$$\prod_{i\in I}(P_i \leq Q_i) \implies P \leq Q \tag{1}$$

is valid in all linear lattices iff the series-parallel graph of Q can be derived from that of P via a finite sequence of deductions *from the list below. Moreover, any lattice satisfying every valid Horn sentence (1) is linear. The list of deductions is:*

(A) Parallel duplication *of an edge. If $e \neq \varepsilon$ is an edge with label $l(e) = a$, replace it by new edges e', e'' with $l(e') = l(e'') = a$, and incident with the same vertices as e was.*

(B) Coalescence *of series edges. Let $e \neq \varepsilon$ and $f \neq \varepsilon$ be edges with label $l(e) = l(f) = a$. Suppose e incident with vertices u, v, f incident with v, w, and no other edge incident with v. Then remove e, f and v, and replace them with a new edge e' incident with u and w, and labelled a.*

(C) Uncontraction *of an edge. Pass from the graph G present before the deduction to a graph H containing a new edge e, having any label $a \in \mathcal{A}$, such that G is the contraction H/e.*

(D) Deletion. *Delete any edge $e \neq \varepsilon$.*

(E) Reversal *of a series-parallel subgraph. Let J be a subgraph of the graph G present before deduction. Suppose J is attached to the rest of G at just two vertices u_0, u_1, i.e., only these vertices are incident with edges of both J and $G \setminus J$. Suppose further that $\varepsilon \notin J$ and that upon deleting $G \setminus J$ and setting $\varepsilon_0 = u_0$, $\varepsilon_1 = u_1$, there results a series-parallel graph J'. Then make each edge $e \in E(J)$ that is incident with u_i incident instead with u_{1-i} $(i = 0, 1)$. The effect of this is to detach J and reattach it with its ends reversed.*

(F) Substitution *of Q_i for P_i. Let J be attached as in deduction (E), and suppose J' is the series-parallel graph associated to P_i, where $P_i \leq Q_i$ is among the hypotheses of the Horn sentence (1). Remove J and replace it with a new subgraph K attached as J was and for which K' is the series-parallel graph associated to Q_i.*

These deductions are illustrated schematically in Figure 1.

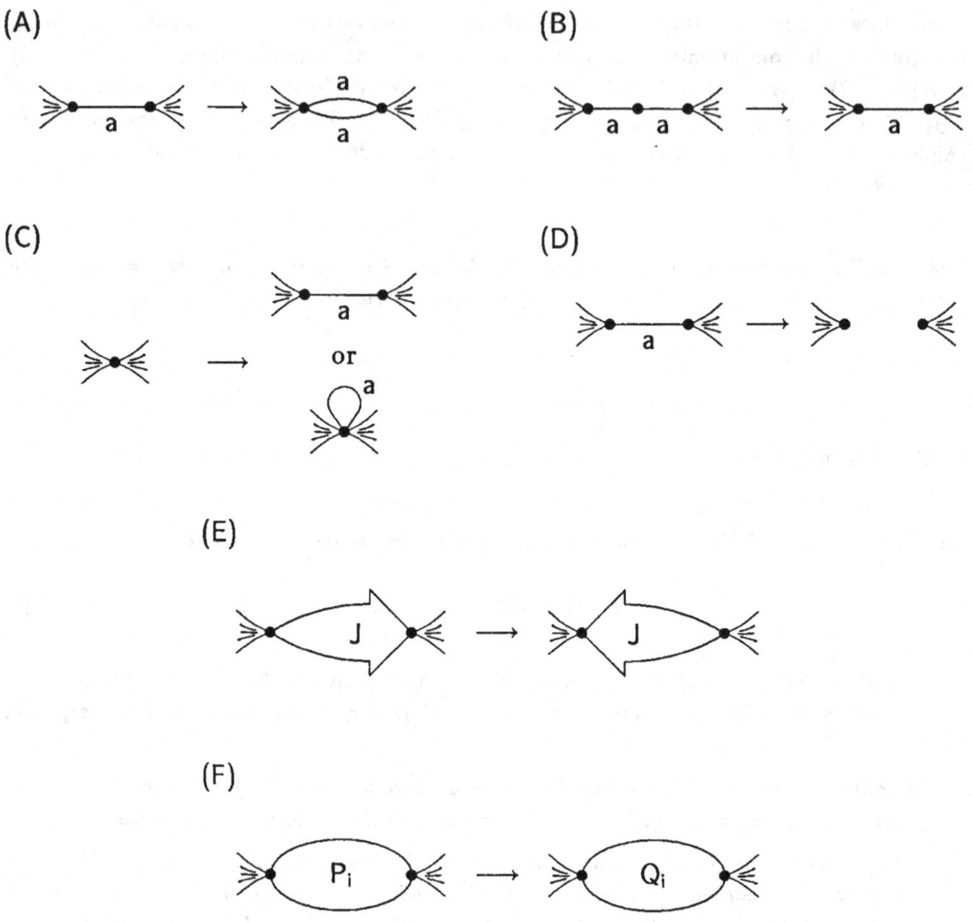

FIGURE I. *The deductions.*

Proof (sketch). The correctness of the deductions is easily checked, for $\jmath(G) \subseteq \jmath(G')$ whenever G' is deduced from G, and this suffices by Lemma 1. The difficulty, of course, is completeness. For this, the deductions are used to build (*via* a transfinite process) from $\underline{P}$ a graph $G_\infty(P)$ whose edge labels define relations on $V(G_\infty(P))$ which generate a linear lattice satisfying $\prod_{i\in I}(P_i \le Q_i)$. If (1) is valid, this linear lattice satisfies $P \le Q$, and Lemma 1 produces $\underline{Q}$ as (almost) a subgraph of $G_\infty(P)$. The construction process for $G_\infty(P)$ possesses a finitary character which makes it possible to extract from it a derivation of $\underline{Q}$ from $\underline{P}$.

Suppose given any lattice L, and take any not necessarily finite presentation of L by generators and relations. Construct $G_\infty(P)$ for every lattice polynomial P in the generators, using the relations of L in place of the hypotheses of (1), so as to insure that the linear lattice on $V(G_\infty(P))$ satisfies all the relations. Then there are homomorphisms from L to these linear lattices, giving a linear representation of L on the disjoint union $\biguplus_P V(G_\infty(P))$. If L satisfies every valid Horn sentence (1), then any inequality $P \le Q$ with left-hand side P true in the lattice on $V(G_\infty(P))$ is true in L also. From this it follows that the above representation of L is faithful.

COROLLARY. *The class of linear lattices is axiomatizable by finitary universal Horn sentences.*

Proof. Since a finite derivation uses only finitely many of the hypotheses $P_i \le Q_i$, any valid infinitary universal Horn sentence (1) is a consequence of a finitary one. The axiomatizability by infinitary universal Horn sentences is contained in Theorem 1.

Duality. Every series-parallel graph is planar. A plane graph G has a *dual* plane graph G^* whose vertices are the regions separated by the edges of the original graph, and whose edges correspond to the edges of the original graph. Each edge is incident in the dual with the vertices

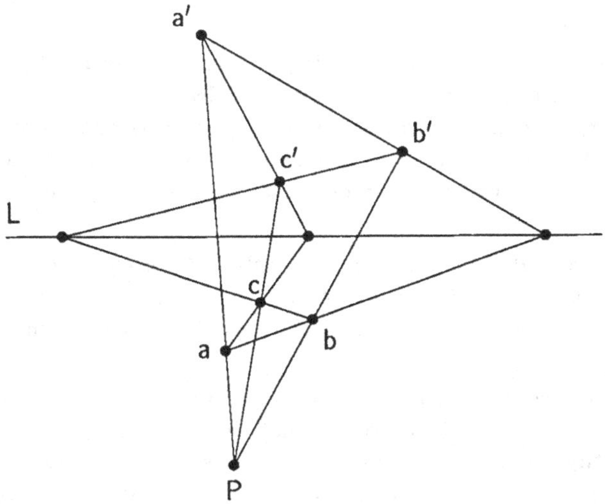

FIGURE II. *The Desargues configuration. P and L are the center and axis of perspectivity, respectively.*

representing the regions on either side of the edge; these regions may be the same. The operations $\underline{\wedge}$ and $\underline{\vee}$ are dual to each other in the sense that for plane graphs G and H, $(G\underline{\wedge}H)^* \cong G^*\underline{\vee}H^*$ and $(G\underline{\vee}H)^* \cong G^*\underline{\wedge}H^*$. Also, atomic graphs are self-dual: $\underline{a}^* \cong \underline{a}$.

Furthermore, there is a duality between the various deductions in Theorem 1 when they are applied to plane graphs in a sufficiently "planar" fashion (e.g., when the parts removed are embedded within a simply connected region, and the parts replacing them go inside that same region; slightly weaker restrictions will also suffice). Specifically, if $G \underset{(A)}{\longrightarrow} G'$ is a valid deduction, then so is $G'^* \underset{(B)}{\longrightarrow} G^*$, and *visa versa*. Similarly, if $G \underset{(C)}{\longrightarrow} G'$ is a valid deduction, then so is $G'^* \underset{(D)}{\longrightarrow} G^*$, and *visa versa*. If $G \underset{(E)}{\longrightarrow} G'$ is a valid deduction, then $G'^* \longrightarrow G^*$ is derivable *via* zero or more applications of *(E)*, depending on how the graphs are embedded in the plane. Finally if $G \underset{(F)}{\longrightarrow} G'$ is valid, using the hypothesis $P_i \leq Q_i$, then so is $G'^* \underset{(F)}{\longrightarrow} G^*$, using the lattice-theoretically dual hypothesis $Q_i^\partial \leq P_i^\partial$.

It follows that if the Horn sentence (1) has a proof using only plane graphs and planar deductions (*i.e.*, obeying the conditions set forth in the previous paragraph), the lattice-theoretic dual $(1)^\partial$ is also valid.

Example 1. The *Arguesian implication* [**18, 19**] is

$$(a \vee a') \wedge (b \vee b') \leq c \vee c' \tag{2}$$

implies

$$(a \vee b) \wedge (a' \vee b') \leq [(a \vee c) \wedge (a' \vee c')] \vee [(c \vee b) \wedge (c' \vee b')]. \tag{3}$$

If a, b, c, a', b', and c' are points in a (coordinate) projective geometry, then (2) expresses that the lines aa', bb' and cc' are concurrent ("central perspectivity"). (3) expresses that the points $ab \cap a'b'$, $bc \cap b'c'$, and $ac \cap a'c'$ are collinear ("axial perspectivity"). That the one implies the other is Desargues' theorem of projective geometry. Hence the term "Arguesian implication." Figure 2 illustrates this plane configuration.

Figure 3 gives a proof of the Arguesian implication. Series-parallel graphs are shown with their associated lattice polynomials. The sequence of deductions is detailed at the bottom. As the proof is planar, the dual to each graph is given just to its right. Reading the dual graphs from (8′) back to (1′), we see a proof of the dual implication. In fact, the Arguesian implication is equivalent to its dual, as can be shown by a suitable lattice computation [**22**].

The next theorem deals with proofs of linear lattice *identities* $P \leq Q$. By Theorem 1, $P \leq Q$ is valid iff it has a proof using deductions *(A)*–*(E)*. It will prove useful to restrict somewhat the identities we wish to prove; please note that the restrictions we shall impose involve no loss of generality.

If x is a variable occurring in $P \leq Q$, then substituting $\mathbf{1}$ for x in P and $\mathbf{0}$ for x in Q and simplifying gives a stronger (perhaps even invalid) identity $P|_{x\leftarrow 1} \leq Q|_{x\leftarrow 0}$. If $P \leq Q$ is a valid linear lattice identity and no such strengthening of it is valid, let us say $P \leq Q$ is *short*. Since every valid identity is a consequence of a short one gotten by strengthening it as above relative to some subset of its variables, to recognize all valid identities it is enough to be able to recognize all short ones.

Before stating Theorem 2, we require one other notion.

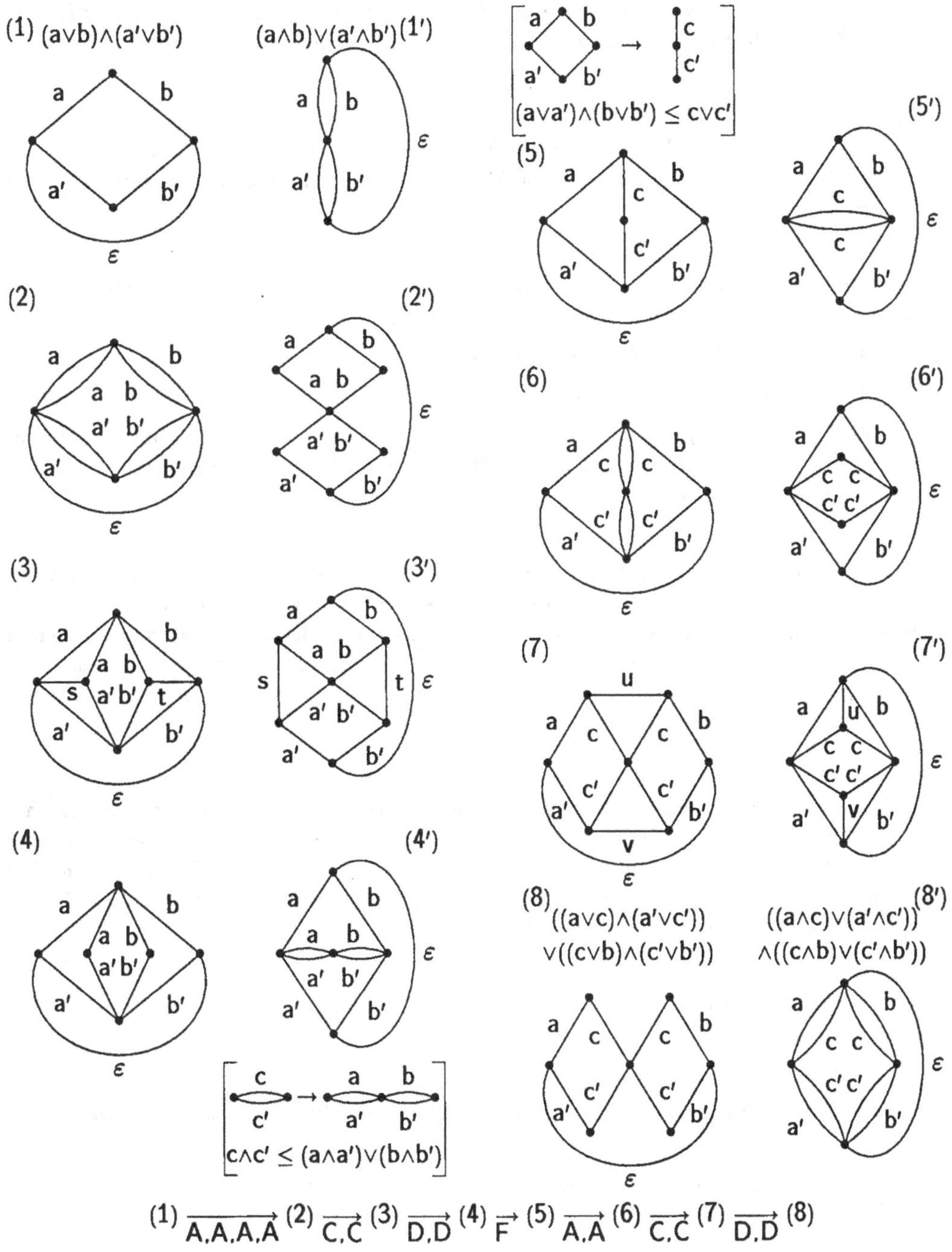

FIGURE III. *Proof of the Arguesian implication and its dual.*

PROPOSITION (Rota). *Every lattice identity $P \leq Q$ can be effectively converted into a lattice-theoretically equivalent one in which every variable occurs exactly once on each side.*

A proof can be found in [**13**]. The idea is to subscript occurrences of a repeated variable x on one side and replace each occurrence on the other side by $x_1 \wedge \ldots \wedge x_n$ (on the left side) or $x_1 \vee \ldots \vee x_n$ (on the right side). Repetition of this gives the required form. Note that strengthening to a short identity preserves this form.

THEOREM 2. *(Normal form). Let $P \leq Q$ be a short linear lattice identity in which each variable occurs exactly once on each side. Then $P \leq Q$ has a derivation in which:*

I. *All deductions (A) occur at the beginning;*

II. *All deductions (B) occur at the end;*

III. *All deductions (C) and (D) occur in consecutive pairs in which the edge uncontracted by (C) is immediately deleted by (D). Thus (C) and (D) can be replaced by a composite deduction (CD) which partitions a vertex.*

Proof (sketch). Parts *I* and *II* follow from a careful analysis of the cases in which the order of two deductions may be reversed. Once a proof satisfying *I* and *II* is found, shortness and unique occurrence of variables can be used to show that every edge uncontracted must later be deleted, and every edge deleted must have earlier been uncontracted. Further analysis of the interchangeability properties of the deductions shows that the deletion can always be made to follow the uncontraction immediately.

It can clearly be decided effectively whether or not one graph can be obtained from another by partitioning vertices and applying deduction *(E)*; for these operations conserve the number of edges in the graph. A recursive bound on the number of deductions *(A)* (necessarily equal to the number of deductions *(B)*) required for a Normal Form proof would thus give an effective solution of the free linear lattice word problem.

CONJECTURE 1. *There is a recursive bound on the number of deductions (A) required in a Normal Form proof.*

CONJECTURE 2. *The number of deductions (A) required in a Normal Form proof is zero.*

While Conjecture 2 may seem somewhat improbable, we do have

PROPOSITION. *If $P \leq Q$ has a Normal Form proof not using deduction (E), then it has one not using (A) and (B).*

[**13**] contains a proof of this Proposition, and also an example showing that deduction *(E)* is sometimes essential.

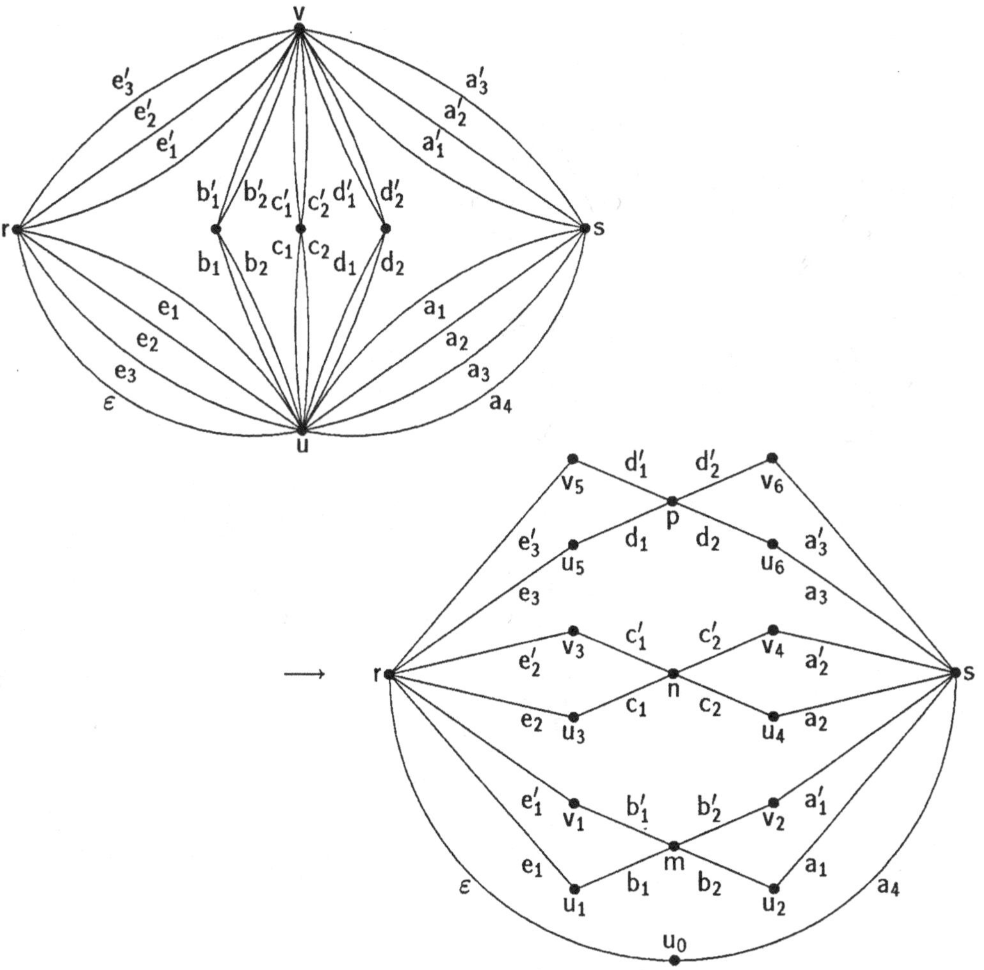

FIGURE IV. *An identity whose natural proof is not planar.*

In this context, we remark that the *generalized* linear lattice word problem is unsolvable because the theorems of Hutchinson [**16**] and Lipshitz [**24**] apply to it. On the other hand, the method used by Freese [**9**] in the modular lattice case is inapplicable to the free linear lattice word problem because linearity highly restricts non-trivial Hall-Dilworth gluings.

As the next example shows, the truth of Conjecture 2 would imply that the variety generated by linear lattices is not self-dual, hence [**22**] not equal to the variety of Arguesian lattices.

Example 2. The lower series-parallel graph in Figure 4 is derived from the upper one by using paired deductions *(CD)* to partition vertices u and v as indicated by the subscripted labels in the lower graph. The corresponding lattice polynomials (which can be read from the graphs, so there is no need to write them out in symbols) therefore are the left- and right-hand sides of a valid linear lattice inequality. This derivation is not planar, so it is not surprising to find (and

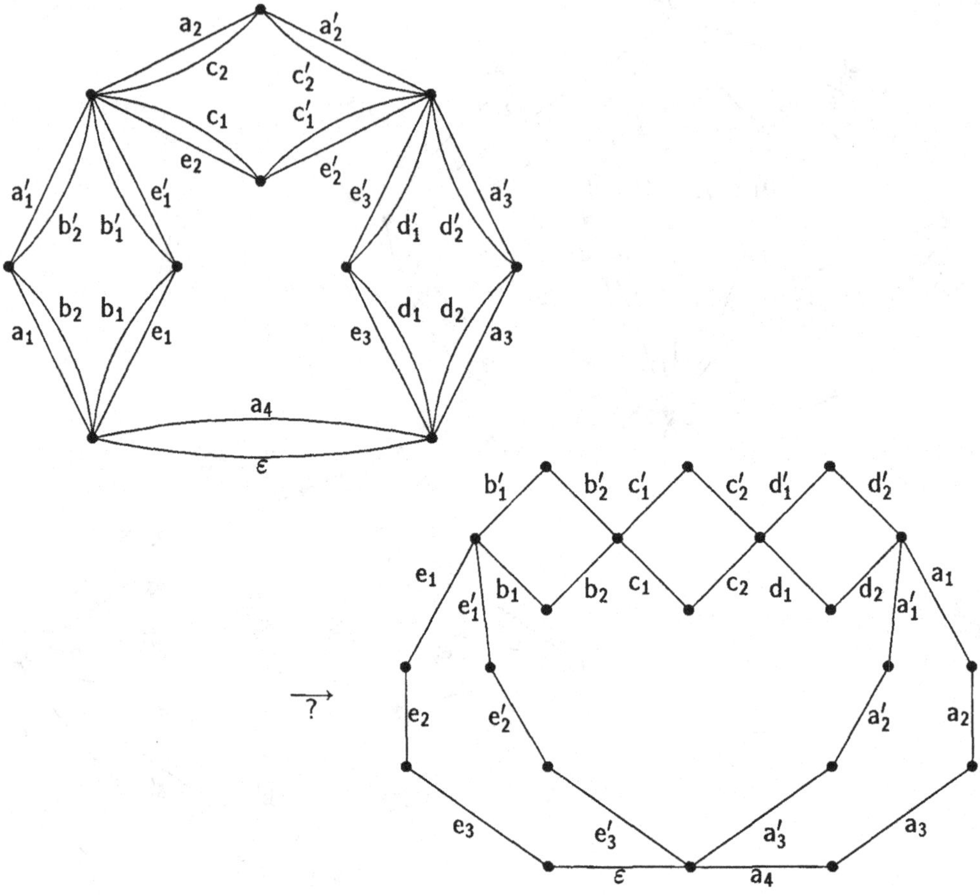

FIGURE V. *The dual to the identity of Figure 4. No proof is possible without deductions (A) and (B).*

routine, though tedious, to verify) that the dual inequality, displayed in Figure 5, has no proof using vertex partitioning and deduction *(E)* alone. Both inequalities are short, so if Conjecture 2 holds, the dual inequality cannot be valid in all linear lattices.

A linear lattice violating the inequality of Figure 5 would of course be of interest, independent of Conjecture 2. In this connection we remark that the inequality is valid for Mal'cev algebra congruence lattices [**12**].

As a final example, we illustrated the methods presented here by proving a sequence of generalizations of the Arguesian identity.

Example 3. The following is a simplified equivalent form of the Arguesian identity [**6, 20, 23**], converted by Rota's Proposition above into a form in which every variable occurs exactly

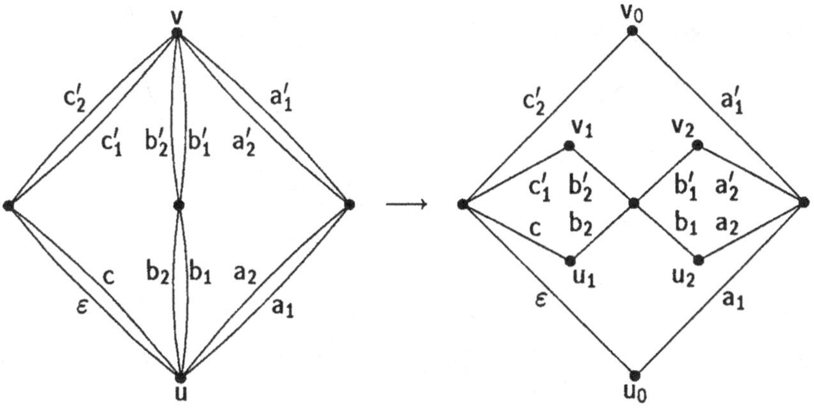

FIGURE VI. *Proof of the Arguesian identity.*

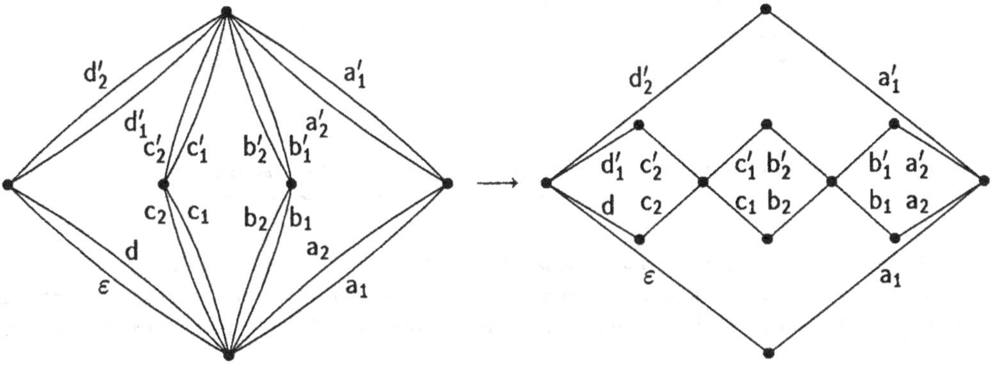

FIGURE VII. *The first higher Arguesian identity.*

once on each side:

$$c \wedge ([((a_1 \wedge a_2) \vee (a_1' \wedge a_2')) \wedge ((b_1 \wedge b_2) \vee (b_1' \wedge b_2'))] \vee (c_1' \wedge c_2'))$$

$$\leq$$

$$a_1 \vee ([((a_2 \vee b_1) \wedge (a_2' \vee b_1')) \vee ((b_2 \vee c) \wedge (b_2' \vee c_1'))] \wedge (a_1' \vee c_2'))$$

Figure 6 gives a Normal Form proof; vertices u and v are partitioned by paired deductions *(CD)* as indicated. The proof is planar and, up to a change of variable names, self-dual.

To interpret the Arguesian identity geometrically, let a_1 and a_2 both stand for the point a in Figure 2, and likewise for the other letters. Then the left-hand side evaluates to c under the assumption of central perspectivity, while the right side evaluates to a line which passes through c only in case of axial perspectivity.

Figure 7 gives the next in a sequence of *higher Arguesian identities* obtained by generalizing Figures 6 and 7 in an evident way. The identity in Figure 7 can be interpreted geometrically as

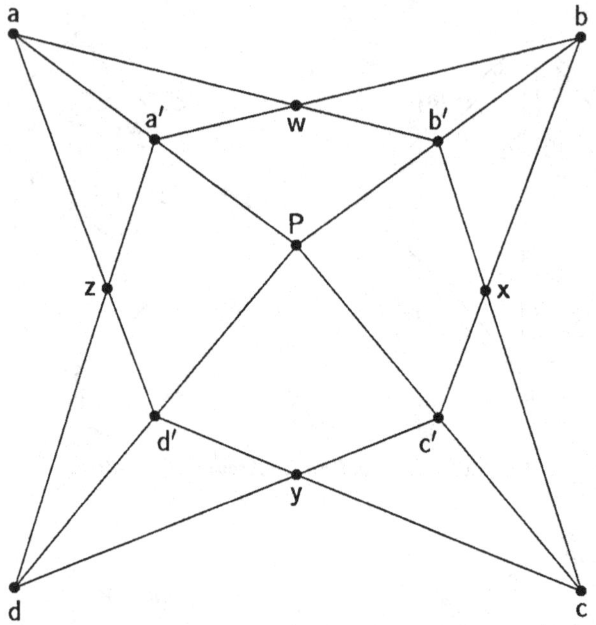

FIGURE VIII. *Geometric interpretation of a higher Arguesian identity. In the 3-dimensional figure, points w,x,y,z are coplanar.*

the theorem that two tetrahedra in 3-space perspective from a point are perspective from a plane. In general, these identities show that two n-dimensional simplices perspective from a point are perspective from an $(n-1)$-dimensional flat. Figure 8 illustrates the situation for the case $n = 3$.

REFERENCES

[1] B. Artmann, "On coordinates in modular lattices." *Illinois J. Math.* **12** (1968) 626-648.

[2] G. Birkhoff, *Lattice Theory, Third Edition.* AMS Colloquium Publications XXV, Providence RI (1967).

[3] T. Brylawski, "A combinatorial model for series-parallel networks." *AMS Transactions* **154** (1967) 1-22.

[4] P. Crawley and R. P. Dilworth, *Algebraic Theory of Lattices.* Prentice-Hall, Englewood Cliffs NJ (1973).

[5] G. Czedli, "On lattice word problems with the help of graphs." Preprint, Szeged (1982).

[6] A. Day, "Geometrical applications in modular lattices." *Universal Algebra and Lattice Theory (Puebla, 1982)*, Springer-Verlag Lecture Notes in Mathematics **1004** (1983) 111-141.

[7] A. Day and D. Pickering, "The coordinatization of Arguesian lattices." *AMS Transactions* **278** (1983) 507-522.

[8] R. J. Duffin, "Topology of series-parallel networks." *J. Math. Analysis and Applications* **10** (1965) 303-318.

[9] R. Freese, "Free modular lattices." *AMS Transactions* **261** (1980) 81-91.

[10] G. Grätzer, *General Lattice Theory*. Birkhauser-Verlag, Basel (1978).

[11] ________, *Universal Algebra, Second Edition*. Springer-Verlag, New York NY (1979).

[12] M. Haiman, "The theory of linear lattices." Ph.D. thesis, M. I. T. (1984).

[13] ________, "Proof theory for linear lattices." *Advances in Math.*, to appear (1985).

[14] Ch. Herrmann, "On the work problem for the modular lattice with four free generators." *Math. Annalen* **265** (1983) 513-527.

[15] W. Hodge and D. Pedoe, *Methods of Algebraic Geometry, Volumes I, II, III.* Cambridge Univ. Press, Cambridge (1953).

[16] G. Hutchinson, "Recursively unsolvable word problems of modular lattices and diagram chasing." *J. Algebra* **26** (1973) 385-399.

[17] ________, "A complete logic for n-permutable congruence lattices." *Alg. Universalis* **13** (1981) 206-224.

[18] B. Jónsson, "On the representation of lattices." *Math. Scand.* **1** (1953) 193-206.

[19] ________, "Modular lattices and Desargues' theorem." *Math. Scand.* **2** (1954) 295-314.

[20] ________, "Arguesian lattices of dimension $n \leq 4$." *Math. Scand.* **7** (1959) 133-145.

[21] ________, "Representation of modular lattices and of relation algebras." *AMS Transactions* **92** (1959) 449-464.

[22] ________, "The class of Arguesian lattices is self-dual." *Alg. Universalis* **2** (1972) 396.

[23] B. Jónsson and G. S. Monk, "Representations of primary Arguesian lattices." *Pacific J. Math.* **30** (1969) 95-139.

[24] L. Lipshitz, "The undecidability of the word problems of modular lattices and projective geometries." *AMS Transactions* **193** (1974) 171-180.

[25] A. I. Mal'cev, "On the general theory of algebraic systems." *Mat. Sbornik (NS)* **35**, no. 7 (1954) 3-20. In Russian.

[26] O. Öre, "Theory of equivalence relations." *Duke Math. J.* **9** (1942) 573-627.

[27] J. D. H. Smith, "Mal'cev varieties." Springer-Verlag Lecture Notes in Mathematics **554**, Berlin-Heidelberg (1976).

[28] R. Smullyan, *First Order Logic*. Springer-Verlag, Berlin-Heidelberg (1968).

[29] J. Von Neumann, *Continuous Geometries*. Princeton Univ. Press, Princeton NJ (1960).

INTERPOLATION ANTICHAINS IN LATTICES

Denis Higgs
Pure Mathematics Department, University of Waterloo
Waterloo, Ontario, Canada N2L 3G1

1. INTRODUCTION. The set $\mathcal{A}$ of bases in a matroid on a finite set E may be characterized as a non-empty antichain in the power set $\mathcal{P}(E)$ of E which enjoys the following interpolation property: if A,B are in $\mathcal{A}$ and $A \subseteq Y$, $Y \supseteq X$, $X \subseteq B$ (where $X,Y \subseteq E$) then $X \subseteq C \subseteq Y$ for some C in $\mathcal{A}$. This characterization (along with two others) has been used by Pezzoli [12] to extend matroid theory to situations in which $\mathcal{P}(E)$ is replaced by some more general lattice, or even poset. The work presented here is a contribution to this program.

The fact that the bases in a specific matroid on E are all on the same level in $\mathcal{P}(E)$ leads to the notion of a *nicely graded* poset (see §3). At least in the case of a lattice L, it appears that the structure of the interpolation antichains in L, and in particular whether L is nicely graded, depends on the two-element interpolation antichains in L; for convenience, we refer to two elements a,b of L as being *adjacent* if {a,b} is an interpolation antichain in L.

In §2 we prove a key result, namely the connectedness of the adjacency graphs of interpolation antichains in lattices in which all bounded chains are finite. In §3 we characterize nicely graded lattices in terms of the adjacency relation and show that two classes of lattices are nicely graded: semimodular lattices and the lattices of independent sets (together with a top element) in finite matroids. §4 contains some miscellaneous examples and questions.

2. INTERPOLATION ANTICHAINS AND ADJACENT ELEMENTS. Let P be an arbitrary poset. A subset A of P will be said to have the *interpolation property* if, given $a \le y$, $y \ge x$, $x \le b$ in P with a,b in A, there exists c in A such that $x \le c \le y$. An *interpolation antichain* is an antichain in P which has the interpolation property.

Two elements a,b of P will be said to be *adjacent*, written $a \sim b$, if {a,b} is a two-element interpolation antichain in P; we write $a \not\sim b$ if a and b are not adjacent. Note that if $a \sim b$ then a and b are incomparable. Since adjacency is an irreflexive and symmetric binary relation on P, $(P,\sim)$ is an undirected graph without loops or multiple edges. For any subset A of P, $(A,\sim)$ denotes the full subgraph of $(P,\sim)$ with A as its vertex set. $(A,\sim)$ depends on the containing poset P as well as on A; which P is being used will always be clear from the context.

A quadruple (a,b,c,d) of elements in a poset P is called an N if $a < b$, $b > c$, $c < d$, and the remaining three pairs of elements are incomparable. The following result is easily verified.

LEMMA 1. *Let* a *and* b *be incomparable elements of a poset* P. *Then* $a \not\sim b$ *in* P *iff there exist* x,y *in* P *such that either* (a,y,x,b) *or* (b,y,x,a) *is an* N.

LEMMA 2. *Let* L *be a lattice, let* $d < e$ *in* L, *and let* A *be an antichain in the interval* $[d,e]$ *of* L. *Then* A *is an interpolation antichain in* $[d,e]$ *iff it is an interpolation antichain in* L *(cf [12, Teorema 5]).*

Proof. It is clear that if A is an interpolation antichain in L then A is an interpolation antichain in $[d,e]$. Suppose that A is an interpolation antichain in $[d,e]$ and let $a \le y$, $y \ge x$, $x \le b$ in L where a,b are in A. Then $a \le y \wedge e$, $y \wedge e \ge x \vee d$, $x \vee d \le b$ in $[d,e]$ and so $x \le x \vee d \le c \le y \wedge e \le y$ for some c in A.

LEMMA 3. *Let* a *and* b *be incomparable elements of a lattice* L. *Then* $a \not\sim b$ *in* L *iff there exist* x,y *in* L *such that either*

(i) $a \wedge b < x < b$, $a < y < a \vee b$, $y = x \vee a$, $x = y \wedge b$,

or

(ii) $a \wedge b < x < a$, $b < y < a \vee b$, $y = x \vee b$, $x = y \wedge a$.

Proof. If such x,y exist then certainly $a \not\sim b$. So let $a \not\sim b$ and let (a,y,x,b), say, be an N in L; by the preceding lemma we can in fact take the N to be in $[a \wedge b, a \vee b]$. Let $x' = y \wedge b$, $y' = x' \vee a$. Then $x \le x'$ and $y' \le y$ so that x' and y' satisfy (i).

As an immediate consequence of Lemma 3 we have:

LEMMA 4. *If* a *and* b *are elements of a lattice* L *such that* $a \vee b$ *covers* a *and* b, *or such that* a *and* b *cover* $a \wedge b$, *then* $a \sim b$.

LEMMA 5. *Let* A *be an interpolation antichain in a lattice* L *and let* a *and* b *be distinct elements of* A *such that* $a \not\sim b$ *in* L. *Then there exists* c *in* A *such that either*

(i) $a \wedge b < b \wedge c < b$, $a < a \vee c < a \vee b$,

or

(ii) $a \wedge b < a \wedge c < a$, $b < b \vee c < a \vee b$.

Proof. We may suppose without loss of generality that x and y are elements of L satisfying (i) in Lemma 3. Then by the interpolation property there exists c in A such that $x \le c \le y$, and necessarily $x = b \wedge c$, $y = a \vee c$.

Let us say that a poset P is a *BCF* poset if all bounded chains in P are finite.

THEOREM 1. *Let* A *be an interpolation antichain in a BCF lattice* L. *Then the adjacency graph* $(A,\sim)$ *is connected.*

Proof. Suppose that a and b are elements of A which lie in different components of $(A,\sim)$ and suppose further that, subject to this condition, a and b are chosen so that $[a \wedge b, a \vee b]$ is minimal. Then $a \not\sim b$ and there exists c in A as in Lemma 5; suppose without loss of generality that c is as in (i) of Lemma 5. Then $[a \wedge c, a \vee c] \subseteq [a \wedge b, a \vee c] \subsetneq [a \wedge b, a \vee b]$ and hence, by the choice of a and b, a and c lie in the same component of $(A,\sim)$. Similarly we see that b and c lie in the same component of A and therefore so also do a and b. This contradiction proves the result.

3. NICELY GRADED LATTICES. A poset P is *graded* if there is an integer-valued function g on P such that $x > y$ implies $g(x) > g(y)$ and x covers y implies $g(x) = g(y) + 1$ (Birkhoff [2, p.5]). Let P be graded; then it is clear that P is BCF and that the grading function g on P is unique to within an additive constant if P is $+$-indecomposable (that is, if P is not the cardinal sum of two non-empty subposets). P will be said to be *nicely graded* if it is $+$-indecomposable, graded with grading function g, and g is constant on every interpolation antichain in P. P will be said to satisfy *condition* Δ_n, where n is an integer ≥ 2, if there do not exist elements $a_0,\ldots,a_n$ in P such that $a_i \sim a_{i+1}$ for $i = 0,\ldots,n-1$ and $a_0 < a_n$.

THEOREM 2. *Every BCF lattice satisfying* Δ_2 *is graded.*

Proof. We show that the given lattice L satisfies the Jordan-Dedekind chain condition; the argument is essentially the same as that given in [2, p.40] for semi-modular posets. Suppose that L does not satisfy the Jordan-Dedekind chain condition and let $[a,b]$ be a minimal interval in L containing maximal chains of different lengths, say $a < x_1 < \ldots < x_{m-1} < b$ of length m and $a < y_1 < \ldots < y_{n-1} < b$ of length $n < m$; necessarily $m \geq 3$. By Lemma 4, $x_1 \sim y_1$ and hence $x_2 \not\sim y_1$ by Δ_2. By Lemmas 1 and 2, there exist u and v in $[a,b]$ such that either (x_2,v,u,y_1) or (y_1,v,u,x_2) is an N. The former is clearly impossible and so (y_1,v,u,x_2) is an N. By the minimality of $[a,b]$, all maximal chains in $[x_2,b]$ have the same length, which we denote by $\ell(x_2,b)$, and similarly for the other proper subintervals of $[a,b]$; note in particular that $\ell(a,x_2) = 2$ and so u covers a and x_2 covers u. Thus we have

$$\begin{aligned} m &= \ell(x_2,b) + 2 = \ell(u,b) + 1 = \ell(u,v) + \ell(v,b) + 1 \\ &= \ell(y_1,v) + \ell(v,b) + 1 = \ell(y_1,b) + 1 = n, \end{aligned}$$

a contradiction.

THEOREM 3. *The following conditions are equivalent for every BCF lattice* L:

(i) L *is nicely graded,*

(ii) L *is graded with grading function* g *and* $a \sim b$ *implies* $g(a) = g(b)$ *for all* a,b *in* L,

(iii) L *satisfies* Δ_n *for all* $n \geq 2$.

Proof. Clearly (i) implies (ii) and (ii) implies (iii) when L is any poset.

The fact that (ii) implies (i) follows immediately from Theorem 1. To see that (iii) implies (ii), let $a \sim b$ and suppose that $g(a) < g(b)$. Let $A = \{x \in L: g(x) = g(b)\}$ and let c be an element of A such that $c > a$. A is certainly an interpolation antichain and hence by Theorem 1 there exist $a_1,\ldots,a_n$ in A such that $a_1 = b$, $a_n = c$, and $a_i \sim a_{i+1}$ for $i = 1,\ldots,n-1$. But then the sequence $a, a_1,\ldots,a_n$ shows that L does not satisfy Δ_n.

THEOREM 4. *Every semimodular BCF lattice is nicely graded.*

Proof. Let L be a semimodular BCF lattice. We first verify that, for all a,b in L, the condition $a \vee b$ covers a and b of Lemma 4 is now necessary, as well as sufficient, for $a \sim b$: suppose that $a \vee b$ does not cover a (say) and let $c \leq b$ cover $a \wedge b$; then $a \vee c$ covers a and so $(a, a \vee c, c, b)$ is an N, whence $a \not\sim b$. (In fact, it is easy to see that, for a BCF lattice, semimodularity is equivalent to the condition that $a \sim b$ iff $a \vee b$ covers a and b.) It is now clear that L satisfies Δ_2 and is thus graded by Theorem 2. Evidently L then satisfies (ii) in Theorem 3 and the theorem is proved.

Since the property of being nicely graded is self-dual, it follows from Theorem 4 that every dually semimodular BCF lattice is nicely graded. Theorem 4 has as a particular case the result of Pezzoli [12, Prop.4] that modular BCF lattices are nicely graded.

A source of nicely graded lattices which are not in general semimodular or dually semimodular is provided by the following result (in which $\oplus$ denotes ordinal sum and 1 is the one-element poset).

THEOREM 5. *Let I be a non-empty order ideal in the power set of a finite set E. Then the lattice $L = I \oplus 1$ is nicely graded iff I is the set of independent sets in some matroid on E.*

Proof. First note that L is graded iff all maximal elements of I have the same cardinality k and that if this is the case then $g(X) = |X|$ for X in I, $g(1_L) = k+1$, defines a grading function on L. Assume that L is nicely graded and let $\mathcal{A}$ be the set of maximal elements of I. We have to show that $\mathcal{A}$ is the set of bases in a matroid on E and this will be done once we show that if A,B are in $\mathcal{A}$ and x is in A then $(A\backslash\{x\})\cup\{y\}$ is in $\mathcal{A}$ for some y in B. So let A,B and x be as stated. Then $A\backslash\{x\} \not\sim B$ since $|A\backslash\{x\}| < |B|$ and there must exist U,V in I such that $(A\backslash\{x\},U,V,B)$ is an N. But then U is in $\mathcal{A}$ and is of the form $(A\backslash\{x\})\cup\{y\}$ for some y in $V \subset B$.

Now suppose that I is given to be the set of independent sets in a matroid on E. Let A and B be incomparable elements of I with $|A| < |B|$. Then there exists y in $B\backslash A$ such that $A\cup\{y\}$ is in I and thus $(A,A\cup\{y\},\{y\},B)$ is an N in L so that $A \not\sim B$ in L. It follows that L satisfies (iii) in Theorem 3 and hence L is nicely graded.

Both Theorem 4 (when applied to the lattice of subsets of a finite set) and

Theorem 5 have as a consequence the invariance of base cardinality in matroids on finite sets.

4. SOME COUNTEREXAMPLES AND QUESTIONS. The assumption that the poset involved is a lattice in various results of the previous two sections is necessary. Let P_1, P_2, and P_3 be the three posets in Figure 1; the dotted lines show the corresponding adjacency relations.

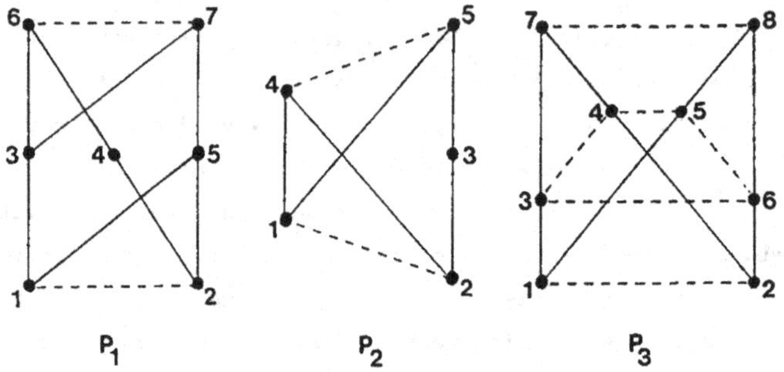

Figure 1

Then, even though it is nicely graded, P_1 fails to satisfy either the conclusion of Lemma 2 or that of Theorem 1: $3 \sim 5$ in [1,7] but not in P_1; {3,4,5} is an antichain in P_1. Since P_2 satisfies Δ_n for all n but is not graded, Theorem 2 and the implication from (iii) to (ii) in Theorem 3 also fail for posets in general. That (ii) implies (i) in Theorem 3 fails in general is shown by P_3, in which {3,4,8} is an interpolation antichain (this also gives another instance of the failure of Theorem 1 for posets).

Theorem 1 provides the only general information at present available on adjacency graphs (except for one easy consequence of Lemma 2, that the only isolated vertices in $(L,\sim)$, where L is a lattice, are those elements of L, if any, which are comparable with every element of L; P_1 shows that the lattice hypothesis is necessary here). Maurer [7], [8], [9], [10] has characterized and studied in detail the graphs $(A,\sim)$ where A is an interpolation antichain in $\mathcal{P}(E)$, E finite; Astie-Vidal [1], Donald, Holzmann, and Tobey [3], Krogdahl [6], and Oubiña [11] have further studied these graphs.

It is clear that Δ_{n+2} implies Δ_n for all $n \geq 2$. However the five-element non-modular lattice satisfies Δ_n just when n is odd; thus for n even, Δ_{n+1} does not imply Δ_n in general, and Δ_2 cannot be replaced by Δ_3 in Theorem 2. The question as to whether Δ_{n+1} implies Δ_n for n odd, and whether Δ_n implies Δ_{n+1} for any n, is more difficult. That Δ_2 implies no other Δ_n is shown by the lattice of Figure 2, whose adjacency graph is given in Figure 3: this lattice satisfies Δ_n for $n = 2$ only.

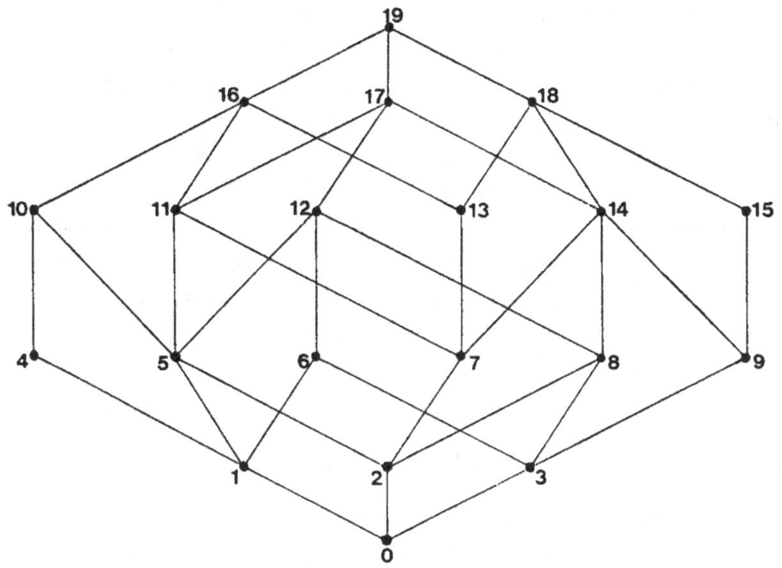

Figure 2

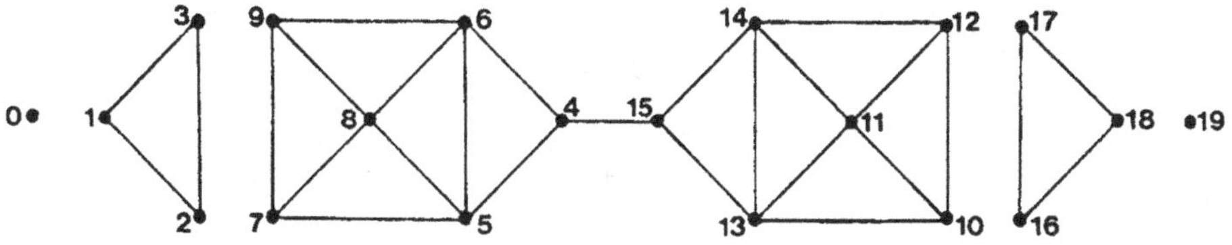

Figure 3

A positive answer to the following question would ease the labour in checking whether a given finite lattice L satisfies Δ_2: If L does not satisfy Δ_2, can one always find a,b,c in L such that $a \sim b$, $a \sim c$, and c covers b?

Condition (iii) in Theorem 3 does not depend on the covering relation or any finiteness conditions and it is thus a reasonable candidate for being the definition of "nicely graded" in the case of arbitrary posets and lattices. As a test case here, one would expect that every modular lattice can be shown to be nicely graded in this sense; this would extend the result of Pezzoli referred to after Theorem 4 and would suggest further that in Theorem 4 the BCF condition could be omitted (one would then interpret semimodularity either as in Birkhoff [2, p.83] or as in Dubreil-Jacotin, Lesieur and Croisot [4, p.85]).

Theorem 4 shows that the property of being nicely graded can be regarded as a generalization of semimodularity and one might hope that various results which are known to be true for semimodular (BCF) lattices, and which in particular are also self-dual, hold more generally for nicely graded lattices. An example of such a result, due to Duffus and Rival [5, Theorem 5.3], states that if a,b,c are elements of a semimodular lattice such that $d(a,b) = d(a,c) + d(c,b)$ then c is in $[a \wedge b, a \vee b]$ (here $d(x,y)$ denotes the distance between x and y in the graph of the lattice under the relation "x covers, or is covered by, y").

A simple example of a nicely graded lattice which is neither semimodular nor dually semimodular is given by Theorem 5 when one takes the matroid on $E = \{1,2,3,4\}$ which has $\{1,3\}$ and $\{2,4\}$ as its circuits; the resulting lattice is just the face lattice of a square and is shown in Figure 4.

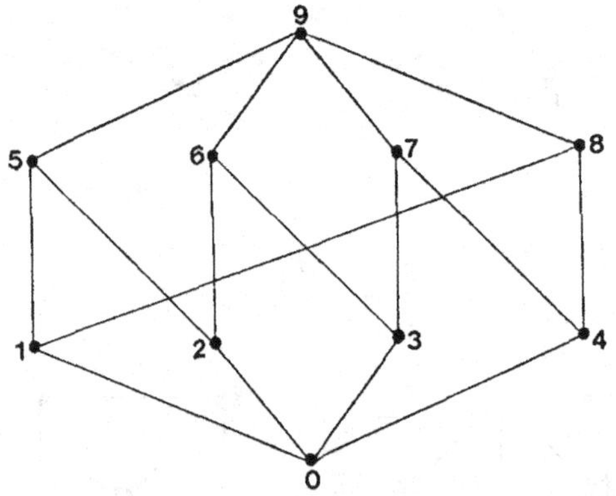

Figure 4

The face lattices of n-gons for $n > 4$ are not nicely graded however. Since these lattices are relatively complemented, it follows that relatively complemented graded lattices are not necessarily nicely graded.

We have remarked that if a lattice is nicely graded then so is its dual. Also, it is clear from Lemma 2 that intervals of nicely graded lattices are themselves nicely graded. Further closure properties of the class of nicely graded lattices will be discussed elsewhere. In this connection, notice that the nicely graded lattice in Figure 4 has a cover-preserving sublattice, namely $\{0,1,3,5,7,9\}$, which is not nicely graded; this contrasts with the situation for semimodular lattices of course.

REFERENCES

[1] A. Astie-Vidal, Factor group of the automorphism group of a matroid basis graph with respect to the automorphism group of the matroid, Discrete Math. 32(1980), 217-224.

[2] G. Birkhoff, Lattice Theory, Third Edition, Amer. Math. Soc., Providence, 1967.

[3] J.D. Donald, C.A. Holzmann, and M.D. Tobey, A characterization of complete matroid basis graphs, J. Combinatorial Theory B 22 (1977), 139-158.

[4] M.L. Dubreil-Jacotin, L. Lesieur and R. Croisot, Leçons sur la théorie des treillis, des structures algebriques, et des treillis geométriques, Gauthier-Villars, Paris, 1953.

[5] D. Duffus and I. Rival, Path length in the covering graph of a lattice, Discrete Math. 19(1977), 139-158.

[6] S. Krogdahl, The dependence graph for bases in matroids, Discrete Math. 19(1977), 47-59.

[7] S.B. Maurer, Matroid basis graphs I, J. Combinatorial Theory B 14(1973), 216-240.

[8] S.B. Maurer, Matroid basis graphs II, J. Combinatorial Theory B 15(1973), 121-145.

[9] S.B. Maurer, Intervals in matroid basis graphs, Discrete Math. 11(1975), 147-159.

[10] S.B. Maurer, A maximal-rank minimum-term-rank theorem for matroids, Linear Algebra and Appl. 10(1975), 129-137.

[11] L. Oubiña, Localisation d'une propriété du graphe des bases d'un matroide binaire, Colloques internationaux C.N.R.S. No. 260 (Problèmes combinatoires et théorie des graphes), Paris-Orsay, 1978, pp.323-324.

[12] L. Pezzoli, Sistemi di indipendenza modulari, Boll. Un. Mat. Ital. B (5) 18(1981), 575-590.

SUBDIRECTLY IRREDUCIBLE AND SIMPLE BOOLEAN ALGEBRAS WITH ENDOMORPHISMS

Jaroslav Ježek
Charles University
Praha, Czechoslovakia

0. INTRODUCTION. Given an interesting variety V of universal algebras, one can also investigate some varieties derived from V in a natural way. For example, for every monoid S we can consider the variety V^S of V-algebras A with operators from S, acting as endomorphisms on A. We shall see in Section 1 that the variety V^S shares many nice properties with V. In particular, if V is residually small then V^S is, too.

The variety $\mathcal{S}$ of semilattices is residually small: its only nontrivial subdirectly irreducible member is the two-element semilattice. It follows that the variety $\mathcal{S}^S$ is residually small for any monoid S and it is only natural to try to describe its subdirectly irreducible or at least simple members. However, in $\mathcal{S}^S$ the situation is much more complicated than in $\mathcal{S}$. In [4], all simple algebras in $\mathcal{S}^S$ are found in the case when S is the free commutative group with two generators, and it turns out that in this case there are uncountable simple algebras in $\mathcal{S}^S$. No more information about simple and subdirectly irreducible algebras in the varieties $\mathcal{S}^S$ is known to the author.

The variety $\mathcal{B}$ of Boolean algebras is residually small, too, and again the two-element Boolean algebra is the only nontrivial subdirectly irreducible algebra in $\mathcal{B}$. The purpose of the present paper is to study subdirectly irreducible and simple algebras in the varieties $\mathcal{B}^S$.

Given a monoid S, we can take the Boolean algebra of all subsets of S and consider it as an algebra from $\mathcal{B}^S$ in a natural way; this algebra is denoted by P_S and it turns out that an algebra from $\mathcal{B}^S$ is subdirectly irreducible iff it is isomorphic to a subalgebra of P_S (see Section 2).

In order to describe simple algebras in $\mathcal{B}^S$, it is then enough to clarify which of the subalgebras of P_S are simple. However, this task seems to be more difficult. It will be considered in the present paper in the special case when S=Z, the group of integers with respect to addition; the variety $\mathcal{B}^Z$ is equivalent to the variety of Boolean algebras with one fixed automorphism. The notion of a weakly periodic subset of Z is introduced and it is proved (in Section 4) that a subalgebra of P_Z is simple iff every subset of Z belonging to it is weakly periodic. It turns out that every simple subalgebra of P_Z is contained in a maximal one, and that a subalgebra of a simple subalgebra is simple. In order to complete the picture of all simple algebras in $\mathcal{B}^Z$, it would thus be sufficient to know the maximal simple subalgebras of P_Z. In particular, it would be desirable to know how many maximal subalgebras formed by weakly periodic sets there are; how many of them are nonisomorphic; and how large they are. However, these questions are left open in this paper.

In Section 6 the notion of an admissible filter is introduced and a correspondence between simple subalgebras of P_Z and admissible filters is established; this correspondence is not very nice.

The paper ends with pointing out a connection between these questions and some questions studied in topological dynamics.

The author is grateful to M. Adams, R. McKenzie, J. Sichler and P. Simon for many interesting discussions and suggestions.

1. EXPANDING VARIETIES OF UNIVERSAL ALGEBRAS BY MONOIDS OF ENDOMORPHISMS. Let V be a variety of algebras of type Δ and let S be a monoid with unit 1. We denote by Δ+S the type which is the disjoint union of Δ and S, where every element of S is considered as a unary operation symbol. (The unary operations corresponding to the symbols $s \in S$ will be denoted by $x \mapsto sx$.) Further, we denote by V^S the class of algebras A of type Δ+S satisfying the following four conditions:

(1) the Δ-algebra $A \upharpoonright \Delta$, obtained from A by forgetting the unary operations corresponding to symbols from S, belongs to V;

(2) for every $s \in S$, the corresponding unary operation $x \mapsto sx$ is an endomorphism of $A \upharpoonright \Delta$;

(3) $st.x=s.tx$ for all $s,t \in S$ and $x \in A$;

(4) $1x=x$ for all $x \in A$.

It is evident that V^S is a variety.

For example, if S is the free monoid (free group, resp.) on n generators, n being an arbitrary cardinal number, than the variety V^S is polynomially equivalent to the variety of algebras from V equipped with n fixed endomorphisms (automorphisms, resp.). If S is the free commutative monoid (group, resp.) over n generators then V^S is equivalent to the variety of algebras from V with n fixed pairwise commuting endomorphisms (automorphisms, resp.).

Let $A \in V$. The algebra B of type Δ+S, defined by $B \upharpoonright \Delta = A$ and $sx=x$ for all $s \in S$ and $x \in A$, belongs evidently to V^S; it is called the V^S-algebra obtained from A by identical expansion. The class of algebras obtained by identical expansion from algebras of V is evidently a subvariety of V^S which is polynomially equivalent to V.

The varieties V^S were introduced and studied in [2]. The followi theorem is a survey of some results from [2].

1.1.THEOREM. Let V be a nontrivial variety and S be a monoid. Then:

(1) The varieties V and V^S have the same Mal'cev properties.

(2) V^S is locally finite iff V is locally finite and S is finite.

(3) If V^S is finitely generated then S is finite and V is finitely generated.

(4) V^S is a discriminator variety iff V is a discriminator variety an S is a finite group.

(5) V^S is an Abelian variety iff V is Abelian.

We are going to add several simple results in this direction.

A variety V is said to have the strong amalgamation property if the following is true for all algebras $A,B,C \in V$: if A is a subalgebra of both B and C and $A=B\cap C$ then there exists an algebra $D \in V$ such tha both A and B are subalgebras of D.

A variety V is said to have the amalgamation property if for eve pair $f:A \rightarrow B$, $g:A \rightarrow C$ of monomorphisms in V there exists an algebra $D \in V$ and a pair of monomorphisms $h:B \rightarrow D$, $k:C \rightarrow D$ such that $hf=kg$.

A variety V is said to have the congruence extension property if the following is true: whenever A,B are algebras from V such that A i a subalgebra of B, then every congruence of A can be extended to a co gruence of B.

A variety V is said to be residually small if there exists a cardinal number k such that every subdirectly irreducible algebra from V is of cardinality $\leq k$.

1.2.THEOREM. The variety V^S has the strong amalgamation property iff V has the strong amalgamation property.

<u>Proof</u>. Let V have the strong amalgamation property. Let $A,B,C \in$ be algebras such that A is a subalgebra of both B and C and $A=B\cap C$. Since V has the property, there exists the free amalgamated product in V of the algebras $B\restriction\Delta$ and $C\restriction\Delta$ over $A\restriction\Delta$, i.e. an algebra $P \in V$ su that $B\restriction\Delta$ and $C\restriction\Delta$ are subalgebras of P and whenever $Q \in V$ and $f:B\restriction\Delta \rightarrow$ $g:C\restriction\Delta \rightarrow Q$ are two homomorphisms coinciding on A then $f \cup g$ can be uniq

ely extended to a homomorphism of P into Q. Let $s \in S$. The mappings $f_s(x)=sx$ $(x \in B)$ and $g_s(x)=sx$ $(x \in C)$ are homomorphisms of $B \upharpoonright \Delta$ into P and of $C \upharpoonright \Delta$ into P coinciding on A; hence there exists a unique endomorphism h_s of P extending both f and g. For all $x \in P$ put $sx=h_s(x)$. We get an algebra D of type $\Delta + S$ with $D \upharpoonright \Delta = P$. If $s,t \in S$ then the mappings $x \mapsto st.x$ and $x \mapsto s.tx$ are endomorphisms of P coinciding on $B \cup C$ and so on P. We see that $D \in V^S$; evidently, both B and C are subalgebras of D. The converse implication is evident.

1.3. THEOREM. The variety V^S has the amalgamation property iff V has the amalgamation property.

<u>Proof</u>. It is analogous to that of 1.2.

1.4. THEOREM. The variety V^S has the congruence extension property iff V has the congruence extension property.

<u>Proof</u>. Let V have the congruence extension property. Let $A,B \in V^S$ where A is a subalgebra of B and let r_1 be a congruence of A. Since V has the property, there exists a congruence r_3 of $B \upharpoonright \Delta$ such that $r_1 = r_3 \cap (A \times A)$. Define a binary relation r_2 on B by $(x,y) \in r_2$ iff $(sx,sy) \in r_3$ for all $s \in S$. It is easy to see that r_2 is a congruence of B and $r_2 \cap (A \times A) = r_1$. The converse is evident.

The following result is due to R. McKenzie.

1.5. THEOREM. The variety V^S is residually small iff V is residually small.

<u>Proof</u>. If V^S is residually small, then it is evident that V is, too. Let V be residually small, so that there exists a cardinal number k such that every subdirectly irreducible algebra from V is of cardinality $\leq k$. Put $n=|S|$. Let $A \in V^S$ be an algebra of cardinality $> k^n$; it is enough to prove that A is not subdirectly irreducible. Since A is not subdirectly irreducible and any algebra is isomorphic to a subdirect product of subdirectly irreducible algebras, there exists a family r_i $(i \in I)$ of nontrivial congruences of $A \upharpoonright \Delta$ with trivial intersection such that all the algebras $(A \upharpoonright \Delta)/r_i$ are subdirectly irreducible. For every $i \in I$ define a binary relation t_i on A by $(x,y) \in t_i$ iff $(s(x),s(y)) \in r_i$ for all $s \in S$. It is easy to verify that t_i is a congruence of A and $t_i \subseteq r_i$, so that t_i $(i \in I)$ is a family of congruences

of A with trivial intersection. It remains to prove that the congruences t_i are nontrivial. Let $i \in I$. The set $(A/r_i)^S$ is of cardinality $\leq k$. For every element $a \in A$ define an element $a^* \in (A/r_i)^S$ by $a^*(s)=s(a)/r_i$ for all $s \in S$. Since $|A| > |(A/r_i)^S|$, there are two distinct elements $a,b \in A$ with $a^*=b^*$. But then $(s(a),s(b)) \in r_i$ for all $s \in S$, i.e. $(a,b) \in t_i$. We have proved that t_i is nontrivial.

2. SUBDIRECTLY IRREDUCIBLE ALGEBRAS IN $\mathcal{B}^S$. We denote by $\mathcal{B}$ the variety of Boolean algebras.

Let S be a monoid. The set S is considered also as a type, consisting of unary operation symbols. By a right S-set we mean an algebra of the type S, with the unary operations denoted by $x \mapsto xs$, satisfying the identities $xs.t=x.st$ $(s,t \in S)$ and $x1=x$.

For every right S-set X we denote by P_X the algebra from $\mathcal{B}^S$ defined as follows: the underlying Boolean algebra of P_X is the Boolean algebra of all subsets of X; if $s \in S$ and $a \in P_X$ then $sa=\{x \in X;\ xs \in a\}$.

2.1. LEMMA. Let $A \in \mathcal{B}^S$. Define a right S-set X as follows: X is the set of ultrafilters of the Boolean algebra A; if $x \in X$ and $s \in S$ then $xs=\{y \in A;\ sy \in x\}$. The mapping $f:A \to P_X$ defined by $f(a)=\{x \in X;\ a \in x\}$ is an embedding of A into P_X.

<u>Proof</u>. It is easy.

2.2. LEMMA. Let A be a subdirectly irreducible algebra from $\mathcal{B}^S$. Then there exists a one-generated right S-set X such that A can be embedded into P_X.

<u>Proof</u>. By 2.1, A can be considered as a subalgebra of P_Y for some right S-set Y. For every one-generated subalgebra U of Y define a mapping g_U of A into P_U by $g_U(a)=a \cap U$. It is easy to verify that g_U is a homomorphism of A into P_U; indeed, g_U evidently preserves the Boolean operations, while $g_U(sa)=sg_U(a)$ means $\{x \in Y;\ xs \in a\} \cap U=\{x \in U;\ xs \in a \cap U\}$ which is evident, since U is a subalgebra. The intersection of the congruences $Ker(g_U)$, where U ranges over all one-generated subalgebras of Y, is easily seen to be equal to id_A. Since A is subdirectly irreducible, we deduce that $Ker(g_X)=id_A$ for some one-generated subalgebra X of Y. But then g_X is an embedding of A into P_X.

A monoid S can be considered as a right S-set in a natural way;

this right S-set is evidently just the free one-generated right S-set, the unit of S being the free generator. Consequently, for every monoid S we can form the algebra P_S, belonging to $\mathcal{B}^S$.

2.3. LEMMA. Let A be a subdirectly irreducible algebra from $\mathcal{B}^S$. Then A can be embedded into P_S.

Proof. By 2.2, A can be embedded into P_X for some one-generated right S-set X. There exists a homomorphism f of the right S-set S onto X. Define a mapping g of P_X into P_S by $g(a)=\{x\in S;\ f(x)\in a\}$. It is easy to see that g is an embedding of P_X into P_S.

Let A be an algebra from $\mathcal{B}^S$. By an S-ideal of A we mean an ideal of the underlying Boolean algebra A which is closed under the unary operations sx $(s\in S)$. Obviously, the restriction of the canonical bijection between congruences and ideals of the underlying Boolean algebra of A is a bijection between congruences and S-ideals of A.

The following lemma was proved by J. Sichler.

2.4. LEMMA. Every subalgebra of P_S is subdirectly irreducible.

Proof. Let Q be a subalgebra of P_S and suppose that Q is not subdirectly irreducible, so that there is a family I_λ $(\lambda\in\Lambda)$ of nontrivial ideals with trivial intersection. Since the ideals are nontrivial, there are nonempty sets $M_\lambda\in I_\lambda$. Take elements $s_\lambda\in M_\lambda$. We have $s_\lambda M_\lambda\in I_\lambda$; the intersection D of the sets $s_\lambda M_\lambda$ $(\lambda\in\Lambda)$ belongs to the (trivial) intersection of the ideals I_λ. Now, D is just the set of the elements x such that $xs_\lambda\in M_\lambda$ for all λ, so that the unit of S belongs to D and consequently $D\neq\emptyset$. This is a contradiction.

Combining Lemmas 2.3 and 2.4, we get:

2.5. THEOREM. Let S be a monoid. An algebra from $\mathcal{B}^S$ is subdirectly irreducible iff it is isomorphic to a subalgebra of the algebra P_S.

3. SIMPLE ALGEBRAS IN $\mathcal{B}^S$.

3.1. PROPOSITION. Let S be a monoid. The algebra P_S is simple iff S is a finite group.

Proof. First, let S be a finite group and r be a nontrivial congruence of P_S. Then $(\emptyset,\{a\}) \in r$ for some $a \in S$. For all $s \in S$ we have $(s\emptyset, s\{a\}) \in r$ and so $(\emptyset,\{x\}) \in r$ for all $x \in S$. (Notice that $s\{a\}=\{as^{-1}\}$. But P_S is finite and so every element of P_S is a finite join of atoms hence $(\emptyset,y) \in r$ for all $y \in P_S$.

Now let P_S be simple. Denote by I the set of elements x of P_S such that there exists a finite number of elements $s_1,\dots,s_n$ of S with $x \subseteq s_1\{1\} \cup \dots \cup s_n\{1\}$. Evidently, I is an S-ideal of P_S; since P_S is simple, we get $S \in I$, i.e. $S=s_1\{1\} \cup \dots \cup s_n\{1\}$ for some $s_1,\dots,s_n \in S$. We get $S=\{x \in S;\ xs_1=1\} \cup \dots \cup \{x \in S;\ xs_n=1\}$. Hence every element of S has a right inverse, and so S is a group; it is evident that it has at most n elements.

3.2. PROPOSITION. Let A be a simple algebra from $\mathcal{B}^S$. Then every subalgebra of A is simple.

Proof. It follows from 1.4 that the variety $\mathcal{B}^S$ has the congruence extension property.

A variety is said to be semisimple if any of its subdirectly irreducible algebras is simple. It follows from 3.1, 3.2 and 2.3 that the variety $\mathcal{B}^S$ is semisimple iff S is a finite group.

Let us remark that if S is a monoid which is not a finite group then it follows from Zorn's lemma that every simple subalgebra of P_S is contained in a maximal simple subalgebra. (Of course, this is true for arbitrary universal algebras.)

4. SIMPLE ALGEBRAS IN $\mathcal{B}^Z$ AND WEAKLY PERIODIC SETS.

We denote by Z the monoid of integers with respect to addition. In this section we shall consider the special case when the monoid S equals Z. The variety $\mathcal{B}^Z$ is equivalent to the variety of Boolean algebras with one fixed automorphism. P_Z is the Boolean algebra of all subsets of Z, together with the family of unary operations $A \mapsto A-c$ $(c \in Z)$.

For every subset $A \subseteq Z$ we denote by $\langle A \rangle$ the subalgebra of P_Z generated by the element A.

If $I=[u,v]$ is an interval (in Z), then the nonnegative integer $v-u$ is called the length of I. (By an interval we mean a closed interval.)

Let $A \subseteq Z$. Two intervals $[a,b]$ and $[c,d]$ are called similar with respect to A if they are of the same length and, for any $i \in [a,b]$, the integer i belongs to A iff $i+c-a \in A$. We then write $[a,b] \underset{A}{\sim} [c,d]$. Evidently, this relation is an equivalence.

By a weakly periodic set we mean a subset A of Z such that for any interval I there exists a positive integer n such that any interval of length n contains a subinterval similar with I with respect to A.

The collection of all weakly periodic sets is denoted by WP.

4.1.PROPOSITION. Let $A \subseteq Z$. Then $\langle A\rangle$ is simple iff A is a weakly periodic set.

Proof. Let $\langle A\rangle$ be simple. Take an interval $[u,u+d]$. For every $i \in [0,d]$ define a set B_i as follows: if $u+i \in A$ then $B_i = A-i = \{a-i;\ a \in A\}$; if $u+i \notin A$ then $B_i = Z \setminus (A-i)$. We have $u \in B_0 \cap \ldots \cap B_d$. Hence the set $B = B_0 \cap \ldots \cap B_d$ is nonempty; it belongs to $\langle A\rangle$. Since $\langle A\rangle$ is simple, the Z-ideal of $\langle A\rangle$ generated by B equals $\langle A\rangle$, so that $Z = (B-m) \cup \ldots \cup (B-1) \cup B \cup (B+1) \cup \ldots \cup (B+m)$ for some $m>0$; but then $Z = B \cup (B+1) \cup \ldots \cup (B+n)$ for some $n>0$. Take any $x \in Z$. We have $x \in B+j$ for some $j \in [0,n]$, i.e. $x-j \in B$; by the definition of B, this means that the intervals $[x-j, x-j+d]$ and $[u,u+d]$ are similar with respect to A. It follows that any interval of length $n+d$ contains a subinterval similar with $[u,u+d]$ with respect to A. Hence A is weakly periodic.

Now let A be weakly periodic. Denote by L the system of sets of the form $e_0(A+u) \cap e_1(A+u+1) \cap \ldots \cap e_d(A+u+d)$ where $u \in Z$, $d>0$ and $e_j \in \{-1,1\}$ for all j; here $e(X)=X$ if $e=1$ and $e(X)=Z\setminus X$ if $e=-1$. Then $\langle A\rangle$ is just the set of unions of finite subfamilies of sets in L. Let r be a nontrivial congruence of $\langle A\rangle$. Then $(\emptyset,B) \in r$ for some nonempty $B \in \langle A\rangle$ and thus for some nonempty $B \in L$. Express B in the form $B = e_0(A+u) \cap e_1(A+u+1) \cap \ldots \cap e_d(A+u+d)$. Take an element $c \in B$. There exists an $n>0$ such that every interval of length $n+d$ contains a subinterval similar with $[c-u-d, c-u]$ with respect to A. Let $x \in Z$ be arbitrary. The interval $[x-n-u-d, x-u]$ is of length $n+d$ and so there exists an $i \in [0,n]$ such that the interval $[x-i-u-d, x-i-u]$ is similar with $[c-u-d, c-u]$ with respect to A. Since $c \in B$, for every $j \in [0,d]$ we have

$c \in e_j(A+u+j)$,

$c \in A+u+j$ iff $e_j=1$,

$c-u-j \in A$ iff $e_j=1$,

$x-i-u-j \in A$ iff $e_j=1$ (because the intervals are similar),

$x-i \in A+u+j$ iff $e_j=1$,

$x-i \in e_j(A+u+j)$

and so $x-i \in e_0(A+u) \cap \ldots \cap e_d(A+u+d) = B$, i.e. $x \in B+i$. Since $x \in Z$ was arbitrary, we have proved $Z = B \cup (B+1) \cup \ldots \cup (B+n)$. But then $(\emptyset,B) \in r$ implies $(\emptyset,Z)\in r$ and r is the greatest congruence of $\langle A\rangle$, so that $\langle A\rangle$ is simple.

4.2.PROPOSITION. Let Q be a subalgebra of P_Z. Then Q is simple iff $Q\subseteq WP$.

Proof. If Q is simple, then $Q\subseteq WP$ by 3.2 and 4.1. Let $Q\subseteq WP$ and let r be a nontrivial congruence of Q. Then $(\emptyset,A)\in r$ for some nonempty $A\in Q$. Since $Q\subseteq WP$, A is weakly periodic. By 4.1, r restricted to $\langle A\rangle$ is the greatest congruence of $\langle A\rangle$, so that $(\emptyset,Z)\in r$; hence r is the greatest congruence of Q and Q is simple.

Let us remark that we shall prove in the following section that WP is not a subalgebra of P_Z.

By a periodic set we mean a subset $A\subseteq Z$ such that there is a positive integer n with the following property: if $a\in Z$ then $a\in A$ iff $a+n\in A$. (The least number n with this property is called the period of A.) Evidently, there are only countably many periodic sets.

By an almost periodic set we mean a subset $A\subseteq Z$ such that for any interval $[a,b]$ there exists an $n>0$ such that for any $k\in Z$, the intervals $[a,b]$ and $[a+kn,b+kn]$ are similar with respect to A.

We denote by PER the set of all periodic sets and by AP the set of almost periodic sets. Evidently,

$$PER\subseteq AP\subseteq WP.$$

4.3.PROPOSITION. PER and AP are subalgebras of P_Z.

Proof. For PER it is obvious. Evidently, AP is closed under complementation and the unary operations and it remains to prove that if A,B are two almost periodic sets then $A\cap B$ is almost periodic, too. Let $[a,b]$ be an interval. There are $n_1,n_2>0$ such that for any $k\in Z$, the intervals $[a,b]$ and $[a+kn_1,b+kn_1]$ are similar with respect to A and the intervals $[a,b]$ and $[a+kn_2,b+kn_2]$ are similar with respect to B. Put $n=n_1n_2$. Then for any $k\in Z$, the intervals $[a,b]$ and $[a+kn,b+kn]$ are similar with respect to both A and B and thus with respect to $A\cap B$, too.

It follows that PER and AP are simple algebras.

5. A CONSTRUCTION OF WEAKLY PERIODIC SETS. By a word we shall in this section mean a word over the alphabet $\{0,1\}$.

Let a,b be two words of the same tength $k\geq 1$. Further, let e,f be two mappings of the set $\{1,2,3,\ldots\}$ into $\{0,1\}$. Put $\bar{0}=1$ and $\bar{1}=0$. Define words $a_0,b_0,a_1,b_1,a_2,b_2,\ldots$ by induction as follows:

$a_0=a$,

$b_0=b$,

$a_n=a_{n-1}e_nb_{n-1}a_{n-1}\bar{e}_nb_{n-1}a_{n-1}$ for $n\geq 1$,

$b_n=b_{n-1}a_{n-1}f_nb_{n-1}a_{n-1}\bar{f}_nb_{n-1}$ for $n\geq 1$.

For $n\geq 0$ we have

$$|a_n|=|b_n|=5^nk+\tfrac{1}{2}(5^n-1),\qquad |a_n0b_n|=|a_n1b_n|=5^n(2k+1).$$

5.1.LEMMA. Let $0\leq m<n$. Then any subword of a_n0b_n of length $5^{m+1}(2k+1)$ contains both a_m0b_m and a_m1b_m as subwords.

Proof. Let m be fixed and proceed by induction on n. Let w be a subword of

$$a_n0b_n=a_{n-1}e_nb_{n-1}a_{n-1}\bar{e}_nb_{n-1}a_{n-1}0b_{n-1}a_{n-1}f_nb_{n-1}a_{n-1}\bar{f}_nb_{n-1}$$

of the given length. If $w=a_n0b_n$, the conclusion is clear. Assume that $w\neq a_n0b_n$; thus $m+1\leq n-1$. If w is a subword of either a_{n-1} or b_{n-1}, we can use the induction assumption. Otherwise some subword w' of w of length $5^{m+1}k+\frac{1}{2}(5^{m+1}-1)$ is either a beginning or an end of either a_{n-1} or b_{n-1}. But then w' is either a_{m+1} or b_{m+1} and everything is evident.

Define a subset $M_{a0b,e,f}$ of Z as follows: if $i\in Z$ then $i\in[-5^nk-\frac{1}{2}(5^n-1),5^nk+\frac{1}{2}(5^n-1)]$ for some $n\geq 0$; denote this interval by $[-c_n,c_n]$ and put $a_n0b_n=s_{-c_n}s_{-c_n+1}\cdots s_{c_n}$; put $i\in M_{a0b,e,f}$ iff $s_i=1$. (This does not depend on the particular n chosen.) From 5.1 it follows that $M_{a0b,e,f}$ is a weakly periodic set.

Quite analogously, we can construct a weakly periodic set $M_{a1b,e,f}$.

5.2.LEMMA. The subalgebra of P_Z generated by the sets $M_{a0b,e,f}$ and $M_{a1b,e,f}$ is not contained in WP.

Proof. We have $M_{a1b,e,f}\setminus M_{a0b,e,f}=\{0\}$ and the set $\{0\}$ is not weakly periodic.

5.3.LEMMA. If $M_{a0b,e,f}=M_{a0b,e',f'}$ then $e=e'$ and $f=f'$.

Proof. It is evident.

5.4.COROLLARY. There are uncountably many weakly periodic sets and the set WP is not a subalgebra of P_Z.

6. SIMPLE SUBALGEBRAS OF P_Z AND ADMISSIBLE FILTERS. By an admissible filter we shall mean a filter F of subsets of Z such that for any $M \in F$ there is an $n>0$ with $[i-n,i+n] \cap M \neq \emptyset$ for any $i \in Z$.

Let F be an admissible filter. We denote by $\mathcal{S}_F$ the system of subsets $A \subseteq Z$ such that for any $n \geq 1$ there exists an $M \in F$ with $[-n,n] \underset{A}{\sim} [-n+i,n+i]$ for all $i \in M$.

Let Q be a subalgebra of P_Z such that $Q \subseteq WP$. Then we denote by $\mathcal{F}_Q$ the set of all $I \subseteq Z$ such that there exists an $n \geq 1$ and an $A \in Q$ with $I \supseteq \{i \in Z;\ [-n,n] \underset{A}{\sim} [-n+i,n+i]\}$.

6.1. PROPOSITION.
(1) If F is an admissible filter then $\mathcal{S}_F$ is a subalgebra of P_Z contained in WP.
(2) If Q is a subalgebra of P_Z contained in WP then $\mathcal{F}_Q$ is an admissible filter.
(3) If Q_1, Q_2 are two subalgebras of P_Z contained in WP and $Q_1 \subseteq Q_2$ then $\mathcal{F}_{Q_1} \subseteq \mathcal{F}_{Q_2}$.
(4) If F_1, F_2 are two admissible filters and $F_1 \subseteq F_2$ then $\mathcal{S}_{F_1} \subseteq \mathcal{S}_{F_2}$.
(5) Let Q be a subalgebra of P_Z contained in WP. Then $Q \subseteq \mathcal{S}_{\mathcal{F}_Q}$.
(6) Let F be an admissible filter. Then $\mathcal{F}_{\mathcal{S}_F} \subseteq F$.
(7) If Q is a maximal subalgebra of P_Z contained in WP then $\mathcal{S}_{\mathcal{F}_Q} = Q$.

<u>Proof</u>. We shall prove only the first two assertions.

(1) Let F be an admissible filter. Evidently, $\mathcal{S}_F$ is contained in WP and is closed under $\cap, \cup$ and $\setminus$. It remains to show that if $A \in \mathcal{S}_F$ then $A+1 \in \mathcal{S}_F$ and $A-1 \in \mathcal{S}_F$. Let $n \geq 1$. There exists an $M \in F$ with $[-n-1,n+1] \underset{A}{\sim} [-n-1+i,n+1+i]$ for all $i \in M$. If $c \in [-n,n]$ and $i \in M$ then $c-1$ and $c+1$ belong to $[-n-1,n+1]$ and so

$$c \in A+1 \Leftrightarrow c-1 \in A \Leftrightarrow c-1+i \in A \Leftrightarrow c+i \in A+1,$$
$$c \in A-1 \Leftrightarrow c+1 \in A \Leftrightarrow c+1+i \in A \Leftrightarrow c+i \in A-1.$$

Hence $[-n,n] \underset{A+1}{\sim} [-n+i,n+i]$ and $[-n,n] \underset{A-1}{\sim} [-n+i,n+i]$ for all $i \in M$. Hence $A+1 \in \mathcal{S}_F$ and $A-1 \in \mathcal{S}_F$.

(2) Let Q be a subalgebra of P_Z contained in WP. It is enough to show that if $I, J \in \mathcal{F}_Q$ then $I \cap J \in \mathcal{F}_Q$. There are sets $A, B \in Q$ and integers $n, m \geq 1$ with

$$I \supseteq \{i \in Z;\ [-n,n] \underset{A}{\sim} [-n+i,n+i]\},$$
$$J \supseteq \{i \in Z;\ [-m,m] \underset{B}{\sim} [-m+i,m+i]\}.$$

It is enough to assume that n=m. For every $c \in [-n,n]$ define two numbers $e_c, f_c \in \{-1,1\}$ as follows: if $c \in A$ then $e_c=1$; if $c \notin A$ then $e_c=-1$; if $c \in B$ then $f_c=1$; if $c \notin B$ then $f_c=-1$. Put $D_{(1)}=D$ and $D_{(-1)}=Z \setminus D$ for all $D \subseteq Z$. Put

$$C=(A+n)_{(e_{-n})} \cap (A+n-1)_{(e_{-n+1})} \cap \dots \cap (A-n)_{(e_n)} \cap (B+n)_{(f_{-n})} \cap \cap (B+n-1)_{(f_{-n+1})} \cap \dots \cap (B-n)_{(f_n)}.$$

We have $C \in Q$ and $0 \in C$. Hence $\{i \in Z;\ [0] \underset{C}{\sim} [i]\} \in \mathcal{F}_Q$, i.e. $\{i \in Z;\ i \in C\} \in \mathcal{F}_Q$, i.e. $C \in \mathcal{F}_Q$. It is enough to prove $C \subseteq I \cap J$. We shall prove $C \subseteq I$ only, since $C \subseteq J$ is analogous. Let $x \in C$. For every $i \in [-n,n]$ we have $x \in (A-i)_{(e_i)}$; hence $x \in A-i$ iff $e_i=1$ iff $i \in A$. Hence $i \in A$ iff $i+x \in A$. Hence $[-n,n] \underset{A}{\sim} [-n+x,n+x]$; this means $x \in I$.

7. CONNECTIONS WITH TOPOLOGICAL DYNAMICS.

The problem of describing simple algebras in the variety $\mathcal{B}^Z$ has been reduced in Section 4 (see Introduction) to the problem of studying maximal subalgebras of the algebra P_Z that are formed by weakly periodic sets.

It is worth mentioning that this problem is related to some questions studied in topological dynamics. For fundamental notions and results in this area see [1] and [3]. In topological dynamics, so called dynamical systems are studied. A dynamical system is a pair (X,f) where X is a compact Hausdorff space and f is a continuous mapping of X into itself. A point $x \in X$ is said to be uniformly recurrent if for any neighborhood U of x, the set $H=\{n;\ f^n(x) \in U\}$ is syndetic, i.e. there exists a positive integer p such that for any $m \geq 0$, the interval $[m,m+p]$ has nonempty intersection with H. Let us remark that the well-known Birkhoff's theorem says that each dynamical system contains at least one uniformly recurrent point.

Now, consider the Cantor set C as the power $\{0,1\}^Z$ with the topology of pointwise convergence; this means that its open basis is given by all the sets $[\varphi]=\{f \in \{0,1\}^Z;\ f \supseteq \varphi\}$, where φ ranges over functions from finite subsets of Z into $\{0,1\}$. Consider the dynamical system (C,S) where the mapping S is defined by $S(g)(n)=g(n+1)$. (S is called the shift.)

It is easy to see that a subset of Z is weakly periodic iff its characteristic function is a uniformly recurrent point in (C,S).

ADDED FEBRUARY 1985: Results of Section 1 are contained in a more general form in the paper [5].

R e f e r e n c e s

[1] G. D. Birkhoff: Dynamical systems. Amer. Math. Soc. Colloq. Publ. Vol. 9, Providence 1927.

[2] S. Burris and M. Valeriote: Expanding varieties by monoids of endomorphisms. Algebra Universalis 17, 1983, 150-169.

[3] R. Ellis: Lectures on topological dynamics. Benjamin, New York 1969.

[4] J. Ježek: Simple semilattices with two commuting automorphisms. Algebra Universalis 15, 1982, 162-175.

[5] W. H. Cornish: Antimorphic action. (Preprint)

A NOTE ON VARIETIES OF GRAPH ALGEBRAS

Emil W. Kiss*
Mathematical Institute of the
Hungarian Academy of Sciences
1364 Budapest, P.O.B. 127, Hungary.

Graph algebras have been invented by C. Shallon [10] to construct examples of non-finitely based varieties (see G. McNulty, C. Shallon [5] for an account). They have also been used to investigate the lattice of subvarieties of finitely generated varieties. There is very little known about the structure of these lattices. Some unexpected restrictions have been revealed by R. McKenzie [4]. Graph algebras seem to provide interesting examples. In [8] and [9] M. Vaughan-Lee and S. Oates-Williams have shown that the lattice L of subvarieties of the variety generated by an important three element graph algebra (Murskii's groupoid, see [6]) satisfies neither the maximum nor the minimum condition. Later, S. Oates-Williams proved (see [7]) that the lattice L actually contains a chain isomorphic to the reals, and hence it is uncountable.

In this note we present an observation stating that varieties of graph algebras correspond to classes of graphs that are closed under forming products, induced subgraphs, disjoint and direct unions. As an application, we obtain the result of S. Oates-Williams [7] mentioned above.

We use the notations and terminology of [5]. Graphs are undirected without multiple edges, but they may contain loops. A rooted graph is a graph with a distinguished vertex. For a graph G = (V,E) the associated *graph algebra*, $\mathfrak{A}(G)$ has universe $V \cup \{\infty\}$, where ∞ is a symbol outside V, and a binary operation * satisfying a*b = ∞ unless (a,b) is in E, when a*b = a. A *homomorphism* of graphs is a mapping of vertices carrying edges to edges. Thus, an edge can be collapsed only to a loop. The *product* of graphs $G_i = (V_i, E_i)$ has vertex set $\prod V_i$, and two vertices in the product are adjacent if so are all their components. A graph G' = (V',E') is an *induced subgraph* of G = (V,E) iff V' is a subset of V and the edges of G' are exactly the edges of G connecting elements of V'.

---

*Research supported by the NSERC of Canada.

THEOREM. *Let G be a class of graphs and let K be the class of the corresponding graph algebras. Then the graph algebra of a finite connected graph* H *is in the variet generated by K if and only if* H *is an induced subgraph of a product of members of* G *If* H *is an arbitrary graph, then* $\mathfrak{A}$(H) *is in* HSP(K) *iff* HSP(K) *contains the graph alg bras of all the connected components of the finite induced subgraphs of* H.

This result can be generalized to directed graphs. The details, and the proof of the last statement of the theorem are left to the reader. The "if" part is easy, and is found in [9]. For the other direction, we relate graphs to terms. Our method is slightly different from that in [5]. We define a rooted graph G(t) for all groupoid terms t. The vertex set V(t) of G(t) is the set of variables occurring in t, and the root of G(t) is the leftmost variable of t. The edges of G(t) are defined by induction on the complexity of t. If t is a variable, then G(t) has no edges. If $t = t_1 * t_2$ then the edges of G(t) are precisely those of $G(t_1)$ and $G(t_2)$ together with a new one connecting the roots of $G(t_1)$ and $G(t_2)$. For example, $G((x_1*x_2)*(x_1*x_1))$ can be drawn as  , where ● is the root. Our definitions immediately yield the following:

LEMMA 1. *Let* G = (V,E) *be a graph,* $t(x_1,\dots,x_n)$ *a term with leftmost variable* x_i *and* f *a map from* V(t) *to* V. *Then the value of* $t(f(x_1),\dots,f(x_n))$ *in* $\mathfrak{A}$(G) *is* $f(x_i)$ *if f is a graph homomorphism from* G(t) *to* G, *and* ∞ *otherwise.*

Since we want to construct equations, the following claim will be useful.

LEMMA 2. *A finite, connected rooted graph* G *is of the form* G(t) *for some term* t.

Proof. We proceed by induction on the number of edges. If G has no edges, then it corresponds to a variable. For the induction step, let r be the root of a finite connected graph G, and (r,g) an edge of G. Throw this edge out, and let G_1 and G_2 be the connected components of r and g, with root r and g, respectively. By the induction hypothesis, $G_1 = G(t_1)$ and $G_2 = G(t_2)$ for some terms t_1 and t_2. Hence $G = G(t_1*t_2)$.

Suppose now that G is a class of graphs, K is the class of the corresponding grap algebras, and H is a finite connected graph such that $\mathfrak{A}$(H) is in the variety generate by K. We may assume that the vertices of H are $x_1,\dots,x_n$. One can see immediately that H is an induced subgraph of a product of members of G if and only if

(1) For every two different vertices x and y of H there is a homomorphism f from H to a member of G such that $f(x) \neq f(y)$;

(2) For every two non-adjacent vertices x and y of H there is a homomorphism f from H to a member of G such that f(x) and g(x) are not adjacent.

Let x be a vertex of H and apply Lemma 2 to find a term t_x such that $G(t_x)$ equals H, and has root x. To prove (1) let x and y be different vertices of H and consider the equation $t_x = t_y$. Lemma 1 shows, with f being the identity of $V(t_x) = V(t_y) = \{x_1,\ldots,x_n\}$ that this equation fails in $\mathfrak{A}(H)$. Hence, it fails in some $\mathfrak{A}(G)$, where $G = (V,E)$ is a member of $\mathcal{G}$. So there exist elements $g_1,\ldots,g_n$ of V such that $t_x(g_1,\ldots,g_n) \neq t_y(g_1,\ldots,g_n)$. Thus Lemma 1 shows that the mapping f sending x_i to g_i is a homomorphism desired in (1). The statement (2) can be verified analogously, by considering the equation $t_x * t_y = t_x$. Thus the Theorem is proved.

Let G_0 be the two element graph . The graph algebra $\mathfrak{A}(G_0)$ is called *Murskii's groupoid.*

COROLLARY 1 [5]. *Let* G *be a finite connected graph. The variety generated by* $\mathfrak{A}(G)$ *contains all loopless graph algebras iff* G_0 *is an induced subgraph of* G.

<u>Proof</u>. By the Theorem, we may consider finite connected graphs only. We use conditions (1) and (2). Since every proper homomorphic image of a complete graph contains a loop, applying (2) with x = y we see that the condition is necessary. Conversely, each loopless graph has the homomorphisms required in (1) and (2) into G_0.

COROLLARY 2 [7]. *The lattice of subvarieties of the variety generated by Murskii's groupoid contains a poset isomorphic to the Boolean lattice of all subsets of a countable set.*

<u>Proof</u>. Let G_i $(i = 1,2,\ldots)$ be a countable family of finite, loopless, connected graphs such that G_i has no homomorphism into G_j if $i \neq j$. The existence of such a family has been proved by Z. Hederlin, P. Vopenka and A. Pultr [2], [3]. Then for all i, $\mathfrak{A}(G_i)$ is not contained in the variety generated by $\{\mathfrak{A}(G_j) : j \neq i\}$ by the Theorem. Hence all subsets of $\{\mathfrak{A}(G_i) : i = 1,2,\ldots\}$ generate different varieties.

We mention that W. Dziobiak [1] seems to be the first, who constructed a finite algebra *A* such that HSP(*A*) has continuum many subvarieties. Note that Murskii's groupoid is a subalgebra of *A*.

ACKNOWLEDGEMENTS. The author expresses his thanks to R. Freese for his helpful and encouraging remarks, to A. Day and J. Sichler for calling his attention to the paper [3] and to R. W. Quackenbush and A. Day for inviting him to Canada where this result has been obtained, and providing excellent circumstances for working and living there.

REFERENCES

[1] W. Dziobiak, *A variety generated by a finite algebra with* $2^{\aleph_0}$ *subvarieties*, Algebra Universalis 13 (1981), 148-156.

[2] Z. Hederlin, P. Vopenka, A. Pultr, *A rigid relation exists on any set*, Comment Math. Univ. Carolinae, 6 (1965), 149-155.

[3] Z. Hederlin, A. Pultr, *Symmetric relations (undirected graphs) with given semi-groups*, Monatsch. Math. 69 (1965), 318-322.

[4] R. McKenzie, *Finite forbidden lattices*, Universal algebra and lattice theory, Lexture notes, Vol. 1004, Springer (1982), 176-205.

[5] G. McNulty, C. Shallon, *Inherently nonfinitely based finite algebras*, Universal algebra and lattice theory, Lecture notes, Vol. 1004, Springer (1982), 206-231.

[6] V. L. Murskii, *The existence in three valued logic of a closed class with finite basis not having a finite complete set identities*, Dokl. Akad. Nauk. SSSR 163 (1965), 815-818.

[7] S. Oates-Williams, *On the variety generated by Murskii's algebra*, Algebra Universalis 18 (1984), 175-177.

[8] S. Oates-MacDonald, M. Vaughan-Lee, *Varieties that make one Cross*, J. Austral. Math. Soc. (Ser. A), 26 (1978), 368-382.

[9] S. Oates-Williams, *Murskii's algebra does not satisfy min*, Bull. Austral. Math. Soc. 22 (1980), 199-203.

[10] C. Shallon, *Nonfinitely based finite algebras derived from lattices*, Ph. D. Dissertation, U.C.L.A. 1979.

How to Construct Finite Algebras Which Are Not Finitely Based

George F. McNulty*
University of South Carolina
Columbia, South Carolina 29208

A variety V of algebras is said to be *finitely based* provided V is the class of all models of some finite set of equations; an algebra is called *finitely based* iff the variety generated by the algebra is finitely based. Roger Lyndon [54] offered the first nonfinitely based finite algebra. It has seven elements and its only (essential) operation is binary. In the ensuing thirty years infinitely many nonfinitely based algebras have been added to the catalog begun by Lyndon. McNulty and Shallon [83] give a fairly complete account of this catalog as of 1982.

How can a nonfinitely based finite algebra be discovered?

One possible procedure for finding such algebras will be given here. The first step is to replace "nonfinitely based" by a property which is considerably stronger and perhaps easier to spot. Recall that a variety is locally finite iff every finitely generated algebra in it is finite. Every variety generated by a finite algebra is locally finite.

A variety V is called *inherently nonfinitely based* iff

(i) V is locally finite, and

(ii) W is not finitely based whenever W is a locally finitely variety such that $V \subseteq W$.

Every inherently nonfinitely based variety fails to be finitely based. An algebra is said to be inherently nonfinitely based iff the variety it generates has this property. Thus a finite algebra $\underset{\sim}{A}$ is inherently nonfinitely based iff W is not finitely based for all locally finite varieties W such that $\underset{\sim}{A} \in W$. This notion emerged around 1979, implicitly in Murskii [79] and explicitly in Perkins [85]. We can modify our original question:

How can an inherently nonfinitely based finite algebra be found?

To determine whether a locally finite variety is inherently nonfinitely based, one is confronted with a host of supervarieties which may or may not be either finitely based or locally finite. Happily, there is no need to consider all these varieties.

*) This research was supported at various times by a Fulbright-Hays Grant from the Philippine-American Educational Foundation, a Fellowship from the Alexander von Humboldt Stiftung, and NSF Grant ISP-11451.

Let V be a variety and n be a natural number. Let T_n denote the set of all equations true in V which involve no variables other than $x_0, x_1, \ldots, x_{n-1}$. Define $V^{(n)}$ to be the class of all models of T_n. So $V^{(n)}$ is a variety. Evidently

$$V^{(0)} \supset V^{(1)} \supset V^{(2)} \supset \ldots V \text{ and } \bigcap_{n\in\omega} V^{(n)} = V.$$

The varieties $V^{(n)}$ along this chain have the following useful description:

* $\underset{\sim}{B} \in V^{(n)}$ iff $\underset{\sim}{C} \in V$ for all subalgebras $\underset{\sim}{C}$ of $\underset{\sim}{B}$ which can be generated by n or fewer elements.

Birkhoff [35] pointed out that if V is locally finite and has only finitely many basic operations, then $V^{(n)}$ is finitely based, for all natural numbers n. B.H. Neumann [37] remarked this result for varieties of groups. This leads to the following theorem:

EASY THEOREM

Let V *be a locally finite variety with only finitely many basic operations,* V *is inherently nonfinitely based iff* $V^{(n)}$ *is not locally finite for infinitely many natural numbers* n.

Here is an outline of the procedure for discovering a finite algebra which is inherently nonfinitely based. For simplicity, we will look for a groupoid (that is, an algebra whose only operation is binary). Call this, so far unknown, finite groupoid $\underset{\sim}{A}$ and let V be $HSP\underset{\sim}{A}$, the variety generated by $\underset{\sim}{A}$.

STEP 0: For each sufficiently large natural number n, construct an infinite groupoid $\underset{\sim}{B}_n$ which is generated by $n+1$ elements such that each subalgebra of $\underset{\sim}{B}_n$ generated by n elements is finite.

STEP 1: Construct an algebra $\underset{\sim}{C}$ such that, for each sufficiently large natural number n, every n-generated subalgebra of $\underset{\sim}{B}_n$ is embeddable in $\underset{\sim}{C}$.

STEP 2: With $\underset{\sim}{C}$ in hand search for a finite groupoid $\underset{\sim}{A}$ satisfying the formula $\underset{\sim}{C} \in HSP\underset{\sim}{A}$.

Thus the algebras $\underset{\sim}{B}_n$ built in Step 0 witness the failure of $V^{(n)}$ to be locally finite, since Steps 1 and 2 place $\underset{\sim}{B}_n$ in $V^{(n)}$ in view of (*). The constraint imposed in Step 0 on the n-generated subalgebras of $\underset{\sim}{B}_n$ is plainly necessary if V is to be locally finite, much less generated by a finite algebra. Once these steps have been successfully completed, the Easy Theorem entails that the finite groupoid $\underset{\sim}{A}$ is inherently nonfinitely based.

STEP 0

$\underset{\sim}{B}_n$ is constructed for each $n > 3$. We only describe $\underset{\sim}{B}_5$ in any detail, since the general case is very similar. Here is a diagram of

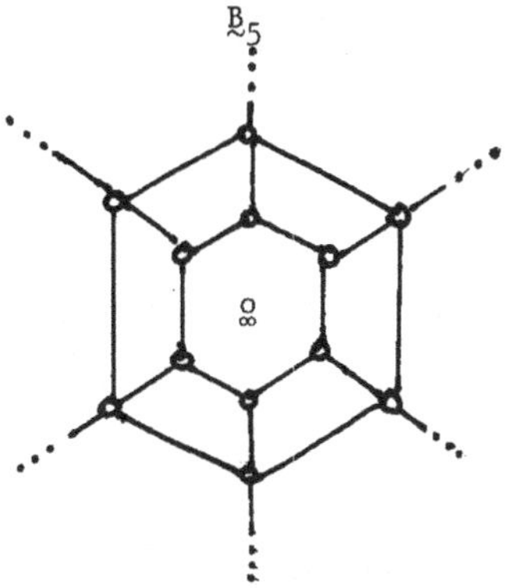

The members of $\underset{\sim}{B}_5$ are rendered here as vertices of an expanding nested collection of hexagons and one additional point labelled ∞. Aside from ∞, each member of $\underset{\sim}{B}_5$ is located on exactly one radial "arm" and on exactly one hexagonal "level". We arrange for the points on the innermost level to generate $\underset{\sim}{B}_5$ by imposing

CONDITION 0: $\alpha\beta = \gamma$ if α is immediately counterclockwise of β and γ is the point one level outward from α on the radial arm through α.

It is easy to verify that the points on the innermost level generate all of $\underset{\sim}{B}_5$ with the possible exception of ∞. As we go through the construction, more conditions will be imposed on the operation, but in the end it will not be defined on all pairs of elements of $\underset{\sim}{B}_5$. One of the destinies of ∞ is to act as a default value.

To insure that all 5-generated subalgebras of B_5 are finite we impose some more conditions:

CONDITION 1: $\alpha\infty = \infty\alpha = \infty$ for all α in $\underset{\sim}{B}_5$.

CONDITION 2: $\beta\alpha \in \{\alpha,\beta,\alpha\beta,\infty\}$ if α is immediately counterclockwise of β.

CONDITION 3: $\alpha\beta \in \{\alpha,\beta,\infty\}$ if neither condition 0 nor condition 2 applies.

These three conditions essentially prohibit the generation of new elements not authorized by Condition 0.

To simplify later steps in the construction we impose three more conditions:

CONDITION 4: $\alpha\beta = \beta\alpha$ for all α and β in B.

CONDITION 5: $\alpha\alpha = \alpha$ for all α in $\underset{\sim}{B}_5$.

CONDITION 6: The map which fixes ∞ and rotates the remainder of $\underset{\sim}{B}_5$ clockwise by 60 degrees is an automorphism of $\underset{\sim}{B}_5$.

Now consider any five given elements of $\underset{\sim}{B}_5$. Pick one of the six radial arms which contains none of the five given elements. Let X be the set of members of $\underset{\sim}{B}_5$ which are not on the radial arm just chosen and which are no further out than the outermost of the five given elements. Then $X \cup \{\infty\}$ generates only 10 new elements. (Only the five outer most elements can play any significant role in the generating process.) Thus the subalgebra of $\underset{\sim}{B}_5$ generated by the five given elements is finite, as required.

A similar construction leads to $\underset{\sim}{B}_n$ for all $n > 3$, so Step 0 is complete.

STEP 1

First consider the algebra $\underset{\sim}{C}_5$ diagrammed as follows:

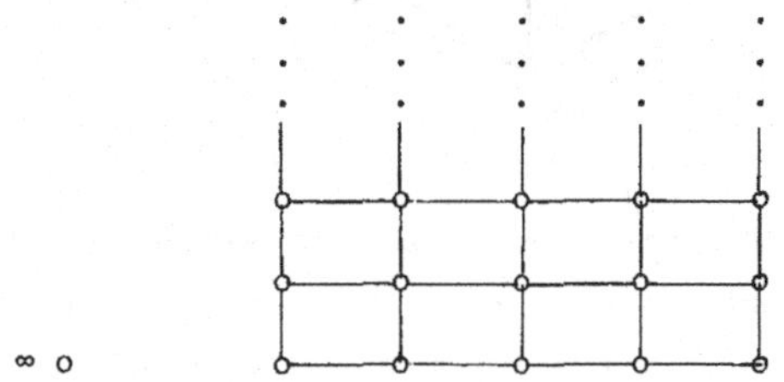

Our intention is that $\underset{\sim}{C}_5$ is made by removing one of the six radial arms of $\underset{\sim}{B}_5$ and then "flattening" the diagram. Thus the five vertical columns used to be radial arms and the horizontal rows are the remnants of the hexagons found in $\underset{\sim}{B}_5$. So $\underset{\sim}{C}_5$ inherits the following conditions from $\underset{\sim}{B}_5$.

CONDITION 0': $\alpha\beta = \gamma$ if α is immediately left of β and γ is immediately above α

CONDITION 1': $\alpha\infty = \infty\alpha = \infty$ for all α.

CONDITION 2': $\beta\alpha \in \{\alpha,\beta,\alpha\beta,\infty\}$ if α is immediately left of β.

CONDITION 3': $\alpha\beta \in \{\alpha,\beta,\infty\}$ if neither Condition 0' nor Condition 2' applies

CONDITION 4': $\alpha\beta = \beta\alpha$ for all α and β .

CONDITION 5': $\alpha\alpha = \alpha$ for all α .

Condition 6 and our analysis of the 5-generated subalgebras of $\underset{\sim}{B}_5$ described in Step 0 imply that every 5-generated subalgebra of $\underset{\sim}{B}_5$ is embeddable in $\underset{\sim}{C}_5$. Now let $\underset{\sim}{C}$ have the diagram below:

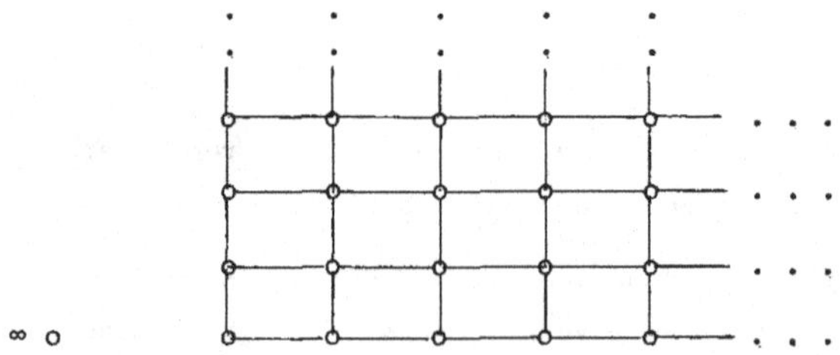

Our intention is that $\underset{\sim}{C}$ be just like $\underset{\sim}{C}_5$ except that it has infinitely many columns arranged to the right. Thus $\underset{\sim}{C}$ looks like the positive integer lattice points in the

plane together with the single additional point ∞. We impose Conditions 0' through 5' on $\underset{\sim}{C}$. Now we really intended that $\underset{\sim}{B}_5, \underset{\sim}{B}_6, \underset{\sim}{B}_7, \ldots$ be defined in a uniform way. So any five consecutive columns of $\underset{\sim}{C}$ should be isomorphic to $\underset{\sim}{C}_5$. To accomplish this we impose the following consition on $\underset{\sim}{C}$:

CONDITION 6': The map which fixes ∞ and translates every other point one unit to the right is an embedding of $\underset{\sim}{C}$ into itself.

This Condition is really our motive for imposing Condition 6 on $\underset{\sim}{B}_5$. Clearly, Condition 6' is much stronger and Condition 6 should really be replaced by some kind of partial embedding condition between $\underset{\sim}{B}_n$ and $\underset{\sim}{B}_m$. In any case, if $\underset{\sim}{C}$ satisfies these seven Conditions, then it is possible to build $\underset{\sim}{B}_n$ as in Step 0 for each $n > 4$ such that every n-generated subalgebra of $\underset{\sim}{B}_n$ is embeddable in $\underset{\sim}{C}$. So Step 1 is finished.

STEP 2

In this final step we want to find a finite groupoid $\underset{\sim}{A}$ such that $\underset{\sim}{C} \in \mathrm{HSP}\underset{\sim}{A}$. Since Lyndon [51] proved that all algebras with two elements are finitely based, $\underset{\sim}{A}$ must have at least three members. So let a, b, and c be three distinct elements of $\underset{\sim}{A}$. Because $\underset{\sim}{A}$ is finite while $\underset{\sim}{C}$ is countably infinite, our simplest hope is to realize $\underset{\sim}{C}$ as a homomorphic image of a subalgebra of $\underset{\sim}{A}^\omega$. Now except for ∞ the elements of $\underset{\sim}{C}$ can be regarded as ordered pairs (m,k) of positive integers. On the other hand, the elements of $\underset{\sim}{A}^\omega$ can be represented as one-way infinite strings of members of $\underset{\sim}{A}$. Thus

aaabccccccbaaaaa.......

belongs to $\underset{\sim}{A}^\omega$ and it is a streamlined representation of

(a, a, a, b, c, c, c, c, c, c, b, a, a, a, a, a, ...)

An even more convenient way to represent the same member of A^ω is

$$a \xrightarrow{3} abc \xrightarrow{6} cba \longrightarrow$$

What we need to do is specify a subset D of $\underset{\sim}{A}^\omega$ and a function F from $\underset{\sim}{D}$ onto $\underset{\sim}{C}$ and then impose an operation on $\underset{\sim}{A}$ in such a way that $\underset{\sim}{D}$ becomes a subalgebra of $\underset{\sim}{A}^\omega$ and F becomes a homomorphism. So the strings in $\underset{\sim}{D}$ will somehow code pairs of positive integers and F will decode the strings in $\underset{\sim}{D}$.

Let m and k be positive integers. We let the code for (m,k) be

$$a \xrightarrow{m} abc \xrightarrow{k} cba \longrightarrow$$

So this string is placed in $\underset{\sim}{D}$ and we set $F(a \xrightarrow{m} abc \xrightarrow{k} cba \longrightarrow) = (m,k)$. Let us hope, for the moment, that $\underset{\sim}{D}$ consists exactly of these strings. So F would be one-to-one. Since F is to be a homomorphism, we can already draw some conclusions about the operation on A. Let us agree to represent the operation on A^ω in a way similar to how addition of real numbers, treated as infinite decimal expansions, is represented. We write one member below the other, draw a horizontal line, and place the result beneath this line, taking care to keep the "digits" in the proper columns. Thus

$$\begin{array}{lcl}
\text{in } \underset{\sim}{A} & \qquad & \text{in } \underset{\sim}{C} \\
a \overset{m}{\text{——}} abc \overset{k}{\text{——}} cba \longrightarrow & & (m, k) \\
a \overset{m+1}{\text{——}} abc \overset{k}{\text{——}} cba \longrightarrow & & (m+1, k) \\
\hline
a \overset{m}{\text{——}} abc \overset{k+1}{\text{——}} cba \longrightarrow & & (m, k+1)
\end{array}$$

represents the result of applying the operation to the codes of (m, k) and (m,k + 1 The code of (m, k + 1) should be obtained if F is a homomorphism since $\underset{\sim}{C}$ satisfies Condition 0'. Condition 5' tells us that the operation in $\underset{\sim}{C}$ is idempotent. Since the operation of $\underset{\sim}{A}$ is defined coordinatewise, from the calculation above and idempotency we can write down part of the operation table of $\underset{\sim}{A}$.

	a	b	c	. . .
a	a	b		
b	b	b	c	
c		c	c	
.				
.				
.				

Using this fragment of the table, we calculate as follows:

$$\begin{array}{l}
a \overset{m+1}{\text{——}} abc \overset{k}{\text{——}} cba \longrightarrow \\
a \overset{m}{\text{——}} abc \overset{k+1}{\text{——}} cba \longrightarrow \\
\hline
a \overset{m}{\text{——}} abc \overset{k+1}{\text{——}} cba \longrightarrow
\end{array}$$

$$\begin{array}{l}
a \overset{m+1}{\text{——}} abc \overset{k}{\text{——}} cba \longrightarrow \\
a \overset{m}{\text{——}} abc \overset{k+2}{\text{——}} cba \longrightarrow \\
\hline
a \overset{m}{\text{——}} abc \overset{k+2}{\text{——}} cba \longrightarrow
\end{array}$$

$$\begin{array}{l}
a \overset{m}{\text{——}} abc \overset{k}{\text{——}} cba \longrightarrow \\
a \overset{m}{\text{——}} abc \overset{k+1}{\text{——}} cba \longrightarrow \\
\hline
a \overset{m}{\text{——}} abc \overset{k+1}{\text{——}} cba \longrightarrow
\end{array}$$

These calculations force the following equalities to hold in $\underset{\sim}{C}$:

$$(m+1,\ k)(m,\ k+1) = (m,\ k+1)$$

$$(m+1,\ k)(m,\ k+2) = (m,\ k+2)$$

$$(m,\ k)(m,\ k+1) = (m,\ k+1)$$

Fortunately, these equalities are in accord with Condition 3'.

At this point it is not possible to determine the product of other members of $\underset{\sim}{D}$ (except those entailed by the commutativity Condition 4'), since the products of a and c are unknown. However, there is no way to define these products so that $\underset{\sim}{D}$ will be a subalgebra of $\underset{\sim}{A}^{\omega}$. We have another problem: F is not yet onto $\underset{\sim}{C}$ since ∞ is not the image of anything in $\underset{\sim}{D}$. We handle these by adding an element u to $\underset{\sim}{A}$ which will be the value of all products of a and c and otherwise to behave like a "zero". This will force us to put certain strings into $\underset{\sim}{D}$ which contain u. For simplicity, put into $\underset{\sim}{D}$ all strings in which u appears. Next, we insist that ∞ be the value assigned by F to any string in which u appears. So $A = \{a,b,c,u\}$ and the operation of $\underset{\sim}{A}$ is given by the following table:

	a	b	c	u
a	a	b	u	u
b	b	b	c	u
c	u	c	c	u
u	u	u	u	u

Evidently, $\underset{\sim}{D}$ is a subalgebra of $\underset{\sim}{A}^{\omega}$. In order to make F into a homomorphism, we must now complete the definition of the operation of $\underset{\sim}{C}$. For all positive integers m, k, p, and q let

$$(m,\ k)(p,\ q) = \begin{cases} (m,\ k+1) & \text{if } p = m+1 \text{ and } k = q \\ (m,\ k) & \text{if } p = m+1 \text{ and } k \in \{q+1,\ q+2\} \\ (m,\ k) & \text{if } p = m \text{ and } k \in \{q,\ q+1\} \\ (p,\ q+1) & \text{if } m = p+1 \text{ and } q = k \\ (p,\ q) & \text{if } m = p+1 \text{ and } q \in \{k+1,\ k+2\} \\ (p,\ q) & \text{if } m = p \text{ and } q \in \{k,\ k+1\} \\ \infty & \text{otherwise} \end{cases}$$

and let $\alpha\infty = \infty\alpha = \infty$ for all α in $\underset{\sim}{C}$. It is easy to check that $\underset{\sim}{C}$ fulfills all the Condition 0' through 6'. Consequently, $\underset{\sim}{C} \in \mathrm{HSP}\underset{\sim}{A}$ and since it is now easy to complete the definition of the operation of each $\underset{\sim}{B}_n$, where $n > 4$, we conclude by the Easy Theorem that $\underset{\sim}{A}$ is inherently nonfinitely based.

SUMMING UP

The algebra $\underset{\sim}{A}$ was discovered by Park [76] (but see also Park [80]),who first showed that it is not finitely based. Kirby Baker observed that $\underset{\sim}{A}$ is inherently nonfinitely based. However, the fact that Park's algebra emerged from our construction seems more or less accidental. Along the way we made a number of arbitrary choices, some of which were made for convenience only and others, which played a more crucial role, had alternatives. In fact, the procedure was manipulated to some extent to arrive at Park's algebra, rather than some other. Much more far-reaching results can be obtained by analyzing the method of construction introduced here. Some of these results will be presented in a forthcoming paper of Baker, McNulty, and Werner.

Even considered only as a demonstration that Park's algebra is not finitely based, the line of reasoning offered here is actually quick and direct. An important advantage that this method enjoys over previous methods of demonstrating an algebra to be nonfinitely based, is that it does not depend on any analysis of what equations are true in the algebra. Instead, the method proceeds simply by constructing the desired algebras.

REFERENCES

Garrett Birkhoff
[35] *On the structure of abstract algebras*, Cambridge Phil. Soc. 31 (1935) 433-454.

Roger Lyndon
[51] *Identities in two-valued calculi* Trans. Amer. Math. Soc. 71 (1951) 457-465.

[54] *Identities in finite algebras*, Proc. Amer. Math. Soc. 5 (1954) 8-9.

G. McNulty and C. Shallon
[83] *Inherently nonfinitely based finite algebras* in Universal Algebra and Lattice Theory, R. Freese and O. Garcia, eds., Lecture Notes in Mathematics, vol. 1004 Springer-Verlag, New York (1983) 206-231.

V.L. Murskii
[79] *On the number of k-element algebras with one binary operation without a finite basis of identities*, (Russian) Problemy Kibernet 35 (1979) 5-27.

B.H. Neumann
[37] *Identical relations in groups*, I.,Math. Ann. 14 (1937) 506-525.

R. Park
[76] *Equational classes of non-associative ordered algebras*, Ph. D. Thesis University of California at Los Angeles, 1976.

[80] *A four-element algebra whose identities are not finitely based*, Algebra Universalis 11 (1980) 255-260.

P. Perkins
[85] *Basic questions for general algebras*, Algebra Universalis, 19 (1984) 16-23.

FINITE INTEGRAL RELATION ALGEBRAS

Roger Maddux
Department of Mathematics
Iowa State University
Ames, Iowa 50011

Given integers n and s, let $F(n,s)$ be the number of isomorphism types of finite integral relation algebras which have n atoms, of which exactly s are symmetric (i.e., satisfy $x^{\smile} = x$). It turns out that $F(n,s) = 0$ iff $1 \leq s \leq n$ and $n-s$ is even. The computation of $F(n,s)$ is time-consuming, but has been done for a few values.

Known values:	$F(1,1) = 1$	$F(3,1) = 3$	$F(4,2) = 37$
	$F(2,2) = 2$	$F(3,3) = 7$	$F(4,4) = 65$
New values:	$F(5,1) = 83$		
	$F(5,3) = 1316$		
	$F(5,5) = 3013$		

THEOREM. $F(n,s)$ approaches $\dfrac{2^{Q(n,s)}}{P(n,s)}$ as n approaches infinity, where $Q(n,s) = (1/6)(n-1)[(n-1)^2+3s-1]$ and

$P(n,s) = (s-1)![(1/2)(n-s)]!2^{(1/2)(n-s)}$.

If E is any set of equations which hold in every representable relation algebra, and $F(E,n,s)$ is the number of isomorphism types of integral relation algebras which have n atoms, satisfy E, and have s symmetric atoms, then the theorem above continues to hold when $F(n,s)$ is replaced by $F(E,n,s)$.

Problem. Does the theorem continue to hold when $F(n,s)$ is replaced by the number of isomorphism types of integral representable relation algebras with n atoms and s symmetric atoms?

Even if the answer is "yes", there are still very many nonrepresentable integral relation algebras with n atoms. In fact, for sufficiently large n, there will be more than $2^{(1/7)n^3}$ isomorphism types of such algebras.

A is a **relation algebra** iff **A** is an algebraic structure of the form $\langle A, +, \cdot, -, 0, 1, ;, {}^{\smile}, 1' \rangle$, where

(Bo) $\langle A, +, \cdot, -, 0, 1 \rangle$ is a Boolean algebra (called the **Boolean part** of **A**).

(Pe) the following formulas are equivalent for all $x,y,z \in A$: $x;y \cdot z = 0$, $x^{\smile};z \cdot y = 0$, $z;y^{\smile} \cdot x = 0$,

(Id) $x;1' = x = x;1'$ for all x in A ,

(As) ; is an associative binary operation.

RA is the class of all relation algebras, and "RA" sometimes serves as an abbreviation of the phrase "relation algebra". An algebra satisfying all the conditions given above except possibly (As) is called a **nonassociative relation algebra.** The class of all such algebras is NA, and "NA" will be an abbreviation of "non-associative relation algebra".

Let **A** be a NA. An element x of A is **symmetric** iff $x^{\smile} = x$. **A** is **symmetric** iff all of its elements are symmetric, **commutative** if $x;y = y;x$ for all x,y in A, and **integral** iff $0 \neq 1$ and for all x,y in A, $x;y = 0$ iff $x = 0$ or $y = 0$. By [5], a RA is integral iff the identity element $1'$ is an atom of **A**, and every integral RA is simple. Given any set U, **Re**U is the algebra $\langle Sb(U\times U), +, \cdot, 0, 1, ;, {}^{\smile}, 1'\rangle$, where $Sb(U\times U)$ is the set of all binary relations on U, $+$ is union, $\cdot$ is intersection, $-$ is complementation (with respect to $U\times U$), 0 is the empty set, 1 is $U\times U$, ; is relative product (or composition) of binary relations, ${}^{\smile}$ produces the converse of a binary relation, and $1'$ is the identity relation on U. It is easy to see that **Re**U is a relation algebra. **Re**U is simple and not integral unless $|U| = 1$. A RA is **representable** iff it is isomorphic to a subalgebra of a direct product of algebras of the form **Re**U. "RRA" is an abbreviation of "representable relation algebra", and also denotes the class of all RRAs. By [5], a simple RA is representable iff it is isomorphic to a subalgebra of **Re**U, for some nonempty set U. A subalgebra **A** of **Re**U is integral iff U is both the range and domain of every element of **A.** A RA is **finitely (infinitely) representable** iff it is isomorphic to a subalgebra of a direct product of algebras of the form **Re**U where U is finite (infinite). Clearly no infinite simple RRA can be finitely representable, but there are finite RRAs which are not finitely representable. The simplest example is the 3-atom integral RA whose atoms are the relations $<$, $>$, and $=$ on the rationals. There are also simple integral RRAs which are finitely representable but not infinitely representable (see [11]). Lyndon was the first to find a RA which is not representable. Such an algebra appears in [9] and has 56 atoms. Another one with only 4 atoms is given by McKenzie in [15]. Other nonrepresentable RAs occur in [6], [11], and [13]. RRA is a variety ([17]) but it is not finitely axiomatizable ([16]).

Integral relation algebras are a very natural class to study. There are 13 RAs having 3 or fewer atoms, and all of them are integral and representable. The smallest nonintegral RA is **Re**2, which has 4 atoms. There is a natural correlation between integral relation algebras and polygroups (see [2] and [3]). Lyndon's first nonrepresentable RA is not integral, but the ones constructed from non-Arguesian projective planes by Jónsson in [6] are integral, and so are all the algebras which arise from projective planes by Lyndon's construction in [11]. The latter algebras are also used by Monk in [16] to show RRA is not finitely axiomatizable. McKenzie's nonrepresentable RA is integral. Any RA arising from a modular lattice by the construction in [12] is integral.

The main tool in this study of finite integral RAs is the correspondence between algebras and their atom structures. This correspondence is discussed in [9] for RAs, in [4] for Boolean algebras with operators, and in [5] for RAs and cylindric algebras. The account here is based on [13], which in turn imitates the development in [8] for cylindric algebras. Other closely related treatments occur in [2], [3], and [7].

Let $\mathbf{A}$ be a NA. Then $At\mathbf{A}$ is the set of atoms of (the Boolean part of) $\mathbf{A}$, and

$$\mathbf{AtA} = \langle At\mathbf{A}, C(\mathbf{A}), {}^{\smile}, I(\mathbf{A})\rangle$$

is the **atom structure** of $\mathbf{A}$, where

$$C(\mathbf{A}) = \{\langle x,y,z\rangle : x,y,z \in At\mathbf{A} \text{ and } x;y \geq z\},$$

and

$$I(\mathbf{A}) = \{x : x \in AtA \text{ and } x \leq 1'\}$$

and "$^{\smile}$" denotes the restriction of $^{\smile}$ to the atoms of $\mathbf{A}$. By [13], Theorem 3.4, the set of atoms of $\mathbf{A}$ is closed under $^{\smile}$.

Given a relational structure $\mathbf{S} = \langle U, C, f, I\rangle$ where C is a ternary relation on U, f is a function from U to U, and I is a subset of U, let

$$\mathbf{CmS} = \langle SbU, \cup, \cap, \sim, \emptyset, U, ;, ^{\smile}, I\rangle$$

where, for any subsets X and Y of U,

$$X;Y = \{z : \text{for some } x \in X \text{ and some } y \in Y, \langle x,y,z\rangle \in C\}$$

and

$$X^{\smile} = \{fx : x \in X\} .$$

CmS is called the **complex algebra** of **S**.

If **A** is a complete atomic NA (for example, if **A** is finite) then **A** is isomorphic to the complex algebra of its atom structure (**A** $\cong$ **CmAtA**). Any two complete atomic NAs are isomorphic iff their atom structures are isomorphic. Hence every complete atomic NA is determined up to isomorphism by its atom structure. So, which structures arise as atom structures of complete atomic NAs?

THEOREM 1. Let **S** = ⟨U, C, f, I⟩ , where C is a subset of UxUxU, f is a function from U to U , and I is a subset of U.

(1) The following conditions (i)-(iii) are equivalent.

(i) **S** is the atom structure of complete atomic NA,

(ii) the complex algebra of **S** is a NA,

(iii) **S** satisfies conditions (a) and (b):

(a) if $\langle x,y,z\rangle \in C$ then $\langle fx,z,y\rangle$, $\langle z,fy,x\rangle \in C$,

(b) for all $x,y \in U$, $x = y$ iff there is some $w \in I$ such that $\langle x,w,y\rangle \in C$.

(2) The complex algebra of **S** is a relation algebra iff it is a NA and satisfies condition (c):

(c) for all $v,w,x,y,z \in U$, if $\langle v,w,x\rangle$, $\langle x,y,z\rangle \in C$, then there is some $u \in U$ such that $\langle w,y,u\rangle$, $\langle v,u,z\rangle \in C$.

Proof. If **S** $\cong$ **AtA** for some complete atomic NA **A** , then **CmS** $\cong$ **CmAtA** $\cong$ **A** and so **CmS** is a NA. Thus (i) implies (ii). Suppose the complex algebra of **S** is a NA. **CmS** is also complete and atomic, and **S** $\cong$ **AtCmS**. Thus (ii) implies (i). The equivalence of (ii) and (iii) is part of Theorem 2.2 of [13], and so is (2).

The conditions (a), (b), and (c) correspond to (Pe), (Id), and (As), respectively.

Suppose **S** = ⟨U, C, f, I⟩ and **CmS** is a NA. Then $ffx = x$ for all $x \in U$. (Use (b), then (a) twice, and (b) again.) For any $x,y,z \in U$, let $[x,y,z] =$

$\{\langle x,y,z\rangle, \langle fx,z,y\rangle, \langle y,fz,fx\rangle, \langle fy,fx,fz\rangle, \langle fz,x,fy\rangle, \langle z,fy,x\rangle\}$.

The set $[x,y,z]$ is called a **cycle** of **S** (or **CmS**, or, more precisely, of $\langle U, f\rangle$). (This term is use in [9] for essentially the same concept.) If $\langle x,y,z\rangle \in C$, then $[x,y,z] \subseteq C$ by (a). Since f has order 2, $[x,y,z]$ is the closure of $\langle x,y,z\rangle$ under the two functions on triples implicit in (a). In fact, it is easy to see that Theorem 1 will still be true if (a) is replaced by (a'):

(a') for all $x,y,z \in U$, either $[x,y,z] \subseteq C$ or $[x,y,z] \cap C = \emptyset$.

Condition (a') simply says that C is a union of cycles.

The complex algebra of **S** satisfies the condition "1' is an atom" just in case I is a singleton, i.e. $I = \{e\}$ for some $e \in U$. Whenever this is the case, (b) is equivalent to (b'):

(b') for all x,y in U, $x = y$ if $\langle x,e,y\rangle$ is in C.

It is not hard to show that Theorem 1 is still true if the hypothesis $I = \{e\}$ is added, (a) is replaced by (a'), and (b) is replaced by (b''):

(b'') $fe = e$, and $C = [(U-\{e\})\times(U-\{e\})\times(U-\{e\})] \cap C \cup \bigcup\{[x,e,x] : x \in U\}$.

A cycle $[x,y,z]$ is called a **diversity cycle**, or simply a **d-cycle**, if all the elements in its triples are in $U-I$. In case $I = \{e\}$, (b'') implies that C is determined by its d-cycles.

The reason for including the condition $fe = e$ in (b'') arises in part of the proof of (b') from (a') and (b''). Suppose $\langle x,e,y\rangle \in C$. Then $\langle x,e,y\rangle \in [z,e,z]$ for some $z \in U$. If $\langle x,e,y\rangle = \langle z,e,z\rangle$ then $x = y$. Three other cases are similar. In the remaining two, namely $\langle x,e,y\rangle = \langle fz,z,e\rangle$ and $\langle x,e,y\rangle = \langle fe,fz,fz\rangle$, use the condition $fe = e$ to deduce that $x = y$.

The conditions in [5], which hold for a RA just in case it is integral, are not equivalent for all NAs.

Theorem 2. Let **A** be a NA.

(1) **A** is integral iff $0 \neq 1$ and for all $x \in A$, if $x \neq 0$, then $x;1 = 1$ (and $1;x = 1$).

(2) If **A** is integral, then 1' is an atom of **A**.

<u>Proof</u>. Suppose **A** is integral. From $x;1.-(x;1) = 0$ it follows that $x^{\smile};-(x;1) = 0$ by (Pe), so $x^{\smile} = 0$ or $-(x;1) = 0$ since **A** is integral. If $x \neq 0$, then $x^{\smile} \neq 0$, so $x;1 = 1$. (Similarly, $1;x = 1$ whenever $x \neq 0$.)

Now assume $x;1 = 1$ whenever $x \neq 0$. If $x \neq 0 \neq y$, then $x^{\smile} \neq 0$ and so $x^{\smile};1 = 1$ by assumption. Hence $0 \neq x^{\smile};1.y$, so $0 \neq x;y$ by (Pe). This completes the proof of (1).

Suppose $1'$ is not an atom. If $1' = 0$ then $1 = 1;1' = 1;0 = 0$ so **A** is not integral. Assume $1' \neq 0$. Then there are $x,y \in A$ such that $0 = x.y$, $0 \neq x$, $0 \neq y$, and $x+y = 1'$. But $x;y = x.y$ by [13], 1.13(15), so $x;y = 0$ and **A** is not integral.

The converse of Theorem 2(2) is false. For example, if $U = \{e,a,b\}$, $|U| = 3$, $C = [a,e,a] \cup [b,e,b]$, and f is the identity on U, then **Cm**$\langle U, C, f, I\rangle$ is a non-integral NA (since $\{a\};\{b\} = \emptyset$), but $1' = \{e\}$ is an atom.

Theorem 2(2) implies that I is a singleton whenever **Cm**$\langle U, C, f, I\rangle$ is an integral NA. So the characterization of those structures whose complex algebras are integral RAs can be restricted to structures of the form $\langle U, C, f, \{e\}\rangle$ where $e \in U$.

THEOREM 3. Let **S** $= \langle U, C, f, \{e\}\rangle$ where C is a subset of $U\times U\times U$, f is a function from U to U, and $e \in U$.

(1) **CmS** is a NA if **S** satisfies (a') and either (b') or (b'').

(2) Let **CmS** be a NA. Then **CmS** is integral iff **S** satisfies (c'):

(c') for all $x,y \in U-\{e\}$, if $x \neq y$ then there is some $w \in U-\{e\}$ such that $\langle x,w,y\rangle \in C$.

(3) Let **CmS** be a NA. Then **CmS** is a RA iff **S** satisfies (c') and also (c''):

(c'') for all $v,w,x,y,z \in U-\{e\}$, if $\langle v,w,x\rangle$ and $\langle x,y,z\rangle$ are in C, and $v \neq z$ or $fw \neq y$, then there is some $u \in U-\{e\}$ such that $\langle w,y,u\rangle$, $\langle v,u,z\rangle \in C$.

<u>Proof</u>. Part (1) follows immediately from Theorem 1 and the remarks following it. Part (2) is easy to prove using Theorem 2(1). For part (3) it suffices to show that if (a') and (b') hold in **S**, then **S** satisfies (c) iff **S** satisfies both (c') and (c''). First suppose **S** satisfies (a'), (b'), and (c).

Assume $x,y \in U-\{e\}$ and $x \neq y$. Then $\langle x,fx,e\rangle$, $\langle e,y,y\rangle \in C$ by (b') and (a'). By (c), there is some $u \in U$ such that $\langle fx,y,u\rangle$, $\langle x,u,y\rangle \in C$. (However, $[fx,y,u] = [x,u,y]$.) Since $x \neq y$, $u \neq e$ by (b'). Thus (c') holds in **S**.

Assume $v,w,x,y,z \in U-\{e\}$, $\langle v,w,x\rangle, \langle x,y,z\rangle \in C$, and either $fw \neq y$ or $v \neq z$. By (c) there is some u in U such that $\langle w,y,u\rangle, \langle v,u,z\rangle \in C$. If $v \neq z$ then $u \neq e$ by (b'), while if $fw \neq y$ then $u \neq e$ by (a') and (b'). Thus (c'') holds in **S**.

Now suppose that **S** satisfies (a'), (b'), (c'), and (c''). Let v,w,x,y,z be in U and assume $\langle v,w,x\rangle, \langle x,y,z\rangle \in C$. We must find some u in U such that $\langle w,y,u\rangle, \langle v,u,z\rangle \in C$. If $v = e$, then $w = x$, so let $u = z$. If $w = e$, then $v = x$, so let $u = y$. If $y = 3$, then $x = z$, so let $u = w$. If $z = e$, then $x = fy$, so let $u = fv$. If $x = e$, then $v = fw$ and $y = z$, so u exists by (c'). If $fw = y$ and $v = z$, then let $u = e$. Finally, if $e \notin \{v,w,x,y,z\}$ and either $fw \neq y$ or $v \neq z$, then u exists by (c'').

Now we present some ways to construct integral relation algebras. The first one shows that every group is isomorphic to the automorphism group of some symmetric integral relation algebra, thus answering a quesion posed by J. D. Monk.

THEOREM 4. Suppose $|U| > 6$, $e \in U$, f is the identity on U, and R is any reflexive binary relation on $U-\{e\}$. Let C be the union of all the following cycles:

$[x,e,x]$ for all $x \in U$,

$[x,y,y]$ for all $x \in U-\{e\}$ where $\langle x,y\rangle \in R$,

$[x,y,z]$ for all $x,y,z \in U-\{e\}$ such that x,y,z are distinct.

Then

(1) $\mathbf{Cm}\langle U, C, F, \{e\}\rangle$ is a symmetric integral relation algebra.

(2) The automorphism groups of $\langle U-\{e\}, R\rangle$, $\langle U, C, f, \{e\}\rangle$, and $\mathbf{Cm}\langle U, C, f, \{e\}\rangle$ are isomorphic.

<u>Proof</u>. For (1) it suffices to show that $\langle U, C, F, \{e\}\rangle$ satisfies (a'), (b''), (c'), and (c''). But (a'), (b''), and (c') obviously hold, so, for (c''), let us assume that $v,w,x,y,z \in U-\{e\}$, $\langle v,w,x\rangle, \langle x,y,z\rangle \in C$, and either $v \neq z$ or $w \neq y$. We want some $u \in U-\{e\}$ such that both $\langle w,y,u\rangle, \langle v,u,z\rangle \in C$. If $v \neq z$ and $w \neq y$, let u be any element in $u-\{e,v,w,y,z\}$. This is possible since $|u| > 6$. So suppose $v = z$ and $w \neq y$. (The other case, in which $v \neq z$ and $w = y$, is handled similarly.) If

$w \neq v = z \neq y$, then it suffices to let $u = v$. If $w = v = z \neq y$ or $w \neq v = z = y$, then let $u = x$.

Every automorphism of a must send atoms to atoms. It is therefore easy to see that $\langle U, C, f, \{e\}\rangle$ and its complex algebra have isomorphic automorphism groups. Notice that if R is the identity on U then all permutations of U which fix e are automorphisms of $\langle U, C, f, \{e\}\rangle$. Also, $\langle x,y\rangle \in R$ iff $x,y \neq e$ and $[x,y,y]$ is included in C. Hence any automorphism of $\langle U, C, f, \{e\}\rangle$ must preserve R, i.e. its restriction to $U-\{e\}$ is an automorphism of $\langle U-\{e\}, R\rangle$. This establishes an isomorphism between the automorphism groups of $\langle U, C, f, \{e\}\rangle$ and $\langle U-\{e\}, R\rangle$. Thus (2) holds.

THEOREM 5. Every group is isomorphic to the automorphism group of some symmetric integral relation algebra.

Proof. Every group is the automorphism group of some reflexive binary relation.

Not all the algebras constructed in Theorem 4 are representable. For if R is the identity on U and U is finite, then $\mathbf{Cm}\langle U, C, f, \{e\}\rangle$ is representable iff there exists a projective plane of order $|U|-2$. (This is proved in [11].) A non-representable relation algebra is thus obtained when $|U| = 8$. On the other hand, if $R = (U-\{e\})\times(U-\{e\})$ then $\mathbf{Cm}\langle U, C, f, \{e\}\rangle$ is representable by Theorem 5.19 of [13]. The proof of 5.19 shows that $\mathbf{Cm}\langle U, C, f, \{e\}\rangle$ is infinitely representable, but it can be shown that $\mathbf{Cm}\langle U, C, f, \{e\}\rangle$ is finitely representable as well.

The next construction produces only RRAs. (Compare Theorem 5.3 of [2].)

THEOREM 6. Assume $e,a \in U$, $a \neq e$, f is a function from U to U, $fe = e$, $ffx = x$ for all $x \in U$, and T is any ternary relation on $U-\{e,a\}$ which is a union of d-cycles. Let C be the union of all the following cycles:

$[x,e,x]$ for all $x \in U$,

$[x,y,a]$ for all $x,y \in U-\{e\}$,

$[x,y,z]$ whenever $[x,y,z] \subseteq T$.

Then $\mathbf{Cm}\langle U, C, f, \{e\}\rangle$ is an infinitely representable integral relation algebra.

Proof. Clearly (a'), (b''), and (c') hold. In the proof of (c''), just let u = a. So **Cm**<U, C, f, {e}> is an integral RA. It satisfies condition 5.13(*) of [13] with x = a. Hence it is infinitely representable by 5.19 of [13].

Let **A** and **B** be finite nonassociative relation algebras in which 1' is an atom. Assume **A** and **B** both have exactly n atoms, and exactly s symmetric atoms. Then $s \geq 1$ since 1' is a symmetric atom. Also n-s is even since the nonsymmetric atoms may be partitioned into two-element sets of the form $\{x, x^{\smile}\}$. Let $m = (1/2)(n-s)$. Suppose h is an isomorphism from **A** and **B**. Then h will map the atoms of **A** onto the atoms of **B**. Let g be the restriction of h to the atoms of **A**. Then g will be a one-to-one correspondence between At**A** and At**B** having the following two properties: g maps 1' of **A** to 1' of **B**, and $g(x^{\smile}) = (gx)^{\smile}$ for every atom x of **A**. The number of functions having these two properties is $(x-1)!m!2^m$. (The symmetric atoms of **A** distinct from 1' may be mapped onto those of **B** in (s-1)! ways. The m two-element sets $\{x, x^{\smile}\}$ of nonsymmetric atoms of **A** can be matched with those of **B** in m! ways. Every such matching of two-element sets gives rise to $m!2^m$ one-to-one correspondences between nonsymmetric atoms of **A** and **B** because there are exactly two one-to-one correspondences between two two-element sets.) Now suppose g is just any one-to-one correspondence between the atoms of **A** and **B** having the two properties mentioned above. Then it is easy to see that g can be extended to an isomorphism between **A** and an algebra **C** which has the same Boolean part as **B**, and the same $\smile$ and 1' as **B**. How many such algebras **C** are there? The answer is computed in the Theorem 7 below.

Until further notice we shall assume:

U is a finite nonempty set of cardinality n,
$e \in U$,
f is a function from U to U,
$fe = e$,
$ffx = x$ for all $x \in U$,
$s = |\{x : x \in U \text{ and } fx = x\}|$.

Let

$$Q(n,s) = (1/6)(n-1)[(n-1)^2+3s-1],$$

and

$$P(n,s) = (s-1)![(1/2)(n-s)]!2^{(1/2)(n-s)}.$$

The significance of $Q(n,s)$ is explained by Theorem 7 below, and $P(n,s)$ is the number of automorphisms of $\langle U, f, \{e\}\rangle$. For any C included in $U\times U\times U$, let

$$\mathbf{A}(C) = \mathbf{Cm}\langle U, C, f, \{e\}\rangle,$$

and

$$K(n,s) = \{\mathbf{A}(C) : C \text{ is included in } U\times U\times U, \text{ and } \mathbf{A}(C) \text{ is a NA}\}.$$

THEOREM 7. $|K(n,s)| = 2^{Q(n,s)}$.

<u>Proof</u>. By Theorem 3(1), for any C included in $U\times U\times U$, $\mathbf{A}(C)$ is a NA iff the structure $S = \langle U, C, f, \{e\}\rangle$ satisfies conditions (a') and (b''). Consequently $\mathbf{A}(C)$ is a NA iff C is the union of all cycles of the form $[x,e,x]$, where x ranges over U, together with some d-cycles, i.e.

$$C = \bigcup_{x \in U}[x,e,x] \cup [x_1,y_1,z_1] \cup \ldots \cup [x_t,y_t,z_t]$$

for some $x_1,\ldots,x_t,y_1,\ldots,y_t,z_1,\ldots,z_t \in Y-\{e\}$. So the number of such ternary relations C is 2 raised to the power $|\{[x,y,z] : x,y,z \in U-\{e\}\}|$. Let $x,y,z \in U-\{e\}$. Then $|[x,y,z]| = 6$ unless there are $u,v \in U-\{e\}$ such that either

$fu = u$ and $[x,y,z] = [u,u,u] = \{\langle u,u,u\rangle\}$,

or $fu \neq u$ and $[x,y,z) = [u,u,fu] = \{\langle u,u,fu\rangle, \langle fu,fu,u\rangle\}$,

or $fu = u$, $u \neq v$, and $[x,y,z] = [u,v,v] = \{\langle u,v,v\rangle, \langle v,fv,u\rangle, \langle fv,u,fv\rangle\}$.

Now use these facts to count the number of triples in $(U-\{e\})\times(U-\{e\})\times(U-\{e\})$ in two ways. On the one hand, there are $(n-1)^3$. Every triple appears in a unique d-cycle. Every d-cycle has cardinality 1, 2, 3, or 6. The number of d-cycles containing just one triple is $s-1$, thus accounting for $s-1$ triples. There are $(1/2)(n-s)$ distinct d-cycles containing two triples, accounting for $n-s$ additional triples. The number of d-cycles containing exactly three triples is $(s-1)(n-2)$. Let N be the number of d-cycles of cardinality 6. Then

$$(n-1)^3 = (s-1) + (n-s) + 3(s-1)(n-2) + 6N,$$

so the total number of d-cycles contained in $(U-\{e\})\times(u-\{e\})\times(U-\{e\})$ is therefore

$$(s-1) + (1/2)(n-s) + (s-1)(n-2) + N = Q(n,s).$$

For any property P of $\mathbf{A}(C)$, let

$$\#P\# = 2^{-Q(n,s)}|\{C : \mathbf{A}(C) \text{ is a NA and has property } P\}|.$$

By Theorem 7, $\#P\#$ is the fraction of all algebras in $\mathbf{K}(n,s)$ which have property P.

An algebra is **rigid** iff it has no nontrivial automorphisms. To estimate the number of isomorphism types of algebras in $\mathbf{K}(n,s)$ we need to estimate the fraction of them which are rigid. It turns out that "almost all" of the algebras in $\mathbf{K}(n,s)$ are rigid.

LEMMA 8. Suppose $n > 11$ and g is a nontrivial isomorphism of $\langle U, f, \{e\}\rangle$, i.e. g is a permutation of U, $ge = e$, and $fgx = gfx$ for all $x \in U$. Then there are at least $(1/2)(n-2)(n-3)$ d-cycles $[x,y,z]$ which are moved by g, i.e. $[x,y,z] \neq [gx,gy,gz]$.

<u>Proof</u>. Suppose $[x,y,z]$ is a d-cycle which is fixed by g, i.e. $[gx,gy,gz] = [x,y,z]$. Then it is easy to see that g maps $\{x,y,z,fx,fy,fz\}$ onto itself. So to know that a cycle $[x,y,z]$ is moved by g, it suffices to know that $\{x,y,z,fx,fy,fz\}$ is not closed under g.

Since g is nontrivial, there is some $x \in U-\{e\}$ such that $gx \neq x$.

Assume $fx \neq x$. Note that if u,v,y,z are in $U-\{e,x,fx\}$, then $[x,u,v] = [x,y,z]$ iff $u = y$ and $v = z$.

Suppose $gx \neq fx$. Take any $u,v \in U-\{e,x,fx,gx,gfx\}$. Then gx is not in $\{x,u,v,fx,fu,fv\}$, so the d-cycle $[x,u,v]$ is moved by g. Hence there are at least $(n-5)(n-5)$ d-cycles which are moved by g, since $|U-\{e,x,fx,gx,gfx\}| = n-5$. However, $(n-5)(n-5) > (1/2)(n-2)(n-3)$ since $n > 11$.

Suppose $gx = fx$. For all $u,v \in U-\{e,x,fx\}$, $[gx,gu,gv] = [fx,gu,gv] = [x,gv,gu]$, so $[gx,gu,gv] = [x,u,v]$ iff $gu = v$ and $gv = u$. There are fewer than $n-3$ pairs $\langle u,v\rangle$ such that $gu = v$, $gv = u$, and u,v are in $U-\{e,x,fx\}$. Consequently the number of d-cycles which are moved by g is at least $(n-3)(n-3) - (n-3)$. But this is more than $(1/2)(n-2)(n-3)$ since $n > 6$.

This completes the proof in case $fx \neq x$.

Assume $fx = x$. Then, for all $u,v \in U-\{e,x,gx\}$, $[x,u,v] = [x,v,u]$, and $[x,u,v]$ is moved by g since gx is not in $\{x,u,v,fu,fv\}$. For all $u,v,y,z \in U-\{e,x,gx\}$, $[x,u,v] = [x,y,z]$ iff $\{u,v\} = \{y,z\}$. Hence at least $(n-3) + (1/2)(n-3)(n-4) = (1/2)(n-2)(n-3)$ d-cycles are moved by g.

Suppose $G(n,s)$ and $H(n,s)$ are real-valued functions of n and s. We say that $G(n,s)$ **approaches** $H(n,s)$ if for every real number $r > 0$ there is some N such that if $n > N$, $n \geq s \geq 1$, and $n-s$ is even, then $|1 - G(n,s)/H(n,s)| < r$.

THEOREM 9.

(1) $\#\mathbf{A}(C)$ is not rigid$\# \leq P(n,s)2^{-(1/4)(n-2)(n-3)}$ if $n \geq 11$.

(2) $\#\mathbf{A}(C)$ is rigid$\#$ approaches 1.

<u>Proof</u>. Suppose g is a nontrivial automorphism of $\langle U, f, \{e\}\rangle$. How many ternary relations C on $U-\{e\}$ are there such that g is also an isomorphism of $\langle U, C, f, \{e\}\rangle$? By Theorem 8, there are at least $(1/2)(n-2)(n-3)$ d-cycles moved by g, so the number of orbits of d-cycles under g can be no more than $Q(n,s)-(1/4)(n-2)(n-3)$. Now if g as an isomorphism of $\langle U, C, f, \{e\}\rangle$, then for every d-cycle $[x,y,z]$, either C contains $[x,y,z]$ and all the cycles in the orbit of $[x,y,z]$ under g, or else C is disjoint from all those cycles. Hence the number of such relations is at most $2^{Q(n,s)-(1/4)(n-2)(n-3)}$. There are only $P(n,s)$ isomorphisms of $\langle U, f, \{e\}\rangle$, so there are no more than $P(n,s)2^{Q(n,s)-(1/4)(n-2)(n-3)}$ algebras in $\mathbf{K}(n,s)$ which have a nontrivial isomorphism. Thus (1) holds, and (2) follows from (1).

If x,y,z are in U, then we use "$Cxyz$" as an abbreviation of "$[x,y,z]$ is included in C". For any integer $t \geq 1$, we define a property (d.t) of algebras in $\mathbf{K}(n,s)$ as follows:

(d.t) for all $x_1,\ldots,x_t,y_1,\ldots,y_t \in U-\{e\}$ there is some $z \in U-\{e\}$ such that Cx_1y_1z and ... and Cx_ty_tz.

It is easy to see that (d.t+1) implies (d.t) for all t. For the results which follow it is convenient to let (d.0) be any property which holds for all algebras in $\mathbf{K}(n,s)$.

THEOREM 10. Let t be any nonnegative integer.

(1) $\#(\text{d.t})$ is false$\# < (n-1)^{2t}(1-2^{-t})^{n-2t-1}$.

(2) $\#(\text{d.t})\#$ approaches 1.

Proof. Note that (1) and (2) hold in case $t = 0$, since # (d.0) holds # = 1 and # (d.0) is false # = 0.

Let $x_1,\ldots,x_t,y_1,\ldots,y_t \in U-\{e\}$. Then, for all $z \in U-\{e\}$, # Cx_1y_1z and ... and $Cx\ y_tz_t\# > 2^{-t}$, so $\#\text{not}(Cx_1y_1z$ and ... and $Cx_ty_tz)\# < 1-2^{-t}$. Furthermore, if $w,z \in U-\{e,x_1,\ldots,x_t,y_1,\ldots,y_t\}$ and $w \neq z$, then

$\#Cx_1y_1w$ and ... Cx_ty_tw and Cx_1y_1z and ... $Cx_ty_tz\# =$

$(\#\ Cx_1y_1w$ and ... and $Cx_ty_tw\ \#)(\#\ Cx_1y_1z$ and ... and$Cx_ty_tz\ \#)$.

It follows that # for all $z \in U-\{e,x_1,\ldots,x_t,y_1,\ldots,y_t\}$, $\text{not}(Cx_1y_1z$ and ... and $Cx_ty_tz)\# < (1-2^{-t})^{n-2t-1}$. There are only $(n-1)^{2t}$ ways to choose $x_1,\ldots,x_t,y_1,\ldots,y_t$ from $U-\{e\}$, so (1) holds, and (2) follows from (1).

Let $D(t,n,s)$ be the number of isomorphism types of algebras in $\mathbf{K}(n,s)$ which satisfy (d.t).

THEOREM 11. Let t be any nonnegative integer. Then $D(t,n,s)$ approaches $2^{Q(n,s)}/P(n,s)$.

Proof. If $\mathbf{A}(C)$ is any algebra in $\mathbf{K}(n,s)$, then the number of algebras in $\mathbf{K}(n,s)$ which are isomorphic to $\mathbf{A}(C)$ is either equal to $P(n,s)$, just in case $\mathbf{A}(C)$ is rigid, or else less than $P(n,s)$. Hence

$2^{Q(n,s)}\#(d.t)\#$

$< P(n,s)D(t,n,s)$

$< |\{C : (d.t) \text{ holds and } \mathbf{A}(C) \text{ is rigid}\}|$

$+ P(n,s)|\{C : (d.t) \text{ holds and } \mathbf{A}(C) \text{ is not rigid}\}|,$

so by Theorem 9(1),

$\#(d.t)\#$

$< D(t,n,s)P(n,s)2^{-Q(n,s)}$

$< \#(d.t) \text{ and } \mathbf{A}(C) \text{ is rigid}\#$

$+ P(n,s)\#(d.t) \text{ and } \mathbf{A}(C) \text{ is not rigid}\#$

$< \#(d.t)\# + P(n,s)\#\mathbf{A}(C) \text{ is not rigid}\#$

$\leq \#(d.t)\# + P(n,s)^2(2^{-(1/4)(n-2)(n-3)})$.

The desired result follows by Theorem 10.

Now comes the theorem stated in the abstract. Let $F(n,s)$ be the number of isomorphism types of algebras in $\mathbf{K}(n,s)$ which are relation algebras, i.e. $F(n,s)$ is the number of isomorphism types of integral relation algebras with n atoms and s symmetric atoms.

THEOREM 12. $F(n,s)$ approaches $2^{Q(n,s)}/P(n,s)$.

<u>Proof</u>. It is easy to see that (d.1) implies (c') and (d.2) implies (c''), so $\mathbf{A}(C)$ is a RA whenever (d.2) holds. Hence

$$D(2,n,s) \leq F(n,s) \leq D(0,n,s) ,$$

so the desired result follows by Theorem 11.

For any set E of equations true in every RRA, let $F(E,n,s)$ be the number of isomorphism types of relation algebras in $\mathbf{K}(n,s)$ which satisfy all equations in E. Note that $F(\emptyset,n,s) = F(n,s)$. The next goal is to extend Theorem 12 by replacing $F(n,s)$ with $F(E,n,s)$ when E is finite. This requires a summary of some results and concepts from [14].

Let $\mathbf{A}$ be an atomic NA. For every ordinal d such that $3 \leq d \leq \omega$ let

$B_d\mathbf{A} = \{x : x$ is a function from $d \times d$ to $At\mathbf{A}$ such that

$x_{i,i} \leq 1'$, $x_{i,j}{}^{\smile} = x_{j,i}$, and $x_{i,j} \leq$

$x_{i,k};x_{k,j}$ for all $i,j,k < d\}$.

A **d-dimensional basis** for $\mathbf{A}$ is a subset M of $B_d\mathbf{A}$ such that the following two statements hold:

(i) if $x \in M$, $i,j,k < d$, $k \neq i,j$, y,z are atoms of $\mathbf{A}$, and $x_{i,j} \leq y;z$, then there is some $w \in M$ such that $w_{i,k} = y$, $w_{k,j} = z$, and $w_{n,m} = x_{n,m}$ whenever $n,m < d$ and $k \neq n,m$,

(ii) for every atom y of $\mathbf{A}$ there is some $x \in M$ such that $x_{0,1} = y$.

$\mathbf{A}$ is a **matrix algebra of degree** d if $\mathbf{A}$ is a subalgebra of some complete atomic NA which has a d-dimensional basis. MA_d is the class of all matrix algebras of degree d.

THEOREM 13. Let $d \geq 3$.

(1) If $d \leq d' \leq \omega$ then MA_d contains $MA_{d'}$.

(2) MA_d is a variety.

(3) Any perfect extension of an algebra in MA_d has a d-dimensional basis and hence is also in MA_d.

(4) Any finite NA is in MA_d iff it has a d-dimensional basis.

(5) Membership in MA_d is decidable for finite NAs.

(6) $RA = MA_4$ and $RRA = MA_\omega$.

(7) RRA is the intersection of all the varieties MA_d where d is finite.

(8) Any finite set of equations which are true in all RRAs must be included in the equational theory of MA_d for some finite d.

<u>Proof</u>. Parts (1), (2), (3), (6), and (7) are Theorems 3, 9, 8, 6, and 10 of [14], respectively. Part (4) follows from (3) since every finite NA is a perfect extension of itself. Part (4) implies (5), and (7) implies (8).

THEOREM 14. Let $\mathbf{A}(C)$ be any algebra in $\mathbf{K}(n,s)$, and let $t > 1$. If $\mathbf{A}(C)$ satisfies (d.t) then

(1) $B_{t+2}\mathbf{A}(C)$ is a $(t+2)$-dimensional basis for $\mathbf{A}(C)$.

(2) $\mathbf{A}(C)$ is in MA_{t+2}.

<u>Proof.</u> It suffices to prove (1).

Suppose y is an atom of $\mathbf{A}(C)$. Define a function x from $(t+2)\times(t+2)$ to the atoms of $\mathbf{A}(C)$ as follows:

$$x_{i,i} = \{e\} \quad \text{for} \quad i < t+2 \ ,$$
$$x_{0,i} = y \quad \text{and} \quad x_{i,0} = y^{\smile} \quad \text{for} \quad 0 < i < t+2 \ ,$$
$$x_{i,j} = \{e\} \quad \text{whenever} \quad 0 \neq i,j < t+2 \ .$$

It is easy to check that $x \in B_{t+2}\mathbf{A}(C)$, so (ii) holds.

Suppose $x \in B_{t+2}\mathbf{A}(C)$, y,z are atoms of $\mathbf{A}$, and $x_{1,2} \leq y;z$. We shall show that there is some w in $B_{t+2}\mathbf{A}(C)$ such that $w_{1,0} = y$, $w_{0,2} = z$, and $w_{k,m} = x_{k,m}$ whenever $k,m < t+2$ and $0 \neq k,m$. (By the symmetry of the definition of $B_{t+2}\mathbf{A}(C)$, this case is sufficient to establish that (i) holds for $i \neq j$. The proof in case $i = j$ is similar.)

If $t = 1$, then w must be defined as follows: $w_{0,0} = w_{1,1} = w_{2,2} = \{e\}$, $w_{1,2} = x_{1,2}$, $w_{2,1} = x_{2,1}$, $w_{1,0} = y$, $w_{0,1} = y^{\smile}$, $w_{0,2} = z$, $w_{2,0} = z^{\smile}$. Clearly w is in $B_{t+1}\mathbf{A}(C)$. Therefore assume $t > 1$.

If $y = \{e\}$ then $x_{1,1} = y$ and $x_{1,2} = z$, so w must be defined as follows: $w_{i,j} = x_{i,j}$, $w_{0,i} = x_{1,i}$, and $w_{i,0} = x_{i,1}$ for $0 < i,j < t+2$. Then w is in $B_{t+2}\mathbf{A}(C)$. So assume $y \neq \{e\}$, and, for a similar reason, $z \neq \{e\}$.

The following parts of the definition of w are forced:

$$w_{i,i} = \{e\} \qquad \text{for} \quad 0 \leq i < t+2 \ ,$$
$$w_{i,j} = x_{i,j} \qquad \text{for} \quad 0 < i,j < t+2$$
$$w_{1,0} = y \ , \quad w_{0,1} = y^{\smile} \ ,$$
$$w_{0,2} = z \ , \quad w_{2,0} = z^{\smile} \ .$$

We must have $w_{i,0} = w_{0,i}{}^{\smile}$ for $2 < i < t+2$, so what remains is to define $w_{0,3}, \ldots, w_{0,t+1}$. It is easy to show that w will be in $B_{t+2}A(C)$ if $w_{0,3}, \ldots, w_{0,t+2}$ are chosen so that for all j, if $2 < j < t+2$, then

$$(*) \quad \{e\} \neq w_{0,j} \leq w_{0,i};w_{i,j} \quad \text{whenever} \quad 0 < i < j.$$

Notice that (*) happens to hold when $j = 2$.

Suppose $2 < k < t+1$ and $w_{0,2}, \ldots, w_{0,k}$ have been determined so that (*) holds whenever $2 < j \leq k$. We wish to define $w_{0,k+1}$ so that (*) holds for $j = k+1$.

We know that $w_{0,1} \neq \{e\}, \ldots, w_{0,k} \neq \{e\}$. Suppose $w_{m,k+1} = \{e\}$ for some m, $0 < m < k+1$. We must let $w_{0,k+1} = w_{0,m}$. Then $w_{0,k+1} \leq w_{0,i};w_{i,m} \leq w_{0,i};w_{i,k+1};w_{k+1,m} = w_{0,i};w_{i,k+1}$. So assume $w_{1,k+1} \neq \{e\}, \ldots, w_{k,k+1} \neq \{e\}$. It follows from (d.t) that there is some u in $U-\{e\}$ such that $\{u\} \leq w_{0,m};w_{m,k+1}$ whenever $0 < m < k+1$. Let $w_{0,k+1} = \{u\}$.

THEOREM 15. Let E be a finite set of equations which are true in every RRA. Then $F(E,n,s)$ approaches $2^{Q(n,s)}/P(n,s)$.

Proof. Choose t so that E is included in the equational theory of MA_{t+2}. By Theorem 14, $D(t+2,n,s) \leq F(E,n,s) \leq D(0,n,s)$, so the conclusion follows by Theorem 11.

It seems reasonable to conjecture that Theorem 15 is still true when E is the equational theory of RRA. It follows from Theorems 6 and 7 that the number of representable relation algebras in $\mathbf{K}(n,s)$ is at least $2^{Q(n-1,n-1)}$, but this is a vanishingly small fraction of all the algebras in $\mathbf{K}(n,s)$ when n is large.

The techniques used so far can be adapted to produce non-representable RAs as well. This time, however, the goal is to show that the number of isomorphism types of nonrepresentable symmetric integral relation algebras is increasing at a rate which is greater than 2 raised to the power of a cubic polynomial in the number of atoms.

Let U be a finite set such that $n = |U| \geq 4$. Let e,a,b,c be distinct elements of U. Let f be the identity on U. Recall that **A**(C) = **Cm**<U, C, f, {e}> for every C included in U×U×U. Let

L(n) = {**A**(C) : C is included in U×U×U, **A**(C) is a NA, [a,a,b], [a,b,b], [a,a,c], [a,c,c], [b,b,c], and [b,c,c] are included in C, [c,c,c] and [a,b,c] are excluded from C, and if x is in U-{e,a,b,c} and [a,c,x] is included in C, then both [a,b,x] and [b,c,x] are excluded from C}.

L(4) contains just 4 algebras, but two of them are isomorphic. The tables below show 3 of these algebras by specifying how the operation ; acts on the atoms distinct from {e}. The fourth algebra in **L**(n) is isomorphic to the one given by the second table. For notational simplicity we shall ignore the distinction between a singleton subset of U and the sole element it contains. The entries in the tables are subsets of U written without brackets or commas to save space, e.g. ebc = {e,b,c}.

;	a	b	c
a	ebc	ab	ac
b	ab	eac	bc
c	ac	bc	eab

;	a	b	c
a	ebc	ab	ac
b	ab	eabc	bc
c	ac	bc	eab

;	a	b	c
a	eabc	ab	ac
b	ab	eabc	bc
c	ac	bc	eab

All of these algebras are nonrepresentable relation algebras. In fact, they are not even in MA_5. (See Theorem 15 of [14] for the first one.) This latter fact and its proof extend to all the algebras in **L**(n).

THEOREM 16.

(1) The following equation fails in every algebra in **L**(n).

$$\text{(e)}\quad t.(u;v.w);(x.y;z) \leq u;[(u^{\smile};t.v;x);z^{\smile}.v;y.u^{\smile};(t;z^{\smile}.w;y)];z.$$

(2) Equation (e) holds in every algebra in MA_5.

(3) **L**(n) and MA_5 are disjoint.

<u>Proof</u>. For every algebra **A**(C) in **L**(n), equation (e) fails when the variables t,u,v,w,x,u,z are assigned to a,c,c,a,b,b,c, respectively. Part (2) follows from Theorems 2 and 15 of [14]. Part (3) follows from parts (1) and (2), but the proof of (1) can be adapted to give the following alternate proof of (3).

Let $\mathbf{A}(C)$ be in $\mathbf{L}(n)$. By Theorem 13(4) we need only show that $\mathbf{A}(C)$ has no 5-dimensional basis. Assume, to the contrary, that M is a 5-dimensional basis for $\mathbf{A}(C)$.

Choose x in M so that $x_{0,1} = a$. Then $x_{0,1} \leq a;b$, so by (i) there is some y in M such that $y_{0,2} = a$, $y_{2,1} = b$, and $y_{0,1} = a$. Now apply (i) twice more, using the conditions $y_{0,2} \leq c;c$ and $y_{2,1} \leq b;c$, to obtain z in M such that

$$z_{0,1} = a, \quad z_{0,2} = a, \quad z_{2,1} = b,$$
$$z_{0,3} = c, \quad z_{3,2} = c,$$
$$z_{2,4} = b, \quad z_{4,1} = c.$$

Here is a diagram intended to express the same information about z:

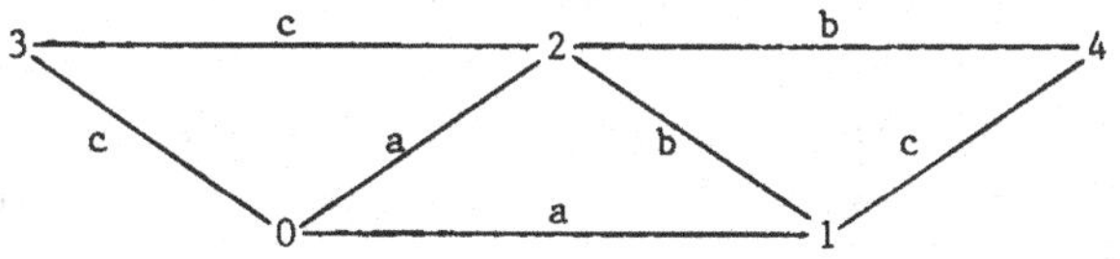

But $z_{3,1} \leq z_{3,0};z_{0,1}\cdot z_{3,2};z_{2,1} = c;a.c;b = c$, and $z_{0,4} \leq z_{0,1};z_{1,4}\cdot z_{0,2};z_{2,4} = a;c.a;b = a$, so $z_{3,4} \leq z_{3,0};z_{0,4}\cdot z_{3,1};z_{1,4}\cdot z_{3,2};z_{2,4} = c;a.c;c.c;b = \emptyset$, a contradiction.

THEOREM 17. $|\mathbf{L}(n)| = 5^{n-4}2^{Q(n,n)-3n+12}$.

<u>Proof</u>. The number of d-cycles of $\langle U, f, \{e\}\rangle$ is $Q(n,n)$. Not every conbination of these d-cycles can occur in a ternary relation C for which $\mathbf{A}(C)$ is in $\mathbf{L}(n)$. In any such C, 6 d-cycles must be included, 2 d-cycles must be excluded, and $3(n-4)$ other d-cycles fall into $n-4$ groups of 3 cycles each, namely $\{[a,b,x], [a,c,x], [b,c,x]\}$, for every x in $U-\{e,a,b,c\}$, such that either $[a,c,x]$ is included in C while $[a,b,x]$ and $[b,c,x]$ are excluded, or else $[a,c,x]$ is excluded from C, and any combination of the other two is included. Thus for each such group there are 5 choices for forming C. The number of remaining d-cycles is $Q(n,n)-3n+12$. Each of them can either be included or excluded, independently. Therefore the number of possible choices for C is $5^{n-4}2^{Q(n,n)-3n+12}$.

THEOREM 18.

(i) The number of isomorphism types of nonrepresentable symmetric integral relation algebras with n atoms which fail to satisfy equation (e), and hence are not in MA_5, is at least

$$((n-4)!)^{-1}[1-(n-1)^4(9/10)^{n-4}]5^{n-4}2^{Q(n,n)-3n+12}.$$

(2) If $0 < r < 1/6$, then, for sufficiently large n, the number of such isomorphism types is at least 2^{rn^3}.

Proof. Part (1) implies (2), so it suffices to prove only (1).

For any property P of algebras in $\mathbf{L}(n)$, let #P# be the fraction of algebras in $\mathbf{L}(n)$ which have property P. Use "Cxyx" as an abbreviation of "[x,y,z] is included in C".

Let v,w,x,y be in U-{e}.

If {v,w,x,y} is included in {a,b,c}, then it is easy to check that #for some z in {a,b,c}, Cvwz and Cxyz# = 1, so #for all z in U-{e}, not (Cvwz and Cxyz)# = 0.

Suppose {v,w,x,y} is not included in {a,b,c}. Then one of the pairs {v,w}, {x,y} must be different from each of the pairs {a,b}, {a,c}, {b,c}. Assume that pair is {x,y}. Let z be in U-{e,a,b,c}. Then #Cxyz# = 1/2, and #Cvwz# is either 1/2, 2/5, or 1/5, since #Cacz# = 1/5 and #Cabz# = #Cbcz# = 2/5. The smallest is 1/5, so #Cvwz and Cxyz# $\geq$ 1/10 and #not (Cvwz and Cxyz)# $\leq$ 9/10. Consequently #for all z in U-{e}, not (Cvwz and Cxyz)# $\leq$ #for all z in U-{e,a,b,c}, not (Cvwz and Cxyz)# $\leq (9/10)^{n-4}$.

From the two preceding paragraphs, plus the fact that there are $(n-1)^4$ ways to choose v,w,x,y from U-{e}, it follows that #(d.2) is false# $\leq$ #there are v,w,x,y in U-{e} such that for all z in U-{e}, not (Cvwz and Cxyz)# $\leq (n-4)^4(9/10)^{(n-4)}$. Consequently

$$|\{\mathbf{A}(C): \mathbf{A}(C) \text{ is in } \mathbf{L}(n) \text{ and (d.2) holds}\}| \geq [1-(n-1)^4(9/10)^{n-4}]5^{n-4}2^{Q(n,n)-3n+12}.$$

The theorem is now a consequence of this inequality and the following facts. Every algebra in $\mathbf{L}(n)$ fails to satisfy (e), is a relation algebra if it satisfies (d.2), and is isomorphic to at most (n-4)! algebras in L(n).

The proof of Theorem 18 is based on the fact that the algebras in L(4) are RAs which fail to be representable because equation (e) fails in each of them in the same way. Other nonrepresentable RAs and other equations could be used in a similar way, with appropriate changes in the definition of L(n), Theorem 17, and Theorem 18(1). But the ultimate goal, namely part (2) of Theorem 18, would be unchanged. An example of such an alternate starting point follows.

Let $U = \{e,a,b,c\}$, and let f be the identity on U. So a cycle consists of all permutations of a triple. Two relation algebras are given in the tables below. The only difference between them is that the second one contains the cycle [c,c,c] while the first one does not. The d-cycles of the first one are [a,a,b], [a,b,b], [b,b,c], [b,c,c], [a,c,c], and [a,b,c].

;	a	b	c
a	eb	abc	bc
b	abc	eac	abc
c	bc	abc	eab

;	a	b	c
a	eb	abc	bc
b	abc	eac	abc
c	bc	abc	eabc

Both of these algebras fail to satisfy equation (f) below when the variables u,v,w,x,y,z are assigned to a,a,a,b,a,c, respectively.

$$\text{(f)}\quad u;v.w;x.y;z \leq u;[u^{\smile};w.v;x^{\smile}.(u^{\smile};y.v;z^{\smile});(y^{\smile};w.z;x^{\smile})];x.$$

Equation (f) is true in every representable relation algebra, so neither of the algebras above is representable. In fact, (f) holds in every algebra in MA_5. This is not hard to prove directly. Alternatively, one can show that the sequent corresponding to (f) has a CUT-free 5-proof, from which the conclusion follows by Theorem 2 of [14]. (Compare Theorem 15 of [14].)

The nonrepresentable relation algebra presented in [9] fails to satisfy (f) (see [1], p. 354). That algebra has 56 atoms. Lyndon mentions that he knows another one with only 52 atoms. Probably that algebra also fails to satisfy (f). Smaller nonrepresentable RAs are obtained from the connection with projective geometries in [11]. From the main result there it follows that a particular RA with 8 atoms is not representable because there is no projective plane of order 6. Whether this algebra satisfies (e) or (f) is not known. The nonrepresentable RA with 4 atoms found by McKenzie ([15]) satisfies (f) but not (e). The RAs in **L**(4) also satisfy (f) but not (e). The two RAs above satisfy (e) but not (f), so (e) and (f) are independent. In fact, out of 102 integral RAs with 4 atoms, those two are the only ones which satisfy (e) but not (f).

Now for some numerical data. The values of $F(n,s)$ for $n \leq 3$ have been computed by several people. (See, for example, footnote 13 of [10].) S. D. Comer found that $F(4,2) = 37$ and $F(4,4) = 65$. The values of $F(5,s)$ were computed using programs written in Pascal and run on a VAX 11/780 and a Zenith Z-100. According to Theorem 12, the number in the rightmost column of the table below should be approaching 1 as n gets large. It is interesting to note that for $n \leq 5$ the values are actually getting smaller as n and s increase.

n	s	Q(n,s)	P(n,s)	$2^{Q(n,s)}/P(n,s)$	F(n,s)	$F(n,s)P(n,s)2^{-Q(n,s)}$
1	1	0	1	1	1	1.0
1	2	1	1	2	2	1.0
3	1	2	2	2	3	1.5
3	3	4	2	8	7	.875
4	2	7	2	64	37	.578
4	4	10	6	170.667	65	.381
5	1	12	8	512	83	.162
5	3	16	4	16384	1316	.080
5	5	20	24	43690.667	3013	.069

References

[1] L. H. Chin and A. Tarski. Distributive and modular laws in the arithmetic of relation algebras, **University of California Publications in Mathemtics** (new series), 1(1951), 341-384.

[2] S. D. Comer, Combinatorial aspects of relations, **Algebra Universalis, 18**(1984), 77-94.

[3] S. D. Comer, A new foundation for the theory of relations, **Notre Dame Journal of Formal Logic, 24**(1983), 181-187.

[4] B. Jónsson and A. Tarski, Boolean algebras with operators, Part 1, **American Journal of Mathematics, 73**(1951), 891-939.

[5] B. Jónsson and A. Tarski, Boolean algebra with operators, Part II, **American Journal of Mathematics, 74**(1952), 127-162.

[6] B. Jónsson, Representation of modular lattices and of relation algebras, **Transactions of the American Mathematical Society, 92**(1959), 449-464.

[7] B. Jónsson, Varieties of relation algebras, **Algebra Universalis, 15**(1982), 273-298.

[8] L. Henkin, J. D. Monk, and A. Tarski, **Cylindric Algebras, Part I.** North-Holland, 1971.

[9] R. C. Lyndon, The representation of relational algebras, **Annals of Mathematics**, series 2, **51**(1950), 707-729.

[10] R. C. Lyndon, The representation of relation algebras, II. **Annals of Mathematics**, series 2, **63**(1956), 294-307.

[11] R. C. Lyndon, Relation algebras and projective geometries, **Michigan Mathematical Journal**, **8**(1961), 21-28.

[12] R. Maddux, Embedding modular lattices into relation algebras, **Algebra Universalis**, **12**(1981), 244-246.

[13] R. Maddux, Some varieties containing relation algebras, **Transactions of the American Mathematical Society**, **272**(1982), 501-526.

[14] R. Maddux, A sequent calculus for relation algebras, **Annals of Pure and Applied Logic**, **25**(1983), 73-101.

[15] R. N. W. McKenzie, The representation of integral relation algebras, **Michigan Mathematical Journal**, **17**(1970), 279-287.

[16] J. D. Monk, On representable relation algebras, **Michigan Mathematical Journal**, **11**(1964), 207-210.

[17] A. Tarski, Contributions to the theory of models, III, **Nederl. Akad. Wetensch. Proc. Ser. A, 58(Indag. Math., 17)** (1955), 56-64.

SOME VARIETIES OF SEMIDISTRIBUTIVE LATTICES

J. B. Nation
University of Hawaii
2565 The Mall
Honolulu, Hawaii 96822

One of the most intriguing questions in lattice theory is this: if a variety of lattices has only finitely many subvarieties, is it necessarily generated by a finite lattice? Indeed, what we would like to prove is that whenever $\mathcal{V}$ is a variety generated by a finite lattice, there are finitely many varieties $\mathcal{W}_1, \ldots \mathcal{W}_k \not\leq \mathcal{V}$ (each generated by a finite lattice) such that every variety $\mathcal{U}$ properly containing $\mathcal{V}$ contains at least one $\mathcal{W}_i$.

Since the general question seems to be quite hard, it is of interest to prove this type of result for some specific varieties of lattices. That is what we will do in this paper, and we will try to do it in such a way that we will encounter some other interesting varieties along the way.

The variety $\mathcal{D}$ of all distributive lattices is the smallest nontrivial lattice variety, and it is generated by the two-element lattice. $\mathcal{D}$ is covered by $V(N_5)$ and $V(M_3)$, the varieties generated by the pentagon and diamond, respectively [2, 8]. We will be concerned with nonmodular lattice varieties; for similar results about modular varieties, see [10, 13, 15].

Ralph McKenzie conjectured in [20] that the lattice variety generated by the pentagon, $V(N_5)$, has exactly sixteen covers. This was subsequently proved by Bjarni Jónsson and Ivan Rival [18]. Fifteen of these covering varieties are join irreducible and generated by a single finite nonmodular subdirectly irreducible lattice, while the sixteenth is $V(N_5) + V(M_3)$.

Jónsson and Rival numbered the lattices generating the join irreducible varieties covering $V(N_5)$ as L_i $(1 \leq i \leq 15)$; see Figure 1. We will adopt their numbering system.

Now M_3 and L_1 - L_5 are not semidistributive lattices. The covers of $V(M_3)$ are known $(V(M_4), V(M_{3,3})$, and $V(M_3) + V(N_5))$ [12, 15], but essentially nothing is known about the covers of $V(L_i)$ for $1 \leq i \leq 5$. On the other hand, the lattices L_6 - L_{15} are all splitting lattices. Henry Rose proved that for $i = 6, 7, 8, 9, 10, 13, 14$, and 15 there is a sequence of lattices L_i^n $(n \in \omega)$ with the property that $L_i^0 = L_i$ and the variety $V(L_i^{n+1})$ is the unique join irreducible cover of $V(L_i^n)$ [23]. Moreover, the lattices L_i^n are all splitting lattices

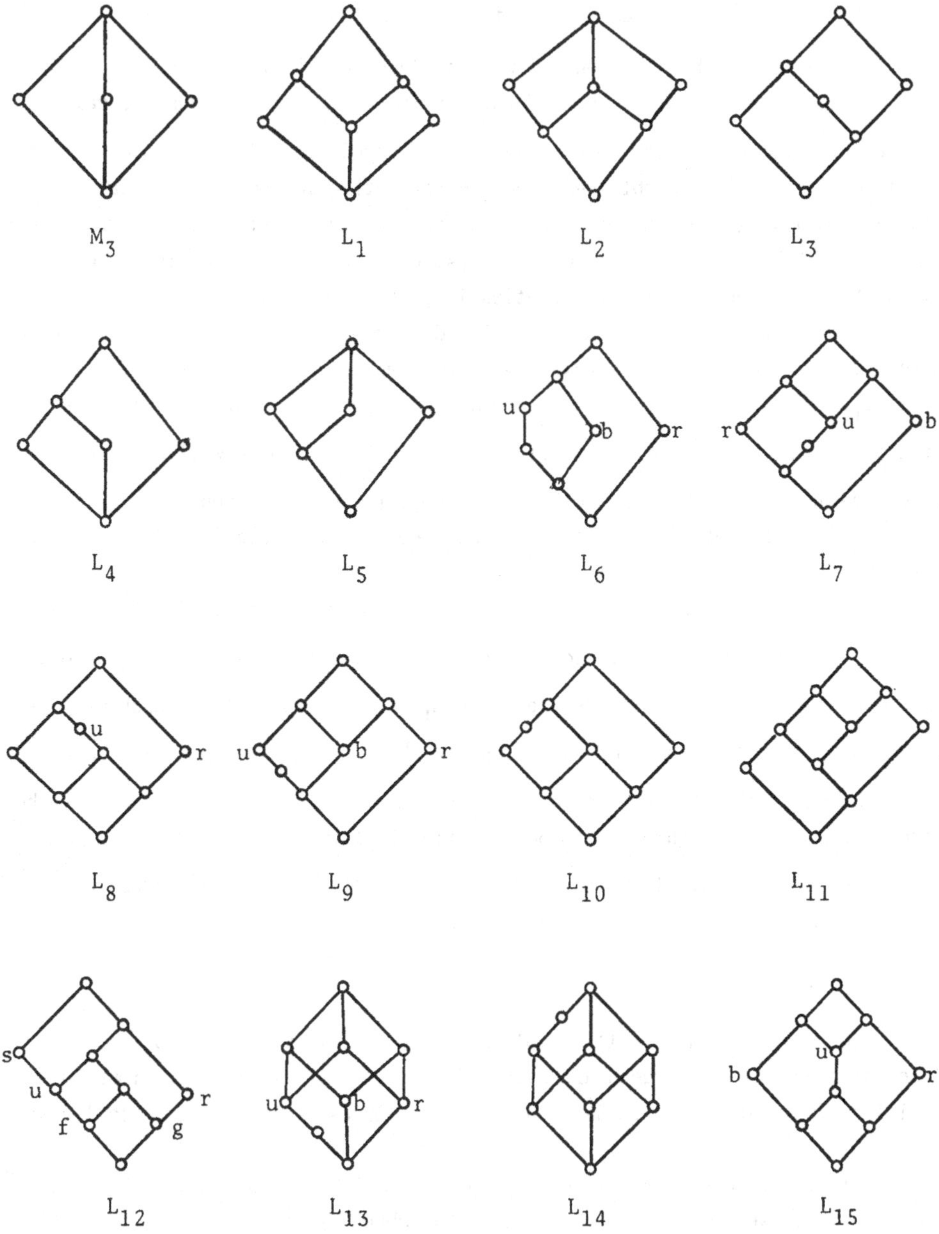

Figure 1.

For $i = 11$ and 12, Rose found two join irreducible covers of $V(L_i)$.

Jeh Gwon Lee, building on the work of Jónsson, Rival, and Rose, considered lattice varieties which contain none of M_3 or $L_1 - L_{12}$ [19]. These he called *almost distributive* varieties. Every subdirectly irreducible lattice in an almost distributive variety can be obtained from a distributive lattice by Alan Day's construction of doubling an element. Lee proved that every almost distributive variety of finite height has only finitely many covers, each generated by a finite lattice, and at most one of these covering varieties is join irreducible.

The techniques used by Rose and Lee depend heavily on the following lemma of Rose [23]. *A subdirectly irreducible semidistributive lattice which does not contain* L_{11} *or* L_{12} *as a sublattice has a unique critical quotient.* In this paper we will be looking at the covers of $V(L_{11})$ and $V(L_{12})$, so we will have to use different methods. In fact, our methods will be derived from techniques used by Ralph Freese and the author in [11] to study the covering relation $(u \succ v)$ in free lattices.

L_{11} and L_{12} are dual lattices, so we can state our results for L_{12} only. We will show that the only join irreducible varieties covering L_{12} are the two found by Rose, which we will denote by $V(L^1_{12})$ and $V(G)$. Above each of these is an infinite ascending chain of varieties, $V(L^n_{12})$ or $V(G^n)$ $(n \in \omega)$, each of which is the only join irreducible variety covering the preceeding one. All of the lattices L^n_{12} and G^n turn out to be splitting lattices.

The author would like to thank the University of Hawaii universal algebra seminar for listening critically to these results.

1. *Preliminaries.*

In this section we will establish some of the terminology and results we will be using here. These ideas are gathered from various papers by various authors; most will be familiar to the reader, and most are discussed in greater detail elsewhere.

First, let us set a little notation. Let L be a lattice. If K is a sublattice of L, we write $K \leq L$. If S is a subset of L, $\langle S \rangle$ denotes the sublattice generated by S. $J(L)$ is the set of all (nonzero) join irreducible elements of L. $\mathcal{I}(L)$ and $\mathcal{F}(L)$ are the ideal lattice and filter lattice of L, respectively. The order on $\mathcal{F}(L)$ is reverse set inclusion, i.e., $F \leq G$ if $F \supseteq G$. $V(L)$ denotes the lattice variety generated by L, and Λ denotes the lattice of all lattice varieties.

If $u \succ v$ in L, there is a unique maximal congruence $\theta \in \mathrm{Con}\, L$ such that $(u, v) \notin \theta$; we use Ψ_{uv} to denote this congruence.

For finite nonempty subsets U, V of a lattice L, we say that U *refines* V (written $U \ll V$) if for every $u \in U$ there exists $v \in V$ such that $u \leq v$. We call V a *join-cover* of $a \in L$ if $a \leq \sum V$. A join-cover is *nontrivial* if $a \nleq v$ for each $v \in V$. V is a *minimal* join-cover of a if whenever $a \leq \sum U$ and $U \ll V$, then $V \subseteq U$.

Let S denote the join closure operator, i.e., if A is a subset of a lattice L then $S(A)$ is the set of joins of all finite, non-empty subsets of A. Dually, P denotes the meet closure operator. If L has a least element, we can also form $S_0(A) = S(A) \cup \{0_L\}$. If A is finite, then $S_0(A)$ will be a lattice, with the join operation inherited from L and the meet defined by

$$b \wedge c = \sum \{a \in A: \quad a \leq bc\}.$$

Similarly, if L is a lattice generated by a finite set X, then for any $k \in \omega$ the subset $S(PS)^k P(X)$ (which already contains $0 = \Pi X$) likewise forms a lattice. This lattice is particularly interesting when L is the free lattice $FL(X)$.

A homomorphism $f: K \to L$ is said to be *lower bounded* if for every $a \in L$, $\{u \in K: f(u) \geq a\}$ is either empty or has a least element. A finitely generated lattice L is called *lower bounded* if every homomorphism $f: K \to L$, where K is finitely generated, is lower bounded. Equivalently, L is lower bounded if there exists a lower bounded epimorphism $f: FL(X) \twoheadrightarrow L$ with X finite. *Upper bounded* is defined dually, and L is *bounded* if it is both lower and upper bounded. When they exist, the least member of $\{u \in K: f(u) \geq a\}$ will be denoted by $\beta(a)$, and the greatest member of $\{v \in K: f(v) \leq a\}$ will be called $\alpha(a)$. These notions were introduced and studied at length by McKenzie [20]; see also [11] and [17].

We will now describe a particularly simple algorithm for determining whether a finitely generated lattice is lower bounded. (Similar algorithms were found independently by Bjarni Jónsson and Ralph McKenzie [20]; what follows is Jónsson's version.) Let $D_0(L)$ denote the set of join prime elements of L, i.e., those elements which have no nontrivial join-cover. For $k > 0$, let $a \in D_k(L)$ if every nontrivial join-cover V of a has a refinement $U \subseteq D_{k-1}(L)$ which is also a join-cover of a. Then a finitely generated lattice L is lower bounded if and only if $\bigcup_{k \geq 0} D_k(L) = L$.

Observe that, from the definition, $D_0(L) \subseteq D_1(L) \subseteq D_2(L) \subseteq \ldots$. If L is lower bounded and $a \in L$, we define the *D-rank* $\rho(a)$ to be the least integer k such that $a \in D_k(L)$. It is easy to see that if U is a finite nonempty subset of $D_k(L)$, then $\sum U \in D_{k+1}(L)$. Thus a finite lattice L will be lower bounded if and only if $J(L) \subseteq D_n(L)$ for some n. The D-rank function is studied in [7].

The proof that Jónsson's algorithm works in fact tells us how to find $\beta(a)$. Let F be a finitely generated lattice, say $F = \langle X \rangle$ with X finite, and let f: $F \to L$; for simplicity, let us assume that L is finite. For every $a \in L$ with $f^{-1}(1/a) \neq \emptyset$, let

$$\beta_0(a) = \Pi \{x \in X: \ f(x) \geq a\}$$

and

$$\beta_{n+1}(a) = \beta_0(a) \cdot \prod_{U \in C(a)} \sum_{b \in U} \beta_n(b),$$

where $C(a)$ denotes the set of all minimal nontrivial join-covers of a in L. If $a \in D_k(L)$, then $\beta_k(a) = \beta_{k+1}(a) = \ldots = \beta(a)$. To find $\alpha(a)$, we define $\alpha_n(a)$ dually.

The following result from [11] connects some of the previous ideas, and provides a starting point for much of what is to follow.

Theorem 1.1. (1) Let $f: F \twoheadrightarrow L$ be a lower bounded epimorphism, where L is a finite lattice and F is finitely generated. Let $B = \{\beta(p): \ p \in J(L)\}$. Then $B \subseteq J(F)$, $L \cong S_0(B)$, and B satisfies the closure condition

(CL) for each $b \in B$, every join-cover of b refines to a join-cover contained in B.

(2) Conversely, let F be finitely generated. If B is a finite subset of $J(F)$ satisfying (CL), then there is a lower bounded epimorphism $f: F \twoheadrightarrow S_0(B)$ with $\beta f(b) = b$ for all $b \in B$, given by

$$f(u) = \sum \{b \in B: \ b \leq u\}$$

for each $u \in F$.

Using this result, it was shown in [11] that to each join irreducible element $w \in J(FL(X))$ there corresponds a finite, lower bounded, subdirectly irreducible lattice $L(w)$ with the property that $L(w) \cong FL(X)/\theta$ where θ is the (unique largest congruence on $FL(X)$ such that w is the least member of w/θ.

A lattice is ***semidistributive*** if it satisfies both of

$(SD_\wedge)$ $u = ab = ac$ implies $u = a(b + c)$

$(SD_\wedge)$ $u = a + b = a + c$ implies $u = a + bc$.

Let u be a completely join irreducible element in a lattice L satisfying $(SD_\wedge)$, and let u_* be the (unique) lower cover of u. If W is any join-cover of u, then $u \leq u_* + w$ for some $w \in W$; for otherwise we would have $u_* = u(u_* + w)$ for each $w \in W$, whence by $(SD_\wedge)$ $u_* = u(u_* + \sum W) = u$, a contradiction. Thus $K(u) = \{v \in L: \ u \nleq u_* + v\}$ is an ideal of L. If $K(u)$ has a largest element, we denote it by $\kappa(u)$.

We note that the ideal or filter lattice of a semidistributive lattice need not be semidistributive [9].

Finally, we will use a special case of Alan Day's interval doubling construction, viz., doubling an element. Given a lattice L and an element $c \in L$, let $L[c] = L\backslash\{c\} \cup \{(c, 0), (c, 1)\}$ with the order given by

$$x \leq y \quad \text{if} \quad \begin{cases} x, y \in L\{c\} \text{ and } x \leq y \\ x \in L\{c\},\ y = (c, j) \text{ and } x \leq c, \\ x = (c, i),\ y \in L\backslash\{c\} \text{ and } c \leq y, \\ x = (c, i),\ y = (c, j) \text{ and } i \leq j. \end{cases}$$

Clearly $L[c]$ is a lattice; for more general uses of this type of construction see [4, 5, 6].

. *Some special varieties*

In this section we will look at some interesting varieties of lattices which will prove useful later.

For a finite lower bounded lattice L, let $\rho(L) = \max\{\rho(p)\colon\ p \in J(L)\}$. Thus $\rho(L(w)) = \rho(w)$ for $w \in J(FL(X))$, and $\rho(L) = k$ implies $D_{k+1}(L) = L$.

For $k \in \omega$ Let $LB(k)$ denote the class of all lattices L such that every finitely generated sublattice $S \leq L$ is a finite lower bounded lattice with $\rho(S) \leq k$.

Theorem 2.1. $L \in LB(k)$ if and only if for every finite set X, for every map $f_0\colon\ X \to L$, there is a homomorphism $f\colon\ S(PS)^kP(X) \to L$ with $f|_X = f_0$.

Thus each $LB(k)$ is a variety, and for X finite the relatively free lattice $F_{LB(k)}(X) \cong S(PS)^kP(X)$.

Proof. If $L \in LB(k)$, X is finite, and $f_0\colon\ X \to L$, then $S = \langle f_0(X)\rangle$ is a finite lower bounded lattice with $\rho(S) \leq k$. Let $h\colon\ FL(X) \to S$ with $h|_X = f_0$. Then by Theorem 1.1 $S \cong S_0(B)$ where $B = \{\beta(p)\colon\ p \in J(S)\} \subseteq (PS)^kP(X)$. Thus by Theorem 3.5 of [11] there is a homomorphism $f\colon\ S(PS)^kP(X) \to S$, and in fact we may use the restriction $f = h|_{S(PS)^kP(X)}$.

Similarly, if L satisfies the condition of Theorem 2.1, we apply the converse portions of Theorems 3.5 of [11] and 1.1 to obtain $L \in LB(k)$.

It is now clear from the first part of the theorem that $LB(k)$ is a variety, and that $F_{LB(k)}(X)$ is a homomorphic image of $S(PS)^kP(X)$ for X finite. It remains to show that $S(PS)^kP(X) \in LB(k)$. This is a consequence of the following.

Claim: If L *is a finite lower bounded lattice with* $\rho(L) = k$, *then for every sublattice* $S \leq L$ *we have* $\rho(S) \leq k$.

For $a \in L$ let $\sigma(a) = \Pi\{s \in S\colon\ s \geq a\}$. It is routine to show that if $a \in D_k(L)$ and $1/a \cap S \neq \emptyset$, then $\sigma(a) \in D_k(S)$. Now let $q \in J(S)$; in order to apply this statement we need to show that there is some $p \in J(L)$ with $\sigma(p) = q$ (so that $\rho_S(q) = \rho_S(\sigma(p)) \leq \rho_L(q) \leq k$). If $q \in J(L)$ we may take $p = q$.

Otherwise, let $q = \sum q_i$ canonically. Then since $q \in J(S)$ there must be some i_0 such that $q/q_{i_0} \cap S = \{q\}$, whence $\sigma(q_{i_0}) = q$ and we may take $p = q_{i_0}$. This proves the claim, and hence the theorem. □

Corollary 2.2. If L is a finite lower bounded lattice with $\rho(L) \le k$, then $V(L) \subseteq \mathcal{LB}(k)$.

The next lemma applies to arbitrary lattices.

Lemma 2.3. If $u \in D_k(L)$ then either u is join irreducible or u is a join of join irreducible elements in $D_{k-1}(L)$.

Proof. If $u \in D_k(L)$ is join irreducible, say $u = \sum U$, then $k > 0$ and there is a refinement $V \ll U$ such that $u = \sum V$ and $V \subseteq D_{k-1}(L)$. Apply induction. □

Let L be a finitely generated lattice. Recall that $D_k(L) \subseteq (\mathcal{PS})^k \mathcal{P}(X)$ for any generating set X of L [17, Theorem 3.1]; in particular, $D_k(L)$ is finite. Thus $J(L) \cap D_k(L)$ satisfies the condition (CL), and there is a homomorphism $d_k\colon L \twoheadrightarrow \mathcal{S}_0(J(L) \cap D_k(L))$ via $d_k(u) = \sum\{v \in J(L) \cap D_k(L)\colon v \le u\}$. This map is the reflection of L into $\mathcal{LB}(k)$, i.e., ker d_k is the smallest congruence on L with $L/\theta \in \mathcal{LB}(k)$.

Lemma 2.4. Let L be finitely generated, and let $u \in J(L)$.

(i) $u \notin D_k(L)$ iff $d_k(u) < u$.

(ii) W is a nontrivial join-cover of u with no refinement in $D_k(L)$ iff $u \le \sum W$, $u \nleq w$ for every $w \in W$, and $u \nleq \sum d_k(W)$.

(iii) If there is a nontrivial join-cover W of u with no refinement in $D_k(L)$, then there is one with two elements.

Proof. (i) and (ii) are straightforward. For (iii), pick an irredundant subcover of W, split it into two parts, and apply (ii). □

Next, a technical lemma.

Lemma 2.5. Let L be a finitely generated sublattice of a lattice L'. Then for every integer $k \ge 0$ we have $D_k(L') \cap L \subseteq D_k(L)$.

Proof. We will prove the following statement. Let $k \ge 1$ and $u \in L$, and let V be a nontrivial join-cover of u with $V \subseteq L$. If V has a refining join-cover $W \subseteq D_{k-1}(L')$, then V has a refining join-cover $S \subseteq D_{k-1}(L)$.

It is clear that $D_0(L') \cap L \subseteq D_0(L)$. So let $k \ge 1$, $u \in L$, $u \le \sum V$ nontrivially with $V \subseteq L$. Assume there is a join-cover $W \subseteq D_{k-1}(L')$ refining V. Let $f\colon FL(X) \twoheadrightarrow L$ with X finite, and for each $w \in W$ let s_w be the least element of $\{t \in FL(X)\colon f(t) \ge w\}$. Note that each $s_w \in D_{k-1}(FL(X))$, and $w \le v$ implies $f(s_w) \le v$ for $w \in W$, $v \in V$. Let $S = \{f(s_w)\colon w \in W\}$. From $W \ll V$ and $u \le \sum W$

we obtain $S >> V$ and $u \leq \sum S$. Now in the proof of Theorem 4.2 of [17] it is shown that $\beta(a) \in D_{k-1}(F)$ implies $a \in D_{k-1}(L)$ when $f\colon FL(X) \twoheadrightarrow L$. Now $s_w = \beta f(s_w)$, so this applies and we obtain $S \subseteq D_{k-1}(L)$, as desired. □

We note in passing that there is a simple construction for the relatively free lattice $F_V(X)$ where V is a variety generated by a finite lower bounded lattice L and X is finite. Let

$$J = \bigcup_S \bigcup_f \{\beta_f(p)\colon\ p \in J(S) \text{ and } p/p_* \text{ is critical in } S\}$$

where S runs over all subdirectly irreducible lattices in $HS(L)$ and f runs over all epimorphisms $f\colon FL(X) \twoheadrightarrow S$. Then $F_V(X) \cong S_0(J)$.

Now if L and X are small, then J is easily constructed by hand. $S_0(J)$ can then be obtained from J by a relatively simple computer program. In particular, the program seems to be at least as simple as the usual subdirect product construction. Ralph Freese has written such a program, and produced the following examples for use.

If $V = V(N_5)$ and $X = \{x, y, z\}$, we obtain

$$\begin{aligned} J = \{&x, y, z, xy, xz, yz, x(xy + z),\\ &x(xz + y), y(xy + z), y(yx + x),\\ &z(xz + y), z(yz + x)\} \end{aligned}$$

and $S_0(J)$ is the well-known 99-element lattice $F_{V(N_5)}(3)$, first constructed by A. G Waterman [24].

If $V = V(L_1)$, then J has three additional elements - $x(y + z)$, $y(x + z)$ and $z(x + y)$ - and $|F_{V(L_1)}(3)| = 178$. If $V = V(L_{15})$, then $|J| = 18$ and $|F_{V(L_{15})}(3)| = 227$.

On the other hand, consider $F_{V(N_5)}(4)$. The reader can check that in this case $|J| = 98$. However, Joel Berman and Barry Wolk [1] have shown that $|F_{V(N_5)}(4)| = 540{,}792{,}672$, so it is not really practical to construct this lattice using our methods.

Our next few results are all variations on a theme of McKenzie, stated in the following theorem.

Theorem 2.6 [20]. Let L be a finite subdirectly irreducible lattice, and let $f\colon FL(X) \twoheadrightarrow L$. Let u/v be a critical prime quotient in L. Form the sequences $\beta_n(u)$ and $\alpha_n(v)$ $(n \in \omega)$. For any lattice variety V: $L \notin V$ if and only if for some n, V satisfies the inclusion $\beta_n(u) \leq \alpha_n(v)$.

As a special case of Theorem 2.6 we get a single splitting equation for L when L is bounded. For example, we will use the following technical lemma later.

Lemma 2.7. Let $B \cong \underset{\sim}{2}^{n+1}$ with $n > 1$, and let c be a coatom of B. The following are equivalent for a lattice variety V.

(1) $B[c] \notin V$

(2) $V \models y(z + \sum_{i=1}^{n} x_i y) \le \sum_{i=1}^{n} (y + \sum_{j=1}^{n} x_j)(z + \sum_{k \ne i} x_k)$

Moreover, (2) implies

AR(n): If u is join irreducible in $L \in V$ and $u \le v + \sum_{i=1}^{n} a_i$ with $a_i < u$ $(1 \le i \le n)$, then $u \le v + \sum_{k \ne i} a_k$ for some i.

Proof. The equivalence of (1) and (2) follows from Theorem 2.6 and the boundedness of $B[c]$, using the map indicated by Figure 2.

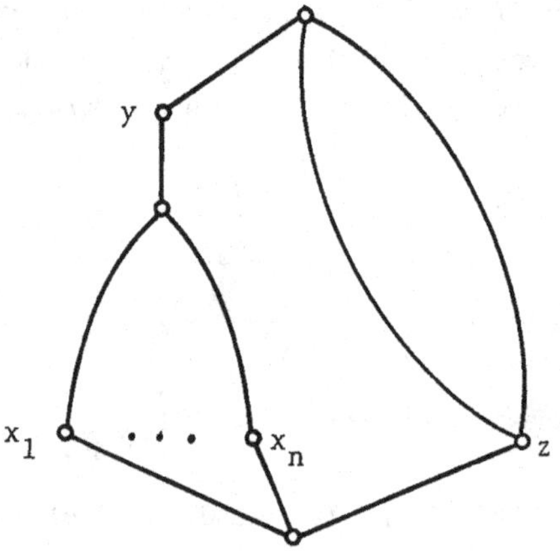

Figure 2.

To see that (2) implies AR(n), set $u \to y$, $v \to z$ and $a_i \to x_i$. Then the left hand side of (2) evaluates to u. while the right hand side is $\sum_{i=1}^{n} u(v + \sum_{k \ne i} a_k)$. If AR(n) fails, then so does (2) by the join irreducibility of u. □

In the same spirit we have the following theorem of Jónsson and Rival.

Theorem 2.8 [18]. For any lattice variety V the following are equivalent

(1) V satisfies (SD_V).

(2) V contains none of M_3, L_2, L_3, L_4 or L_5.

(3) Let

$$y_0 = y \qquad z_0 = z$$
$$y_{n+1} = y(x + z_n) \qquad z_{n+1} = z(x + y_n)$$

Then V satisfies $y_n \le x + yz$ for some n.

Proof. The proof of the corresponding result in [18] uses L_1 in only one place. This occurs about two-thirds of the way down page 467, where we have the following situation.

$$u = x + y = x + z \neq x + yz$$
$$x \prec u$$
$$xy = xz = yz \quad y, z$$
$$v = x(y + z)$$
$$y + v \neq y + z \quad \text{and} \quad z + v \neq y + z.$$

If $v = 0$ then x, y, and z generate L_4. If $(y + v)(z + v) = v > 0$ then x, $y + v$ and $z + v$ generate L_4. Otherwise x, $(y + v)(z + v)$ and z generate L_3.

3. *Reflections*

Let L be any lattice, and let $p \in J(L)$. Then p has a unique lower cover p_* in $I(L)$, viz., $p_* = \{x \in L\colon\ x < p\}$. Of course p_* may or may not be a principal ideal, but we can always regard p_* as a lattice element by enlarging L to $L' = \langle L \cup \{p_*\}\rangle \leq I(L)$. Most of the properties we are considering in this paper are unaffected by this construction; recall in particular Lemma 2.5.

Now we define two binary relations on $J(L)$.

$p\ A\ q$ if $q < p$ and there exists $x \in L$ such that $p \leq q + x$ but $p \nleq q_* + x$.

$p\ B\ q$ if $p \neq q$ and $p \leq p_* + q$ but $p \nleq p_* + q_*$

Let $E = A \cup B$. These relations, due to Jónsson, have been used for some time to study finite semidistributive lattices [17, 21, 22]. We want to use them in finitely generated semidistributive lattices. First of all, we need to extend the main lemma about these relations [17] to the finitely generated case.

Lemma 3.1. Let L be a lattice in a variety V such that

(1) V is semidistributive, and

(2) $B_{n+1}[c] \notin V$ for some n,

where $B_{n+1} \cong 2^{n+1}$ and c is a coatom of B_{n+1}. If $u \in J(L)\backslash D_k(L)$, then there exist an extension $L' \geq L$ with $L' \in V$ and a finite set of elements $\{p, q_1, \ldots, q_m\} \subseteq J(L')$ such that $u\ B\ p$ and $u\ A\ q_i$ for $1 \leq i \leq m$, and $\{p, q_1, \ldots, q_m\}$ is a minimal nontrivial join-cover of u in L'. Moreover, if $k > 0$ then at least one of $p, q_1, \ldots, q_m$ is not in $D_{k-1}(L')$.

Proof. Let $\{a, b\}$ be a nontrivial join-cover of u in L. If $k = 0$ then any nontrivial join-cover will do, but for $k > 0$ we choose $\{a, b\}$ so that it has no refinement in $D_{k-1}(L)$ (by Lemma 2.4). We may assume $u_* \in L$. By

$(SD_\wedge)$ we get $u \le u_* + a$ or $u \le u_* + b$, say the latter. Let p be minimal in $F(L)$ with respect to $p \le b$ and $u \le u_* + p$, and let $L_1 = \langle L, p\rangle$. By $(SD_\wedge)$ we have $p \in J(L_1)$, and clearly $u\ B\ p$.

Let $s = u_* + p$. Let $\{e_1, \ldots, e_m\}$ be a maximal-sized set such that $s = e_1 + \ldots + e_m + p$ irredundantly with each $e_i \le u_*$; by AR(n) $m < n$. There exist minimal filters $q_1, \ldots, q_m$ with $q_1 \le e_i$ such that $q_1 + \ldots + q_m + p = s$. Let $L' = \langle L_1 \cup \{q_1, \ldots, q_m\}\rangle$.

Suppose some q_i were join reducible, say $q_1 = q_1' + q_1''$. Then we could find elements $x_1 \in q_1' - q_1$, $y_1 \in q_1'' - q_1$, $x_2 \in q_2, \ldots, x_m \in q_m$ with $x_i \le e_i$ and $y_1 \le e_1$ and $x_1 + y_1 + x_2 + \ldots + x_m + p = s$. By the choice of m, this is redundant, and no x_i with $i \ge 2$ can be omitted since $\sum e_i + p$ is irredundant. So either $\sum x_i + p = s$ or $y_1 + \sum_{i=2}^{m} x_i + p = s$.

Now fix x_1 and y_1, and let $x_2, \ldots, x_m$ vary. More particularly, let $A_i = \{x \in L_1\colon\ x \le e_i$ and $x \in q_i\}$ for $2 \le i \le m$; we will partition $A_2 \times \ldots \times A_m$ into two sets X and Y. Let $(x_2, \ldots, x_m) \in X$ if $\sum x_i + p = s$, and $(x_2, \ldots, x_m) \in Y$ otherwise. If $\bar{x} \in X$ and $\bar{y} \ge \bar{x}$ (componentwise) then $\bar{y} \in X$; if $\bar{x} \in Y$ and $\bar{y} \le \bar{x}$, then $\bar{y} \in Y$. Hence one of these two sets is cofinal downward in $A_2 \times \ldots A_m$ - say X. This means that $\sum x_1 + p = x$ for every choice of $x_2, \ldots, x_m$. Hence $x_1\ \ q_2 + \ldots + q_m + p = s$, and by $(SD_\vee)$ $x_1 q_1 + q_2 + \ldots + q_m + p = s$. Since $x_1 \nleq q_1$, this contradicts the choice of the q_i's. We conclude that each q_i is irreducible.

One easily checks that $u\ A\ q_i$ for $1 \le i \le m$. It remains to show that at least one of $p, q_1, \ldots, q_m \notin D_{k-1}(L')$ if $k > 0$.

Suppose $p, q_1, \ldots, q_m \in D_{k-1}(L')$, so that $d_{k-1}(p) = p$ and $d_{k-1}(q_i) = q_i$ for $1 \le i \le m$. Then

$$\begin{aligned} u \le \textstyle\sum q_i + p &= d_{k-1}(\textstyle\sum q_i + p) \\ &\le \textstyle\sum d_{k-1}(u) + d_{k-1}(p) \le d_{k-1}(a + b) \end{aligned}$$

which contradicts the choice of $\{a, b\}$. (We are using Lemma 2.4 and 2.5 here.) Hence at least one of $p, q_1, \ldots, q_m \notin D_{k-1}(L')$. □

Lemma 3.2. Let V be a lattice variety such that $V \not\subseteq LB(k)$. Then there is a finitely generated lattice $L \in V$ with a join irreducible element $u \in J(L)\backslash D_k(L)$.

Proof. To begin with, we can find a lattice $L_0 \in V\backslash LB(k)$ which is finitely generated and subdirectly irreducible. Let a/b be critical in L_0. Let u be a minimal in $F(L)$ such that $u \leq a$, $u \nleq b$. Note that u is completely join irreducible in $F(L_0)$, $u_* = ub$, and $u/u_* \nearrow_w a/b$. Let $L = \langle L_0, u\rangle/\Psi_{uu_*}$. Then $L_0 \leq L$ (as $(a, b) \notin \Psi_{uu_*}$). Moreover, $u \notin D_k(L)$ or else we would have $L_0 \leq L \in LB(k)$ as follows.

Suppose $u \in D_k(L)$, and let $f\colon FL(X) \twoheadrightarrow L$ where X is finite. There is a unique completely meet irreducible congruence $\Psi_{\beta(u)} \in \mathrm{Con}\,(FL(X))$ separating $\beta(u)$ from everything below it but with $\Psi^*_{\beta(u)}$ not doing so [11, Theorem 4.1], and $FL(X)/\Psi_{\beta(u)} \cong L(\beta(u))$ via the standard epimorphism. Since ker f has this property, $\ker f = \Psi_{\beta(u)}$ and $L \cong L(\beta(u)) \in LB(k)$. This contradiction gives us $u \notin D_k(L)$. □

We also want to recall the following lemma from [21]. Let A^d and B^d be the relations defined dually to A and B, respectively, on M(L).

Lemma 3.3. Let L be a semidistributive lattice. Let $p, q \in J(L)$ and assume that $\kappa(p)$ and $\kappa(q)$ exist in L.

(i) If $p\,A\,q$, then $\kappa(p)\,B^d\,\kappa(q)$.

(ii) If $p\,B\,q$, then $\kappa(p)\,A^d\,\kappa(q)$.

We will say that a finite subdirectly irreducible semidistributive lattice L is *reflected* in the semidistributive lattice K, written $L \sqsubset K$, if there exists a one-to-one map $h\colon J(L) \rightarrowtail J(K)$ such that

(1) $x \leq y$ in $J(L)$ iff $h(x) \leq h(y)$ in $J(K)$;

(2) if $x\,B\,z$ in L, then $h(x)\,B\,h(z)$ in K and $h(x) \leq h(z) + \sum\{h(y)\colon y < x\}$;

(3) if $w \in \langle h(J(L))\rangle$ and $h(x) \leq h(x)_* + w$, then either $h(x) \leq w$ or $w \geq h(z)$ for some $z \in J(L)$ with $x\,B\,z$.

In condition (3) we need not assume that $h(x)_*$ exists in K, for we can always use the ideal $h(x)_* = \{v \in K\colon v < h(x)\}$. For practical purposes this means that we will want to assume that $V(K)$ is semidistributive.

This definition is designed to make the following theorem work.

Theorem 3.4. Let L be a finitely subdirectly irreducible semidistributive lattice. For any lattice K in a semidistributive variety, if $L \sqsubset K$ then L is a homomorphic image of a sublattice of K.

Proof. To simplify notation, let $X = J(L)$. Assume $L \sqsubset K$ via the map $h\colon X \rightarrowtail J(K)$. We will show first that $S_0(h(X)) \cong S_0(X) = L$, and then that $S_0(h(X))$ is a homomorphic image of $\langle h(X)\rangle$.

To prove that $S_0(h(X)) \cong S_0(X)$, we need to show that for $x \in X$ and $V \subseteq X$, $x \le \sum V$ if and only if $h(x) \le \sum h(V)$. By (1) of the definition of $L \sqsubset K$, we have $x \le v$ for some $v \in V$ if and only if $h(x) \le h(v)$. So first assume $x \le \sum V$ nontrivially. By induction we may assume that this implies $h(y) \le \sum h(V)$ for all $y \in X$ with $y < x$. By $(SD_\wedge)$ we have $x \le x_* + v_0$ for some $v_0 \in V$. Let z be a minimal in L such that $z \le v_0$ and $x \le x_* + z$. Then $x \mathrel{B} z$, so by condition (2) we have $h(x) \le h(z) + \sum\{h(y)\colon\ y < x\} \le \sum h(V)$, as desired.

Conversely, assume $h(x) \le \sum h(V)$ nontrivially. By induction we may assume that $y \le \sum V$ for each $y < x$. By $(SD_\wedge)$ we have $h(x) \le h(x)_* + h(v_0)$ for some $v_0 \in V$, whence condition (3) gives us $h(v_0) \ge h(z)$ for some $z \in X$ with $x \mathrel{B} z$. By (1), $v_0 \ge z$, so $x \le z + \sum\{y \in X\colon\ y < x\} \le \sum V$. We conclude that $S_0(h(X)) \cong S_0(X) = L$.

Let $H = \langle h(X)\rangle$. We will show that $h(X)$ satisfies the closure condition (CL) of Theorem 1.1 in H, from which it follows that $S_0(h(X))$ is a homomorphic image of H. So let W be a (w.l.o.g.) nontrivial join-cover of $h(x)$ in H. By induction we may assume that for each $y < x$ there is a set $V_y \subseteq h(X)$ with $V_y \ll W$ and $h(y) \le \sum V_y$. As before there is a $w_0 \in W$ such that $h(x) \le h(x)_* + w_0$, whence by condition (3) there is a $z \in X$ with $w_0 \ge h(z)$ and $x \mathrel{B} z$. Let $V = \{h(z)\} \cup \bigcup_{y<x} V_y$. Then $V \subseteq h(X)$, $V \ll W$, and $h(x) \le h(z) + \sum_{y<x} h(y) \le \sum V$, so (CL) holds. □

The same sort of argument gives us the following result.

Lemma 3.5. Let L be a finite subdirectly irreducible semidistributive lattice, and let $u \in J(L)$ be such that $\mathrm{con}(u, u_*)$ is critical. Then $J(L) = \{u\} \cup \{x \in L\colon\ u = t_0 \mathrel{E} t_i \mathrel{E} \dots \mathrel{E} t_n = x \text{ for some } n\}$.

Proof. Let H be the smallest subset of $J(L)$ such that $u \in H$ and if $v \in H$ and $v \mathrel{E} x$, then $x \in H$. We will show that H satisfies the condition (CL) of Theorem 1.1. It then follows that there is a standard homomorphism $f\colon L \twoheadrightarrow S_0(H)$ with $\beta f(u) = u$. By the choice of u, f is the identity, and hence $H = J(L)$.

So let $v \in H$, and let W be a (w.l.o.g.) nontrivial join-cover of v in L. Then as in the proof of Theorem 3.1 there is a set $\{a_1, \dots, a_m, b\} \subseteq J(L)$ such that $v \mathrel{A} a_i$ $(1 \le i \le m)$, $v \mathrel{B} b$ and $b \le w$ for some $w \in W$, and $v \le a_1 + \dots + a_m + b$. For each i we have $a_i < v \le \sum W$, and $a_i \in H$, so we may assume by induction that there are sets $T_i \subseteq H$ with $T_i \ll W$ and $a_i \le \sum T_i$. (If $a_i \le \sum W$ trivially, then $T_i = \{a_i\}$.) Let $T = \{b\} \cup \bigcup_{i=1}^{m} T_i$. Then $T \subseteq H$, $T \ll W$, and $v \leq \sum T$, as desired. □

4. *Onlione varieties.*

Let V be a lattice variety. We will call V an *onlione* variety if V is semidistributive $(M_3, L_1 - L_5 \notin V)$ and $L_6 - L_{10}, L_{13} - L_{15} \notin V$.

The property makes onlione varieties interesitng is contained in the following lemma.

Lemma 4.1. A semidistributive lattice variety V is an onlion variety if and only if V satisfies

> (O) for all $L \in V$, for all $u \in L$, if u is completely join irreducible in L and $u \leq u_* + x$ and $u \leq u_* + y$, but $u \nleq x$ and $u \nleq y$, then $u \leq u_* + xy$,

and its dual. In particular, if $L \in V$ where V is an onlione variety, and $u \in J(L)$, and $u\ B\ b$ and $u\ B\ r$, then $b = r$.

Proof. First we note that the property of being an onlione variety is self-dual, since the excluded lattices are either self-dual or else come in dual pairs. Moreover, it is clear that a variety satisfying the condition (O) cannot contain any of $L_6 - L_{10}$ or $L_{13} - L_{15}$, since each of these lattices or its dual contains distinct join irreducible elements u, b, r with $u\ B\ b$ and $u\ B\ r$ (hence violating (O)), as labeled in Figure 1.

Conversely, let V be a semidistributive variety failing (O), and let u be a completely join irreducible element in $L_0 \in V$ with $u \leq u_* + x$ and $u \leq u_* + y$, but $u \nleq u_* + xy$. Then by standard arguments we have in $F(L_0)$ join irreducible elements $b \leq x$ and $r \leq y$ such that $u\ B\ b$ and $u\ B\ r$. Clearly b and r are incomparable. Let $L = \langle L_0 \cup \{b, r\}\rangle$. After a couple of preliminaries, we will show that $L_i \sqsubset L$ for some $i \in \{6, 7, 8, 9, 13, 15\}$.

Let $u' = u + b_* + r_*$. We claim $u' \in J(L')$ where $L' = \langle u', u_* + r_* + b_*, r, b\rangle$. First of all, it is easy to show that for every $w \in L'$ we have $w \leq b$ or $w \geq r_*$, and likewise $w \leq r$ or $w \geq b_*$. The generators of L' have these properties, and the set of elements with the properties is closed under meets and joins. Suppose $u' = c + d$ with $c, d \in L'$. Then, working in L, we have

$$
\begin{aligned}
u' = c + d &= u + b_* + r_* \\
&= uc + ud + b_*c + b_*d + r_*c + r_*d \qquad \text{(by } SD_\wedge) \\
&= uc + ud + b_* + r_*.
\end{aligned}
$$

Since $u \nleq u_* + b_* + r_*$, this implies $u \leq c$ or $u \leq d$, say $u \leq c$. Because $u \nleq b$ or r, we also have $c \nleq b$ or r, so by the above remark $b_* + r_* \leq c$. Hence $u' \leq c$ and $u' = c$, so $u' \in J(L')$. The relations $u'\ B\ b$ and $u'\ B\ r$ also hold in L': for clearly $u' \leq u'_* + b$ and $u' \leq u'_* + r$, while if

$u' \le u'_* + b_*$ say, then by $(SD_\wedge)$ $u \le u_* + u'_* = u'_*$ (since $u_* + b_* + r_* < u'$) or $u \le u_* + b_*$, both of which are contradictions. Thus, replacing L by L' and u by u', we may assume that $r_* + b_* \le u$ in L.

Now we may assume that $L = \langle u, u_*, r, b\rangle$. Then $1 = u_* + b + r$ while $1 \ne u_* + b_* + r_*$ ($\nleq u$), so at least one of b and r, is a canonical joinand of 1, and hence join-prime. Hence we assume $r \in D_0(L)$.

Case 1. Assume $u \nleq b + r$. Then we have

$$(\forall w \in L)\ w \ge u_* \text{ or } w \le b + r \text{ exclusively.}$$

This property holds for the generators of L. and the set of elements with the property is closed under joins and meets. Thus $u_* \in D_0(L)$.

Subcase 1a. Assume also $b \le u + r$. Now $b \nleq b_* + u$ since $u\,B\,b$, so $(SD_\wedge)$ implies $b \le b_* + r$. Arguing as above, we have

$$(\forall w \in L)\ w \ge b_* \text{ or } w \le r \text{ exclusively.}$$

so $b_* \in D_0(L)$.

Note $L = \langle u, u_*, b, b_*, r\rangle$. We want to show that $L_9 \sqsubset L$ using these elements. Conditions (1) and (2) are immediate, and we have already shown that $\{u_*, b_*\} \subseteq D_0(L) \subseteq J(L)$. Since also $r \in D_0(L)$ by assumption, it remains to verify condition (3) for b and u.

If $b \le b_* + w$ nontrivially, then $w \nleq b_*$ so $w \le r$, and since $b_* + r_* \le u_*$ in fact $w = r$. Next we claim that

$$(\forall w \in L)\ w \le u + b \text{ implies } w \ge b \text{ or } w \le u \text{ exclusively.}$$

Certainly the set of elements with this property contains the generators of L and is closed under joins. Let s and t have the property, and assume $st \le u + b$. The interesting case is when $b \le s \le u + b$ and $t \ge r$. If $t = r$ then $st \le r(u + b) = r_* \le u$, while if $t > r$ then $t \ge b_* + r \ge b$ and hence $st \ge b$. Thus the set is closed under meets.

Now let $u \le u_* + w$ nontrivially. If $w \nleq r$ then $w \le u + b$, and inasmuch as $w \nleq u$ we get $w \ge b$ by the claim, as desired. This finishes (3), and we conclude that $L_9 \sqsubset L$.

Subcase 1b. If we assume instead that $b \nleq u + r$, then we have $\{r, u_*, b\} \subseteq D_0(L)$. We need to show first in this case that we can also assume w.l.o.g. that $u \nleq u_* + (b + r)(u + b)(u + r)$.

So suppose $u \le u_* + (b + r)(u + b)(u + r)$, and let v be a minimal in $F(L)$ such that $v \le (b + r)(u + b)(u + r)$ and $u \le u_* + v$. Then $u\,B\,v$, $r \nleq v$ as $v \le u + b$, and $v \nleq r$ as $u \nleq u_* + r_*$. Consider $L' = \langle u, u_*, v, r\rangle$. In this lattice we have $r \in D_0(L')$, $u \nleq v + r$ since $v + r \le b + r$, and $v \le u + r$. Moreover, these relations still hold if we make the change of variables

$u' = u + v_*$ $(+r_*)$ as in the preliminaries. Thus the assumption leads to the conditions of Subcase 1a, so we may assume $u \not\leq u_* + (b + r)(u + b)(u + r)$.

Now $L = \langle u, u_*, b, r\rangle$, and we want to show that $L_{13} \sqsubset L$ using these elements. Since u_*, b, and r are join-prime, we only have to check condition (3) for u, i.e., we must show that if $u \leq u_* + w$ nontrivially then $w \geq b$ or $w \geq r$. If $u \leq u_* + w$ nontrivially, then $w \not\leq u_*$ so $w \leq b + r$. If $w \not\leq b$ then $w \leq u + r$, and if $w \not\leq r$ then $w \leq u + b$. But then $w \leq (b + r)(u + b)(u + r)$, contrary to our assumption. Hence either $w \geq b$ or $w \geq r$. Conditions (1) and (2) are easy to verify, and we can conclude that $L_{13} \sqsubset L$.

Case 2. Assume $u \leq b + r$ and $b \leq u + r$. Then $u + r = b + r = ub + r = b_* + r$. As in an earlier argument we have

$$(\forall w \in L)\ w \geq b_* \text{ or } w \leq r \text{ exclusively,}$$

and

$$(\forall w \in L)\ w \geq r \text{ or } w \leq u + b \text{ exclusively.}$$

Thus $\{b_*, r\} \subseteq D_0(L)$.

We claim that

$$(\forall w \in L)\ w \leq u + b \text{ implies } w \geq b \text{ or } w \leq u \text{ exclusively,}$$

so that $b \in D_0(u + b/0)$. The proof of this is the same as in Subcase 1a.

Subcase 2a. Now also assume $u \leq r_* + b$. We may take $L' = \langle u, b, b_*, r, r_*\rangle$, and we want to show that $L_8 \sqsubset L'$ using these elements. Conditions (1) and (2) are immediate, so we need to verify condition (3). Also, we already have $\{u, b, b_*, r\} \subseteq J(L) \cap L' \subseteq J(L')$, but we must show $r_* \in J(L')$.

Using $u \leq r_* + b$ we easily see that

$$(\forall w \in L')\ w \geq r_* \text{ or } w \leq b \text{ exclusively,}$$

whence $r_* \in D_0(L') \subseteq J(L')$.

Condition (3) is nontrivial for b and u, as the other elements are join-prime. If $b \leq b_* + w$ nontrivially, then $w \geq r$, as otherwise $w \leq u + b$ and $b \in D_0(u + b/0)$. If $u \leq u_* + w$ nontrivially and $w \not\leq r$, then $w \leq u + b$, whence $w \geq b$ as $w \not\leq u$. Thus (3) holds, and we get $L_8 \sqsubset L'$.

Subcase 2b. If however $u \not\leq r_* + b$, note that $L = \langle u, u_*, b, b_*, r\rangle$, and we will show that $L_6 \sqsubset L$ using these elements. Again (1) and (2) are easy, and we have $\{b_*, r\} \subseteq D_0(L)$. So we need to show that $u_* \in J(L)$, and we have to check (3) for u, u_* and b.

It is easy to prove that

$$(\forall w \in L)\ w \leq u + b \text{ implies } w \geq u_* \text{ or } w \leq r_* + b \text{ exclusively.}$$

(If $u_* \leq r_* + b$ then we would have $u \leq u_* + b \leq r_* + b$, contrary to assumption.)

Thus $u_* \in D_0(u + b/0) \subseteq J(L)$. It follows that if $b \le b_* + w$ or $u_* \le u_{**} + w$ nontrivially, then $w \nleq u + b$ so $w \ge r$. The argument that u satisfies (3) is exactly as in the first subcase. Hence $L_6 \sqsubset L$.

Case 3. Assume $b \nleq u + r$ (so $b \in D_0(L)$) and either $u \le b_* + r$ or $u \le r_* + b$.

Subcase 3a. If $u \le b_* + r$ and $u \le r_* + b$, then we may take $L' = \langle u, b, b_*, r, r_* \rangle$ and we want to show that $L_{15} \sqsubset L'$. Conditions (1) and (2) are immediate, so we need to verify condition (3), and to show that $b_*, r_* \in J(L')$. We have $r, b \in D_0(L) \cap L' \subseteq D_0(L')$ by assumption, and it is easy to show that $b_*, r_* \in D_0(L')$.

We need to show that if $w \in L'$ and $u \le u_* + w$ nontrivially, then $w \ge b$ or $w \ge r$. Equivalently, we can show that if $u \le u_* + w$ and $w \le \kappa(b)\kappa(r) = (u + r)(u + b)$, then $u \le w$. This in turn is a consequence of the following claim:

$$(\forall w \in L')\ w \le (u + r)(u + b) \text{ implies}$$
$$w \ge u \text{ or } w \le b_* + r_* \text{ exclusively.}$$

The generators of L' have this property, and the set of elements having the property is closed under joins. So let s and t satisfy the property of the claim, and assume $st \le (u + r)(u + b)$. First suppose $s \ge b$, in which event $t \ngeq b$ so $t \le u + r$. If $s = b$ then $st \le b(u + r) = b_* \le b_* + r_*$, and the conclusion holds. If $s > b$ then $s \ge r_* + b \ge u$, and in this case each of the possibilities $t = r$, $t > r$, or $t \le (u + r)(u + b)$ leads easily to either $st \ge u$ or $st \le b_* + r_*$. Thus we may assume $s \ngeq b$, i.e., $s \le u + r$ and hence by symmetry $s, t \le (u + r)(u + b)$. In that case either $st \ge u$, or else one of the elements is below $b_* + r_*$, and hence so is their product. This proves the claim, and thus we get $L_{15} \sqsubset L'$.

Subcase 3b. Now assume $u \le r_* + b$ and $u \nleq b_* + r$ (the other case is symmetric). Let us first show that in this case we may assume without loss of generality that $u \nleq u_* + (u + b)(b_* + r)$. Assuming $u \le u_* + (u + b)(b_* + r)$, let v be a minimal in $F(L)$ such that $v \le (u + b)(b_* + r)$ and $u \le u_* + v$. Then $u\, B\, v$, and we claim that b and v are incomparable. For $b \nleq v$ as $v \le b_* + r$ and $b \in D_0(L)$, while $v \nleq b$ since $u \le u_* + v$ but $u \nleq u_* + b_*$.

Note that

$$(\forall w \in L)\ w \ge r_* \text{ or } w \le b \text{ exclusively.}$$

This property extends to $F(L)$: for if $F \in F(L)$ and $F \nleq b$, then for every $f \in F$ we have $f \nleq b$ and hence $f \ge r_*$, so $F \ge r_*$. In particular, this gives us $v \ge r_*$.

Now we have $u\, B\, b$ and $u\, B\, v$ with b and v incomparable, $b \nleq u + v$ as $u + v \le u + r$ and $b \nleq u + r$, $v \le u + b$, and $u \le r_* + b \le v + b$. Moreover, these conditions remain valid if we make the change of variables $u' = u + v_* \ (+b_*)$ as done in the preliminaries. Thus the assumption $u \le u_* + (u + b)(b_* + r)$ leads to Case 2 with u, v, and b replacing u, b, and r respectively. Hence we may assume $u \nleq u_* + (u + b)(b_* + r)$.

Now $L = \langle u, u_*, b, r, r_* \rangle$, and we want to show that $L_7 \sqsubset L$ using these elements. Again (1) and (2) are easy, and we need to check condition (3) and to show $u_*, r_* \in J(L)$.

Since

$$(\forall w \in L)\ w \ge r_* \ \text{ or } \ w \le b \ \text{ exclusively.}$$

we have $r_*, b, r \in D_0(L)$. Next we claim that

$$(\forall w \in L)\ w \le u + r \ \text{ implies } \ w \ge u_*$$
$$\text{or } \ w \le r + b(u + r) \ \text{ exclusively.}$$

Certainly this holds for the generators of L, and the property is preserved by joins. Suppose $st \le u + r$. Then at least one of s and t, say s, is below $u + r$, and the interesting case is when $s \ge u_*$ and $t \ge b$. If $t = b$ then $st \le b(u + r)$; otherwise $t \ge r_* + b \ge u_*$. Hence the property is preserved by meets. Also $u_* \nleq r + b(u + r) = r + b_*$ else $u \le u_* + r \le r + b_*$, a contradiction. This proves the claim. Thus $u_* \in D_0(u + r/0)$. In particular, $u_* \in J(L)$, and if $u_* \le u_{**} + w$ nontrivially then $w \ge b$, as desired.

It remains to show that

$$(\forall w \in L)\ w \le (u + r)(u + b) \ \text{ implies } \ w \ge u$$
$$\text{or } \ w \le u_* + (u + b)(b_* + r) \ \text{ exclusively.}$$

This property holds for the generators of L and is preserved by joins. Assume s and t have the property and $st \le (u + r)(u + b)$. If $s = b$ then $st \le b(u + r) = b_* \le u_*$ so the conclusion holds. If $s > b$ then $s \ge r_* + b \ge u$; in this case w.l.o.g. $s \nleq r$ (i.e., $s \ne 1$) so $s = u + b$, and $t \le u + r$ as $t \nleq b$. If $t \ge r$ then by the previous claim either $t \ge u_*$, so $t \ge u_* + r \ge u$ and $st \ge u$, or $t \le b_* + r$ whence $st \le (u + b)(b_* + r)$. Otherwise $t \nleq r$, implying $t \le (u + b)(u + r)$, and the desired conclusion follows easily since $s \ge u$. Therefore we may assume that $s, t \nleq b$. The case where both elements are below $(u + b)(u + r)$ is easy, so we may now assume w.l.o.g. that $r \le s \le u + r$ and $u \le t \le (u + b)(u + r)$. By the previous claim either $s \ge u_* + r \ge u$, so $st \ge u$, or $s \le b_* + r$ whence $st \le (u + b)(b_* + r)$. Thus the property of the claim is preserved by meets.

Now let $u \le u_* + w$ nontrivially. We have just shown that $u \in D_0((u + r)(u + b)/0)$, so $w \nleq (u + r)(u + b)$, i.e., either $w \ge b$ or $w \ge r$, as desired. This finishes (3), and we conclude that $L_7 \sqsubset L$.

Case 4. This leaves the possibility that $u \le b + r$ and $b, r \in D_0(L)$ with $u \nleq b_* + r$ and $u \nleq r_* + b$. Let $b_0 = b$ and $r_0 = r$.

Suppose we have b_n and r_n such that

(i) $b_0 \le b_1 \le \ldots \le b_n$ and $r_0 \le r_1 \le \ldots \le r_n$

(ii) $u \mathrel{B} b_n$ and $u \mathrel{B} r_n$ in $1/ ub_n r_n$

(iii) $u \le b_n + r_n$, $u \nleq ub_n + r_n$ and $u \nleq ur_n + b_n$.

Assume n is odd; for n even we will do the same thing with b_n and r_n interchanged.

Suppose there is some t with $ub_n \le t < r_n + ub_n$ such that $u \le u_* + t$. Then we may choose a filter $f \le t$ minimal in $1/ub_n$ such that $u \le u_* + f$. Consider $L' = \langle u, u_*, f, r_n \rangle$ which satisfies

$$(\forall w \in L')\ w \ge ub_n \quad \text{or} \quad w \le r_n$$

We claim that $f \in J(L')$. By $(SD_\wedge)$ f is join irreducible in $1/ub_n$, so if f were join reducible we would have $f = c + d$ where $ub_n \le c < f$ and $ub_n r_n \le d < r_n$ (since $t \ge f$ and $t \nleq r_n$). But then $u \le u_* + f = u_* + c + d$ implies either $u \le u_* + c$ or $u \le u_* + d$, both contradictions. Hence $f \in J(L')$, and in fact $f_* \ge ub_n$. It now follows easily that $u \mathrel{B} f$ in L'. Thus we have $u \mathrel{B} r_n$ and $u \mathrel{B} f$ in L', while $r_n + f = r_n + ub_n \nleq u$. This is preserved when we replace u by $u' = u + r_{n*} + f_*$, putting us back into Case 1. Hence we may assume that $u \nleq u_* + t$ for $ub_n \le t < r_n + ub_n$.

Let $b_{n+1} = b_n$ and $r_{n+1} = r_n + ub_n$. Clearly inductive hypothesis (i) holds. Since $b_{n+1} = b_n$ and $ub_{n+1}r_{n+1} = ub_n \ge ub_n r_n$ we still have $u \mathrel{B} b_{n+1}$ in $1/ub_{n+1}r_{n+1}$, while the assumption from the previous paragraph gives us $u \mathrel{B} r_{n+1}$ in $1/ub_{n+1}r_{n+1}$. Thus (ii) holds. Moreover, $b_{n+1} + r_{n+1} = b_n + r_n \ge u$. Likewise $u \nleq ub_{n+1} + r_{n+1} = ub_n + r_n$. If $u \le ur_{n+1} + b_{n+1}$, then we make the replacement $u' = u + b_{n+1*} + r_{n+1*}$ as in the preliminaries. It is easy to check that $u' \le u'r_{n+1} + b_{n+1}$ and $u' \le b_{n+1} + r_{n+1}$, so this puts us back in Case 2 or 3. Thus we may assume that $u \nleq ur_{n+1} + b_{n+1}$, and the induction hypotheses are satisfied.

Let B be the ideal generated by $\{b_n\colon n \in \omega\}$ and let R be the ideal generated by $\{r_n\colon n \in \omega\}$. Now by (iii) $u \nleq B$ and $u \nleq R$, while by construction $u B = u R$ and $u \le b_0 + r_0 = B + R$. Using $(SD_\wedge)$ we get

$$u\,B = u\,R = u\,(B + R) = u$$

which is a contradiction. We conclude that the process stopped by leading to a previous case somewhere along the way. This proves the lemma. □

Our next observation is simple but useful.

Lemma 4.2. If L is a lattice in an onlione variety, and $u \mathrel{B} r \mathrel{B} s$ in L, then $u < s$.

Proof. From $u \mathrel{B} r \mathrel{B} s$ we have $u \le u_* + r \le u_* + r_* + s$ whence by $(SD_\wedge)$ $u \le u_* + s$. Now $u \ne s$ since $\kappa(u) < \kappa(r) < \kappa(s)$ by Lemma 3.3. If $u \nleq s$ then we can find a filter $f \le s$ such that $u \mathrel{B} f$. Then Lemma 4.1 implies $r = f \le s$, a contradiction. Hence $u < s$. □

Lemma 4.3. If V is an onlione variety and $L \in V$ and $u \mathrel{B} r \mathrel{B} s$ in L, then $L_{12} \in V$. Dually, if $u \mathrel{A} r' \mathrel{A} s'$ in L, then $L_{11} \in V$.

Proof. Assume $u \mathrel{B} r \mathrel{B} s$ in L. Then arguing as in the proof of Lemma 4.1 (with $n = 2$ since $L_{14} \notin V$) we see that there is a unique filter $f < u$ such that $u \mathrel{A} f$ and $\{r, f\}$ is a minimal nontrivial join-cover of u. Likewise there is a unique filter $g < r$ such that $r \mathrel{A} g$ and $\{s, g\}$ is a minimal nontrivial join-cover of r. Let $L' = \langle u, r, s, f, g\rangle$. We want to show that $L_{12} \sqsubseteq L'$ using these elements.

For condition (1), we have $u < s$ by Lemma 4.2, so $f < u < s$ and $g < r$. Since $\{r, f\}$ is a nontrivial join-cover of u, $f \nleq r$, and similarly $g \nleq s$. Thus (1) holds. Condition (2) is immediate.

For condition (3), we first note that $\{s, f, g\} \subseteq D_0(L')$ as

$(\forall w \in L')\ w \ge s$ or $w \le u + r$ exclusively,

$(\forall w \in L')\ w \ge f$ or $w \le r$ exclusively,

$(\forall w \in L')\ w \ge g$ or $w \le s$ exclusively.

Next, we can show that

$$(\forall w \in L')\ w \le u + r \text{ implies } w \ge r \text{ or } w \le g + s(u + r) \text{ exclusively.}$$

We leave the details to the reader, with the observation that $w > s$ implies $w \ge g + s \ge r$ and hence $w = 1$. It follows that $r \in D_0(u + r/0)$, and thus if $r \le r_* + w$ nontrivially then $w \ge s$. Similarly,

$$(\forall w \in L')\ w \le g + s(u + r) \text{ implies } w \ge u \text{ or } w \le f + r(g + s(u + r)) \text{ exclusively.}$$

This time we observe that $w > r$ implies $w \ge r + f \ge u$, and leave the rest to the reader. In particular, $u \in D_0(g + s(u + r)/0)$. Hence if $u \le u_* + w$ nontrivially, then $w \nleq g + s(u + r)$ so $w \ge r$. This finishes (3), and we conclude that $L_{12} \sqsubseteq L'$, and thus $L \in V$. □

We are now in a position to prove some results about the lattice of lattice varieties. We start with the classic theorem of Jónsson and Rival.

Theorem 4.4 [18]. If V is a lattice variety and $V \supset V(N_5)$, then V contains at least one of $M_3, L_1, \ldots, L_{15}$.

Proof. If V is not semidistributive, then V contains M_3 or some L_i $(1 \le i \le 5)$. If in addition V is not onlione, then V contains some L_i with $i \in \{6, 7, 8, 9, 10, 13, 14, 15\}$. Thus we may assume that V is an onlione variety.

We want to apply Lemmas 3.1 and 3.2. First we note that V contains a finitely generated subdirectly irreducible lattice $L \in V \backslash V(N_5)$. Suppose $V \subseteq LB(1)$. Now every finitely generated subdirectly irreducible lattice L in $LB(1)$ is isomorphic to $L(w)$ for some $w \in J(FL(X))$ with $\rho(w) \le 1$. Since V is onlione and $L \not\cong 2$, this means that $w \mathrel{A} a$ and $L \cong S_\cap(J(L)) \cong N_5$. Hence $V \not\subseteq LB(1)$. Hence applying Lemma 3.2 and then Lemma 3.1 twice, we can find a lattice $L' \in V$ and elements $u, r, s \in L'$ with either $u \mathrel{A} r \mathrel{A} s$ or $u \mathrel{B} r \mathrel{B} s$. By Lemma 4.3, we conclude that either L_{11} or $L_{12} \in V$, as desired. □

Next, let us show that $V(L_{12})$ has only two join irreducible covers, as conjectured by Rose [23]. The dual result of course holds for $V(L_{11})$.

Theorem 4.5. Let V be a lattice variety such that $V \supset V(L_{12})$. Then V contains at least one of M_3, L_i for $i \in \{1, \ldots, 11, 13, 14, 15\}$, L^1_{12} or G.

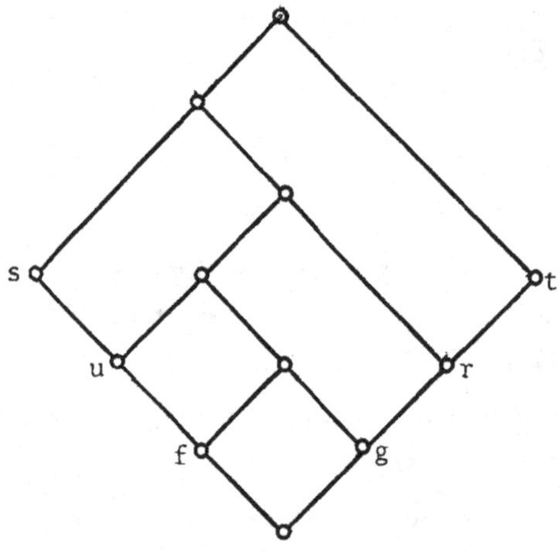

L^1_{12}

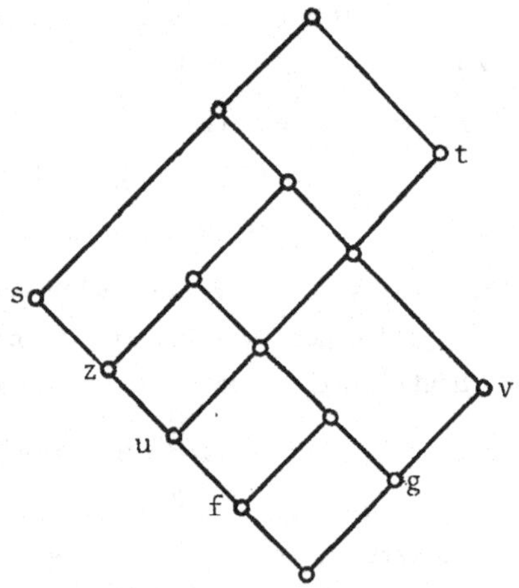

G

Figure 3.

Proof. As before we may assume that V is onlione. Moreover, assuming $L_{11} \notin V$, if $x \mathrel{A} y$ in $L \in V$, then y is join-prime.

Suppose $V \subseteq LB(2)$. Every finitely generated subdirectly irreducible lattice L in $LB(2)$ is isomorphic to $L(w)$ for some $w \in J(FL(X))$ with $\rho(w) \le 2$.

If $\rho(w) \le 1$ we get $L \cong \mathcal{2}$ or N_5 as in the proof of Theorem 4.4. If $\rho(w) = 2$ then we have $w\ A\ a$ and $w\ b\ B$ for a unique pair $\{a, b\}$. Now a is join-prime, so since $\rho(w) = 2$ we have $b\ A\ c$ and $b\ B\ d$ for a unique pair $\{c, d\}$ with c and d join-prime. By Lemma 4.2, $w < d$. It follows from Lemma 3.5 that $J(L) = \{w, a, b, c, d\}$ and $L \cong S_0(J(L)) \cong L_{12}$. Thus $V \not\leq LB(2)$.

Applying Lemmas 3.1 and 3.2 to this situation, we find that there are a lattice $L \in V$ and elements $\{u, r, s, t, f, g, z\} \subseteq J(L)$ satisfying the following properties.

(1) $u\ B\ r\ B\ s\ B\ t$.

(2) $u\ A\ f$, $f \in D_0(L)$ and $\{r, f\}$ is a minimal nontrivial join-cover of u.

(3) $r\ A\ g$, $g \in D_0(L)$ and $\{s, g\}$ is a minimal nontrivial join-cover of r.

(4) $s\ A\ z$, $z \in D_0(L)$ and $\{t, z\}$ is a minimal nontrivial join-cover of s.

(5) $f < u < s$, $g < r < t$ and $z < s$.

We want to show that either $L_{12}^1 \sqsubset L$ or $G \sqsubset L$.

First assume $z \le u$. Working in $L' = \langle u, r, s, t, f, g,\rangle$ we find that

$$1 = s + t = u + t = f + t$$

so in particular $f \nleq t$ and $s \le f + t$. Now we easily have

$$(\forall w \in L')\ w \ge f \text{ or } w \le t \text{ exclusively,}$$

so in L' $f_* = ft < t$. On the other hand $s \nleq s_* + t_*$ implies $s \nleq f + t_*$. Thus in L' $\{t, f\}$ is a minimal nontrivial join-cover of s with $s\ A\ f$.

We are now in a position to prove the following series of statements for L'.

$(\forall w \in L')$ $w \ge g$ or $w \le s$ exclusively.
$(\forall w \in L')$ $w \ge t$ or $w \le r + s$ exclusively.
$(\forall w \in L')$ $w \le r + s$ implies $w \ge s$ or
$w \le u + t(r + s)$ exclusively.
$(\forall w \in L')$ $w \le u + t(r + s)$ implies $w \ge r$ or
$w \le g + s(u + t(r + s))$ exclusively.
$(\forall w \in L')$ $w \le g + s(u + t(r + s))$ implies $w \ge u$ or
$w \le f + t(g + s(u + t(r + s)))$ exclusively.

The proofs of these statements are tedious but straightforward, so we will omit them with the exception of the following detail, which differs somewhat from the arguments used thus far. In the last statement we need to show that $u \nleq f + t(g + s(u + t(r + s)))$. Supposing otherwise, we would have $u \le f + t(g + s(u + t(r + s))) \le u_* + t(g + s(u + t(r + s)))$. By Lemma 4.1 this implies $r \le t(g + s(u + t(r + s))) \le g + s(u + t(r + s))$, contradicting the previous statement.

It is now easy to check that $L_{12}^{1} \sqsubset L'$. Conditions (1) and (2) are immediate, and our list of statements is exactly what is needed to verify condition (3).

On the other hand, assume $z \nleq u$. In this case put $t' = t + u$, $z' = z + u$ and $L' = \langle u, r, s, t', f, g, z' \rangle$. Then we have the following order relations in L': $f < u < z' < s$, $g < r < t'$ and $u < t'$, but $g \nleq s$, $z' \nleq t'$ (since z is join-prime) and $u \nleq r$. Moreover $s \leq z' + t$; that this is a minimal nontrivial join-cover of s in L' will follow from some of the statements below.

The series of statements which must be proved in this case is as follows.

$(\forall w \in L')$ $w \geq f$ or $w \leq r$ exclusively.

$(\forall w \in L')$ $w \geq g$ or $w \leq s$ exclusively.

$(\forall w \in L')$ $w \geq z'$ or $w \leq t'$ exclusively.

$(\forall w \in L')$ $w \geq t'$ or $w \leq r + s$ exclusively.

$(\forall w \in L')$ $w \leq r + s$ implies $w \geq s$ or
$w \leq z' + t'(r + s)$ exclusively.

$(\forall w \in L')$ $w \leq z' + t'(r + s)$ implies $w \geq r$ or
$w \leq g + s(z' + t'(r + s))$ exclusively.

$(\forall w \in L')$ $w \leq g + s(z' + t'(r + s))$ implies $w \geq u$ or
$w \leq f + r(g + s(z' + t'(r + s)))$ exclusively.

Note in particular that we get $\{t', z'\} \subseteq D_0(L') \subseteq J(L')$.

With these statements, it is routine to check that $G \sqsubset L'$, as desired. □

We have $L_{12}^{0} = L_{12}$ and L_{12}^{1}. If we let L_{12}^{2} be the lattice

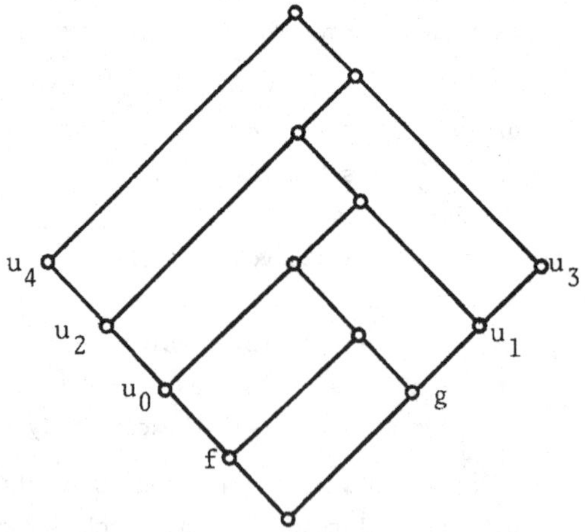

Figure 4.

and so forth for $n \in \omega$, then the varieties $V(L_{12}^{n})$ form a covering chain in Λ. Likewise, if we let $G = G^1$ and let G^2 be the lattice

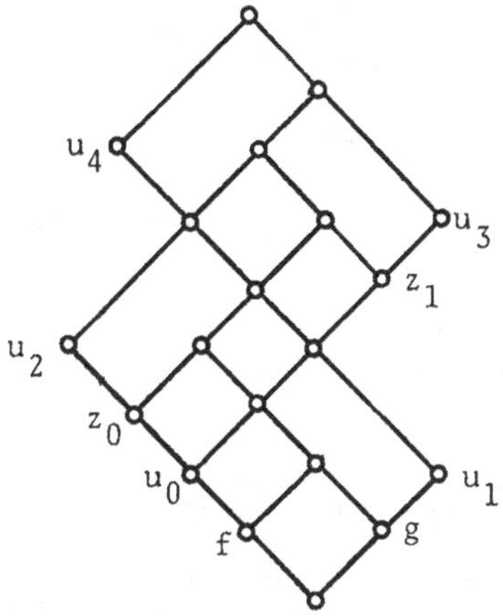

Figure 5.

and so forth for $n \in \omega$, then the varieties $V(G^n)$ form a covering chain in Λ.

Theorem 4.6. For $n \geq 1$, $V(L_{12}^{n+1})$ is the unique join irreducible variety covering $V(L_{12}^n)$, and $V(G^{n+1})$ is the unique join irreducible variety covering $V(G^n)$.

Proof. We give here only the part of the argument which finds the lattice L' for which we must check that $L_{12}^{n+1} \sqsubset L'$ or $G^{n+1} \sqsubset L'$. As before, starting with a lattice variety V properly containing $V(L_{12}^n)$ or $V(G^n)$, but not $V(G)$ in the former case or $V(L_{12}^1)$ in the latter, we may assume that V is an onlione variety, $V \not\subseteq LB(n+2)$, and $x \, A \, y$ in $L \in V$ implies $y \in D_0(L)$. Then there are a lattice $L \in V$ and elements $\{u_0, u_1, \ldots, u_{n+3}, f, g, z_0, \ldots, z_n\} \subseteq J(L)$ satisfying the following properties.

(1) $u_0 \, B \, u_1 \, B \ldots B \, u_{n+3}$

(2) $u_0 \, A \, f$, $f \in D_0(L)$, and $\{u_1, f\}$ is a minimal nontrivial join-cover of u_0.

(3) $u_1 \, A \, g$, $g \in D_0(L)$, and $\{u_2, g\}$ is a minimal nontrivial join-cover of u_1.

(4) For $0 \leq i \leq n$, $u_{i+2} \, A \, z_i$, $z_i \in D_0(L)$, and $\{u_{i+3}, z_i\}$ is a minimal nontrivial join-cover of u_{i+2}.

(5) $f < u_0 < u_2 < u_4 < \ldots$ and $g < u_1 < u_3 < u_5 < \ldots$.

The proof of Theorem 4.5 shows that if $z_i \leq u_i$ for some i $(0 \leq i \leq n)$ then $L_{12}^1 \in V$, while if $z_i \not\leq u_i$ for some i then $G \in V$. Hence we may assume in one case that $z_i \leq u_i$ for all i, and in the other that $z_i \not\leq u_i$ for all i. In the former case, let $L' = \langle u_0, \ldots, u_{n+3}, f, g \rangle$. It is easy to show then that $u_i \leq u_{i+1} + f$ for i even, and $u_i \leq u_{i+1} + g$ for i odd. This case is finished by showing that $L_{12}^{n+1} \sqsubset L'$.

Now assume $z_i \nleq u_i$ for all i. Let $z_0' = z_0 + u_0$ and $z_i' = z_i + u_i + u_{i-1}$ for $1 \leq i \leq n$. Let $u_i' = u_i$ for $0 \leq i \leq 2$ and $u_i' = u_i + u_{i-3}$ for $3 \leq i \leq n + 3$, and let $L' = \langle u_0', u_1', \ldots, u_{n+3}', f, g, z_0', \ldots, z_n' \rangle$. Note that $u_i' < z_i' < u_{i+2}'$, but $u_{i+3}' \nleq u_{i+2}'$ as $u_{i+3}' \nleq z_i$. Thus we can show that for $0 \leq i \leq n$,

$$(\forall w \in L')\ w \geq z_i' \text{ or } w \leq u_{i+3}' \text{ exclusively,}$$

so that $z_i' \in D_0(L')$.

Next we claim that for $3 \leq i \leq n + 3$,

$$(\forall w \in L')\ w \geq u_i \text{ implies } w \geq u_i'.$$

As this is trivial for $i \leq 2$, assume $i \geq 3$. The set of elements in L' having this property contains the generators of L' and is clearly closed under meets. To see that it is closed under joins, assume $u_i \leq a + b$ where a and b are in L' and satisfy the property. If $u_i \leq a$ or $u_i \leq b$ then the conclusion holds. Otherwise $u_i \leq u_{i*} + b$ say (by $(SD_\wedge)$), so by Lemma 4.1 $u_{i+1} \leq b$. Thus

$$u_i' = u_i + u_{i-3} \leq u_i + u_{i+1} \leq a + b$$

as desired. The same argument shows that $u_i' \in J(L')$.

It follows from the above two claims that $\{z_{i-2}', u_{i+1}'\}$ is a minimal nontrivial join-cover of u_i' in L' for $2 \leq i \leq n + 2$. The proof is completed by checking that $G^{n+1} \sqsubseteq L'$ in this case. □

REFERENCES

1. Berman, J. and Wolk, B., Free lattices in some small varieties, Algebra Universalis, 10(1980), 269-289.
2. Birkhoff, G., On the lattice theory of ideals, Bull. Amer. Math. Soc., 40(1934), 613-619.
3. Birkhoff, G., Lattice Theory, 3rd ed. (1967), Providence, R.I., American Mathematical Society.
4. Day, A., A simple solution to the word problem for lattices, Canadian Math. Bull., 13(1970), 253-254.
5. Day, A., Splitting lattices generate all lattices, Algebra Universalis, 7(1977) 163-169.
6. Day, A., Characterizations of lattices that are bounded-homomorphic images or sublattices of free lattices, Canadian J. Math., 31(1979), 69-78.
7. Day, A. and Nation, J.B., A note on finite sublattice of free lattices, Algebra Universalis, 15(1982), 90-94.
8. Dedekind, R., Über die von drei Moduln erzeugte Dualgruppe, Math. Ann., 53(1900 371-403.

9. Freese, R., Ideal lattices of lattices, Pacific J. Math., 57(1975), 125-133.

10. Freese, R., The structure of modular lattices of width four with applications to varieties of lattices, Memoirs Amer. Math. Soc., no. 181(1977), Providence, R.I., American Mathematical Society.

11. Freese, R. and Nation, J. B., Covers in free lattices, to appear in Trans. Amer. Math. Soc.

12. Grätzer, G., Equational classes of lattices, Duke Math. J., 33(1966), 613-622.

13. Hong, D. X., Covering relations among lattice varieties, Pacific J. Math., 40(1972), 575-603.

14. Jónsson, B., Algebras whose congruence lattices are distributive, Math. Scand., 21(1967), 110-121.

15. Jónsson, B., Equational classes of lattices, Math. Scand., 22(1968), 187-196.

16. Jónsson, B., Varieties of lattices: some open problems, Coll. Math. Soc. János Bolyai, 29(1982), Contributions to Universal Algebra (Esztergom), North Holland, 421-436.

17. Jónsson, B. and Nation, J. B., A report on sublattices of a free lattice, Coll. Math. Soc. János Bolyai, 17(1977), Contributions to Universal Algebra (Szeged), North Holland, 233-257.

18. Jónsson and Rival, I., Lattice varieties covering the smallest nonmodular variety, Pacific J. Math., 82(1979), 463-478.

19. Lee. J. G., Almost distributive lattice varieties, Ph.D. Dissertation, Vanderbilt University, 1983.

20. McKenzie, R., Equational bases and nonmodular lattice varieties, Trans. Amer. Math. Soc., 174(1972), 1-43.

21. Nation, J. B., Bounded finite lattices, Coll. Math. Soc. János Bolyai, 29(1982), Contributions to Universal Algebra (Esztergom), North Holland, 531-533.

22. Nation, J. B., Finite sublattices of a free lattice, Trans. Amer. Math. Soc., 269(1982), 311-337.

23. Rose, H., Nonmodular lattice varieties, Memoirs Amer. Math. Soc., no. 292(1984), Providence, R.I., American Mathematical Society.

24. Waterman, A. G., The free lattice with 3 generators over N_5, Portugal. Math. 26(1967), 285-288.

HOMOMORPHISMS OF PARTIAL AND OF COMPLETE STEINER TRIPLE SYSTEMS AND QUASIGROUPS

Don Pigozzi
Iowa State University
Ames, Iowa 50011

and

J. Sichler*
University of Manitoba
Winnipeg, Manitoba R3T 2N2

ABSTRACT Infinite Steiner quasigroups, that is, commutative idempotent groupoids satisfying the identity $(xy)x = y$, form an almost universal category. As a consequence, every monoid is representable by all nonconstant endomorphisms of a Steiner quasigroup, and there is a proper class of nonisomorphic representations of this kind. Similar results are obtained for partial Steiner quasigroups as well as for partial and complete Steiner triple systems naturally corresponding to these algebras.

1980 Mathematics Subject Classification: Primary 08A35, 18B15; Secondary 51E10.

1. INTRODUCTION. A pair (X,T) in which T is a set of three-element subsets of the set X such that every pair of distinct $x,y \in X$ lies in exactly one member T will, as usual, be called a Steiner triple system (STS). Every triple in T is uniquely determined by any of its two-element subsets; it is thus possible to define a binary operation on X by requiring that $xy = z$ just when $\{x,y,z\} \in T$ and $xx = x$ for all $x \in X$. It is clear that $xy = yx$ and $(xy)x = y$ hold in any algebra thus defined; the algebras defined by these three identities are called Steiner quasigroups. Conversely, it is easy to see that any Steiner quasigroup determines a unique STS on its underlying set: to any two-element set $\{a,b\}$ simply assign the triple $\{a,b,ab\}$ and note that each element of such a triple is a product of the remaining two.

A mapping $f : X \to X'$ for which $\{f(x),f(y),f(z)\} \in T'$ whenever $\{x,y,z\} \in T$ will be called an S-homomorphism of the Steiner triple system (X,T) into (X',T'). Observe that all S-homomorphisms must be one-to-one mappings.

---

*The support of the NSERC is gratefully acknowledged.

It is clear that any S-homomorphism is also a homomorphism of the quasigroups Q, Q' corresponding to (X,T), (X',T') respectively; quasigroup homomorphisms will be referred to as Q-homomorphisms. If $f : Q \to Q'$ is a Q-homomorphism with $f(x) = f(y)$ then $f(xy) = f(x)f(y) = f(x) = f(y)$; this enables us to define Q-homomorphisms for STS by requiring that for every $\{x,y,z\} \in T$ either $\{f(x),f(y),f(z)\} \in T'$ or $f(x) = f(y) = f(z)$. With these definitions in place homomorphisms of either type may be considered for both Steiner quasigroups and their corresponding triple systems. Note that the class of Q-homomorphisms includes all constant mappings and that S-homomorphisms are exactly those Q-homomorphisms that are one-to-one.

A category **C** of algebras or relational systems is binding (or universal) if every full category of algebras is isomorphic to a full subcategory of **C**. Any universal category contains a proper class of nonisomorphic rigid objects, that is, objects whose endomorphism monoid consists of the identity mapping alone. Also, every monoid occurs as the endomorphism monoid of an object of any universal category. An extensive list of binding categories can be found in A. Pultr and V. Trnková [7], together with a thorough exposition of methods used to establish universality.

J. Ježek, T. Kepka and P. Němec [3] showed that the variety of all completely symmetric quasigroups (that is, commutative quasigroups satisfying $(xy)x = y$) is binding; their construction employs unary polynomials to control quasigroup homomorphisms. Since Steiner quasigroups possess no nontrivial unary polynomials, this method cannot be applied here and, in fact, Steiner quasigroups do not form a universal category: the existence of constant Q-homomorphisms implies that all rigid Steiner quasigroups are trivial. E. Mendelsohn [5], [6], however, proved that every group occurs as the automorphism group of some STS and hence also as the automorphism group of a Steiner quasigroup.

In the present article we aim to complement these results. A category **C** of algebras is almost universal if the class of all nonconstant homomorphisms between algebras from its subclass **K** forms a universal category. This definition requires that, in particular, any composite of nonconstant homomorphisms between members of **K** be nonconstant again.

THEOREM 1.1. Infinite Steiner quasigroups form an almost universal category.

The morphisms in Theorem 1.1 are, of course, quasigroup homomorphisms. An analogous statement for S-homomorphisms is false in view of the fact that no monoid with more than one idempotent element can be represented by one-to-one mappings.

A *partial Steiner triple system* (PSTS) will be any pair (X,T) in which T is a set of three-element subsets of X satisfying $X = \bigcup X$ and $|t \cap u| \neq 2$ for all $t,u \in T$. A *partial Steiner quasigroup* (PSQ) is a maximal partial quasigroup corresponding to a PSTS; in particular, the partial operation satisfies $x^2 = x$ for all $x \in X$, and whenever its domain contains (x,y) it must also contain pairs (y,x), (x,z), (z,x), (y,z), and (z,y) where $z = xy$; furthermore, we must have $yx = z$, $xz = zx = y$, and $yz = zy = x$. It is easily seen that every PSQ determines a unique PSTS. To avoid the somewhat cumbersome concept of a PSQ we will often consider its corresponding PSTS instead.

The concepts of S-homomorphism and that of Q-homomorphism naturally extend to these partial structures and there is no need to define PS-homomorphisms and PQ-homomorphisms formally. Note that a PS-homomorphism need not be one-to-one; however, like an S-homomorphism, it does not collapse any triple. Following are results concerning these partial structures.

THEOREM 1.2. *Partial Steiner quasigroups form an almost universal category. Furthermore, any finite monoid can be represented by nonconstant endomorphisms of a finite partial Steiner quasigroup.*

THEOREM 1.3. *The category of all partial Steiner triple systems is universal, and every finite monoid occurs as the endomorphism monoid of a finite partial Steiner triple system.*

Partial Steiner quasigroups have a natural interpretation in a category dual to the class of pseudocomplemented distributive lattices - see T. Katriňák [4], or G. Grätzer, H. Lakser and R. Quackenbush [2]. The idea underlying Theorem 1.2. is used again in [1] where pseudocomplemented distributive lattices determined by their endomorphism monoids are investigated.

We begin by exhibiting a simple rigid PSTS that will be subsequently used to construct triple systems required in a proof of the latter two results. Free extensions of PSTS are then described and used to obtain Theorem 1.1 as a consequence of Theorem 1.2.

It is a pleasure to acknowledge a discussion with V. Koubek on free Steiner quasigroups and his suggestions that helped improve the presentation of this note.

2. A SIMPLE RIGID PSTS. An equivalence Θ on the underlying set X of a PSTS (X,T) is a Q-congruence if $\{x,y,z\},\{x',y',z'\} \in T$, $x \Theta x'$ and $y \Theta y'$ imply $z \Theta z'$, and if $x \Theta y$ implies $y \Theta z$ whenever $\{x,y,z\} \in T$; that is, if and only if Θ is a congruence of the PSQ $(X,\cdot)$ that corresponds to (X,T). A Q-congruence that does not collapse any triples from T will be called a S-congruence. A Q-simple PSTS or PSQ has no nontrivial Q-congruences; S-simplicity is analogously defined. Any Q-simple PSTS is clearly also S-simple.

Any finite Q-simple automorphism-free PSTS has no nontrivial S-endomorphisms, while the constant maps are its only nontrivial Q-endomorphisms. We construct such a PSTS (A,B) as follows.

Set $A = \{c,d\} \cup \{v_i : i \in 6\} \cup \{w_j : j \in 6\}$ and let $+$ denote addition modulo six. Then B consists of all triples $\{v_i, v_{i+1}, w_{i+1}\}$, $\{w_i, v_{i+1}, c\}$ for $i \in 6$, of all $\{v_j, v_{j+3}, d\}$ with $j \in 3$, and of the triple $\{w_2, w_4, w_5\}$. Figure 1 illustrates these triples as lines connecting elements of A.

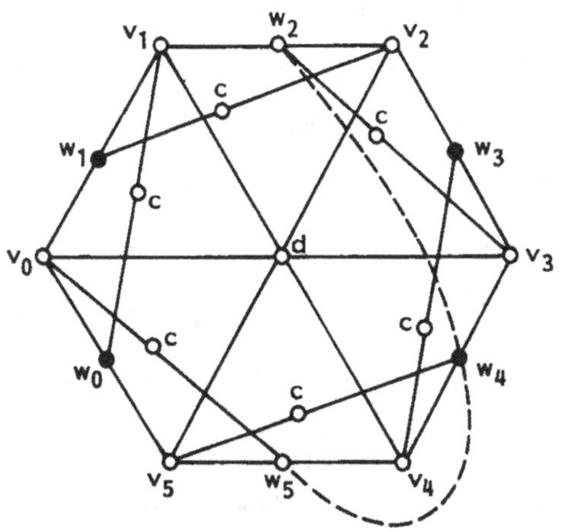

Figure 1

The following properties of (A,B) are immediately apparent:

$$|\{w_0,w_1,w_3,w_4\} \cap b| \leq 1 \quad \text{for every} \quad b \in B \ ; \tag{1}$$

if $d(a)$ denotes the number of distinct triples containing a, then

$$2 \leq d(a) \leq 6 \quad \text{for all} \quad a \in A \ , \tag{2}$$

$$d(a) = 6 \quad \text{if and only if} \quad a = c \ . \tag{3}$$

LEMMA 2.1. (A,B) <u>is</u> Q-<u>simple</u>.

<u>Proof</u>. We will consider Q-congruences of (A,B') where B' consists of all triples from B other than the triple $\{w_2,w_4,w_5\}$. Note that the mapping f that fixes c and d and satisfies $f(v_i) = v_{i+1}$, $f(w_i) = w_{i+1}$ is an automorphism of (A,B').

Let Θ be a Q-congruence of (A,B'). Since for every $i \in 6$, B' contains the triples $\{c,w_i,v_{i+1}\}$, $\{v_i,w_{i+1},v_{i+1}\}$, and also the triple $\{c,w_{i+1},v_{i+2}\}$, we see that

$$c \,\Theta\, v_i \quad \text{iff} \quad w_i \,\Theta\, w_{i+1} \quad \text{iff} \quad v_{i+1} \,\Theta\, v_{i+2} \quad \text{for all} \quad i \in 6 \ . \tag{4}$$

Assume $v_j \,\Theta\, v_{j+1}$ for some $j \in 6$. The triple $\{v_j,w_{j+1},v_{j+1}\} \in B'$ is thus contained in a class of Θ, so $w_{j+1} \,\Theta\, v_{j+1}$. Then $c \,\Theta\, w_{j+2}$ follows because $\{w_{j+1},c,v_{j+2}\}$ and $\{v_{j+1},w_{j+2},v_{j+2}\}$ lie in B'. Then using the triple $\{c,w_{j+2},v_{j+3}\}$ we obtain $c \,\Theta\, v_{j+3}$. Simultaneously, $v_j \,\Theta\, v_{j+1}$ applied to triples $\{v_j,d,v_{j+3}\}$, $\{v_{j+1},d,v_{j+4}\}$ yields $v_{j+3} \,\Theta\, v_{j+4}$. By transitivity of Θ,

$$v_j \,\Theta\, v_{j+1} \quad \text{implies} \quad c \,\Theta\, v_{j+4} \quad \text{for all} \quad j \in 6. \tag{5}$$

If $c \,\Theta\, w_k$ for some $k \in 6$, then the triple $\{w_k,c,v_{k+1}\}$ is contained in the class $[c]\Theta$ of Θ. Using (4) and (5) alternately, we conclude that $A \setminus \{d\} \subseteq [c]\Theta$ whenever Θ identifies two elements of any triple $t \in B'$ not containing d. If this is the case, then $d \,\Theta\, c$ follows from $\{v_0,d,v_3\} \in B'$, so that $\Theta = \iota$.

Assume Θ to be trivial on every triple $t \in B'$ not containing d. If $c \, \Theta \, d$ then $w_i \, \Theta \, v_{i+4}$ since $\{c, w_i, v_{i+1}\}$, $\{d, v_{i+4}, v_{i+1}\}$ lie in B' for every $i \in 6$. Hence the triples $\{w_0, v_0, v_5\}$, $\{v_4, w_5, v_5\}$, for instance, show that $\{v_0, c, w_5\}$ is collapsed by Θ, a contradiction, Hence $[c]\Theta = \{c\}$; also, v_j, w_{j+2} are not Θ-congruent for any j. Thus $v_i \, \Theta \, w_{i+k}$ only if $k = 3$ or $k = 4$. In either of these cases triples containing c may be used to show that Θ must collapse the triple $\{v_{i+4}, v_{i+5}, w_{i+5}\}$, contradicting the hypothesis. Furthermore, above arguments extended by the existence of f to any $i \in 6$ combine to prove that $v_i \, \Theta \, v_{i+k}$ only if $k = 3$. Note that, in this case $d \, \Theta \, v_{i+3}$ as well, and recall that triples $\{d, v_{i+1}, v_{i+4}\}$, $\{v_{i+3}, w_{i+4}, v_{i+4}\}$ belong to B'. Therefore $v_{i+1} \, \Theta \, w_{i+4}$ contrary to the immediately preceding argument. Altogether, $[v_i]\Theta = \{v_i\}$ for all $i \in 6$; since distinct w_i and w_j are linked by triples to c together with distinct v_{i+1} and v_{j+1}, respectively, we find also that $[w_j]\Theta = \{w_j\}$ for all $j \in 6$. The remaining vertex d is connected by a triple to each v_i from which it follows that $[d]\Theta = \{d\}$; this concludes the proof of simplicity of (A,B'). Since $B' \subseteq B$ the PSTS (A,B) is also simple.

LEMMA 2.2 (A,B) _has no nontrivial automorphisms_.

Proof. Clearly, $d(\alpha(a)) = d(a)$ for any automorphism α of (A,B); thus $\alpha(c) = c$ follows immeditely from (3). Since $\{v_i : i \in 6\} = \{a \in A : d(a) = 4\}$ there is a permutation $p : 6 \to 6$ such that $\alpha(v_i) = v_{p(i)}$, and it easily follows that $\alpha(d) = d$, and $\alpha(w_i) = w_{q(i)}$ for another permutation q of 6. Thus $\{w_{q(i)}, c, v_{p(i+1)}\} \in B$ for every $i \in 6$, so that $p(i+1) = q(i)+1$ for all $i \in 6$.

Consider the α-images $\{v_{p(j)}, w_{q(j+1)}, v_{p(j+1)}\}$, $\{v_{p(j+1)}, q(j+2), v_{p(j+2)}\}$ of consecutive triples in B. These triples intersect in $\{v_{p(j+1)}\}$ and lie in B. If $p(j+2) \neq q(j+2)$ then $p(j+2)+1 = q(j+2) = p(j+1)$; since q is one-to-one, $q(j+1) \neq p(j+1)$ and thus also $p(j) = q(j+1) = p(j+1)+1$. Altogther, the contradictory $q(j+2) = p((j+1)+1)+1 = q(j+1)+2 = p(j+1)+3 = q(j+2)+3$ is obtained; thus $p(j+2) = q(j+2)$ for all $j \in 6$. Consequently, $p(i) = q(i) = i+p(0) = i+m$ for all $i \in 6$; hence $\{w_{2+m}, w_{4+m}, w_{5+m}\} \in B$, and this is true only if $m = 0$. Therefore $p = q$ is the identity mapping and $\alpha = id_A$ follows.

COROLLARY 2.3. The only nonconstant Q-endomorphism of (A,B) is the identity mapping.

Proof. Immediate from 2.1 and 2.2.

3. THE CONSTRUCTION. Any pair (X,R) with $R \subseteq X^2$ will be called a digraph; a mapping $f: X \to X'$ is a compatible map of (X,R) into (X',R') if $(f(x),f(y)) \in R'$ for all $(x,y) \in R$. Let G denote the category of all compatible mappings between connected digraphs satisfying $p_0(r) \neq p_1(r)$ for all $r \in R$ and such that $p_0(R) = p_1(R) = X$; here p_i denotes the restriction to R of the i-th projection $X^2 \to X$. The category **G** is universal [7].

A one-to-one functor $\Phi : \mathbf{G} \to \mathbf{P}$ is a full embedding if for every morphism $p : \Phi(G) \to \Phi(G')$ in the category **P** there is a **G**-morphism $g : G \to G'$ such that $p = \Phi(g)$. To prove that the category **P** of all partial Steiner quasigroups and their Q-homomorphisms is almost universal we will construct a functor $\Phi : \mathbf{G} \to \mathbf{P}$ such that $\Phi(g)$ is a nonconstant Q-homomorphism for every compatible g, while each Q-homomorphism $p : \Phi(G) \to \Phi(G')$ not of the form $\Phi(g)$ will be constant. In other words, a class of PSTS will be exhibited for which all nonconstant Q-homormophisms form a universal category.

Let (A,B) be the Q-simple rigid PSTS constructed in the previous section. For any digraph $G = (X,R) \in \mathbf{G}$ set $E_r = (A\times\{r\}, B_r)$, where for every $r \in R$ the set B_r consists of all triples $\{(x,r),(y,r),(z,r)\}$ such that $\{x,y,z\} \in B$. Thus E_r is isomorphic to (A,B), and the disjoint union $\bigcup(E_r\colon r \in R)$ is a PSTS.

Let $\Theta = \Theta(G)$ denote the least Q-congruence on $E(G) = \bigcup(E_r\colon r \in R)$ such that $(w_{4i},r)\ \Theta(G)\ (w_{4j},s)$ and $(w_{2i+1},r)\ \Theta(G)\ (w_{2j+1},s)$ whenever $i,j \in 2$ and $p_i(r) = p_j(s)$. Thus, for instance, $(w_4,(x,y))\ \Theta(G)\ (w_0,(y,z))$ and $(w_3,(x,y))\ \Theta(G)\ (w_1,(y,z))$ if $(x,y),(y,z) \in R$.

For every $x \in X$ set $E(x) = \{(w_{4i},r)\colon i \in 2,\ p_i(r) = x\}$ and $O(x) = \{(w_{2i+1},r)\colon i \in 2,\ p_i(r) = x\}$. From (1) and the definition of $\mathbf{G}$ it follows that $\Theta \restriction E_r = \omega$, and that

$$E(x),\ O(x) \text{ are the only nontrivial classes of } \Theta(G). \tag{6}$$

It is clear that $\Theta(G)$ is also an S-congruence. Furthermore, $\Phi(G) = \bigcup(E_r\colon r \in R)/\Theta(G)$ is a PSTS corresponding naturally to a PSQ. For every triple t of $\Phi(G)$ there is a unique $r \in R$ and a unique triple $\{x,y,z\} \in B$ such that $t = \{\phi(x,r),\phi(y,r),\phi(z,r)\}$, where $\phi = \phi_G$ is the S-homomorphism with kernel $\Theta(G)$.

If $g : G \to G' = (X',R')$ is compatible then ${}^2g(x,y) = (g(x),g(y)) \in R'$ for every $(x,y) \in R$. Therefore the mapping $g^+ : E(G) \to E(G')$ defined by $g^+(a,r) = (a,{}^2g(r))$ is an S-homomorphism. If $p_i(r) = p_j(s)$ then $\phi' g^+(w_{4i},r) = \phi'(w_{4i},{}^2g(r)) = \phi'(w_{4j},{}^2g(s)) = \phi' g^+(w_{4j},s)$ follows from $p_i({}^2g(r)) = g(p_i(r)) = g(p_j(s)) = p_j({}^2g(s))$; similarly, $\phi' g^+(w_{2i+i},r) = \phi' g^+(w_{2j+1},s)$. The kernel of $\phi' \circ g^+$ thus must contain $\Theta(G)$, and there is a unique PS-homomorphism $\Phi(g)$ for which $\Phi(g)\circ\phi = \phi'\circ g^+$. It is easily verified that Φ is a faithful functor and that no $\Phi(g)$ is a constant mapping.

Assume $h : \Phi(G) \to \Phi(G')$ to be a Q-homomorphism and consider the restriction h_r of h to some $\phi(E_r) \cong (A,B)$. Since ϕ is one-to-one on E_r, without loss of generality we may take (A,B) as the domain of h_r.

If $\mathrm{Ker}(h_r) \neq \omega$ then h_r is constant by Lemma 2.1; in particular, $h\phi(w_4,r) = h\phi(w_3,r) = h\phi(w_0,r) = h\phi(w_1,r)$. For $s \in R$ with $p_i(r) = p_j(s)$ then $h\phi(w_{4j},s) = h\phi(w_{4i},r) = h\phi(w_{2i+1},r) = h\phi(w_{2j+1},s)$, so that h_s is not one-to-one and, consequently, $h_r \cup h_s$ is a constant mapping. Since all digraphs $G \in \mathbf{G}$ are connected, the mapping h is constant.

Any nonconstant h must therefore be one-to-one on every copy $\phi(E_r) \subseteq \Phi(G)$ of (A,B). The description of $\Theta(G)$ together with (2) and (3) imply that $d(q) > 6$ in $\Phi(G)$ only if either $q \in \phi(\{c\}\times R)$ or $q \in \phi(\{w_0,w_1,w_3,w_4\}\times R)$ for some $i \in 2$. If $f : (A,B) \to \Phi(G')$ is one-to-one then $d(f(c)) > 6$; suppose that $f(c) = \phi(w_{4i},s)$ first.

For every triple $\{w_k,c,v_{k+1}\}$ there are uniquely determined $s(k) \in R'$ and $\{x,y,z\} \in B$ such that $t_k = \{f(w_k),f(c),f(v_{k+1})\} = \{\phi'(x,s(k)),\phi'(y,s(k)),\phi'(z,s(k))\}$. From (1) it now follows that $\{f(w_k),f(v_{k+1})\} \subseteq \phi'(E_{s(k)} \setminus \{w_0,w_1,w_3,w_4\}\times\{s(k)\}) = A(k)$ and similarly $\{f(w_{k-1}),f(v_k)\} \subseteq A(k-1)$. The triple $\{f(v_{k-1}),f(w_k),f(v_k)\}$ of $\Phi(G')$, however, intersects both $A(k)$ and $A(k-1)$; this is possible only if $s(k) = s(k-1)$. Hence there is a unique $s \in R'$ for which $f(A\setminus\{d\}) \subseteq \phi'(E_s)$; since $\phi'(E_s)$ contains the first two members of $\{f(v_0),f(v_3),f(d)\}$, we conclude that $f(A) \subseteq \phi'(E_s) \cong (A,B)$. This, however, contradicts Corollary 2.3 because $f(c) = \phi'(w_{4i},s)$. An analogous argument shows that $f(c) \neq \phi'(w_{2i+1},s)$.

The remaining case is that of $f(c) = \phi'(c,s)$; the triples $\{f(w_k),\phi'(c,s),f(v_{k+1})\}$ of $\Phi(G')$ are entirely contained in $\phi'(E_s)$ so that $f(A\setminus\{d\}) \subseteq \phi'(E_s)$, and $f(A) \subseteq \phi'(E_s)$ follows as in the above paragraph. Lemma 2.2 now yields $f(a) = \phi'(a,s)$ for every $a \in A$.

As a result, for every $r \in R$ there exists a unique $H(r) \in R'$ such that $h\phi(a,r) = \phi'(a,H(r))$ for all $a \in A$. Consequently, from (6) it follows that the mapping $H : R \to R'$ satisfies

$$\text{if} \quad i,j \in 2 \quad \text{and} \quad p_i(r) = p_j(s) \quad \text{then} \quad p_iH(r) = p_jH(s). \qquad (7)$$

Recall that for every $x \in X$ and $i \in 2$ there exists some $r \in R$ with $p_i(r) = x$. If also $p_i(s) = x$, then $p_iH(r) = p_iH(s)$ by (7); thus, for $i \in 2$, there are mappings $g_i : X \to X'$ satisfying $g_i \circ p_i = p_i \circ H$. If $p_0(r) = x = p_1(r')$, then $g_0(x) = g_0p_0(r) = p_0H(r) = p_1H(r') = g_1p_1(r') = g_1(x)$, by (7) again. Therefore $g_0 = g_1 = g$ is a compatible mapping of G into G', and $H = {}^2g$. Finally, $\Phi(g) = h$ since ϕ is onto.

This concludes the proof of Theorem 1.2 in view of the representability of any finite monoid as the endomorphism monoid of a finite digraph from **G**, see [7], and because Φ preserves finiteness. Since each PQ-homomorphism $\Phi(g)$ is one-to-one on every triple of its domain, Theorem 1.3 has also been verified.

Observe that all triples of $\Phi(G')$ intersecting the image of $\Phi(g)$ are, in fact, $\Phi(g)$-images of triples from $\Phi(G)$. Thus $\Phi(g)$ is an onto mapping whenever ${}^2g$ maps R onto R'. Similarly, $\Phi(g)$ is one-to-one whenever g is.

4. FREE EXTENSIONS. Recall that a partial Steiner quasigroup may be viewed as a partial Steiner triple system $(X,T) = S$ in which $X = \bigcup T$ and $|t \cap u| \neq 2$ for all $t,u \in T$.

Let $N(S)$ denote the set of all two-element subsets of X not contained in any triple $t \in T$, and let $n : N(S) \to N$ be a bijection onto a set N disjoint from X; for a pair $p = \{x,y\}$ we will often write $n(x,y)$ instead of $n(p)$.

Set $X^+ = X \cup N$ and $T^+ = T \cup M$ where $M = \{\{x,y,n(x,y)\}: \{x,y\} \in N(S)\}$. The pair $S^+ = (X^+,T^+)$ clearly is a PSTS whose restriction to X is the original system S; the inclusion mapping $X \subseteq X^+$ is a partial S-homomorphism. Furthermore, for every $u \in X^+ \setminus X$ there is unique triple $t = \{x,y,u\} \in T^+$, namely $t = \{x,y,n(x,y)\}$; note that $t \cap X = \{x,y\}$.

Let $f : S \to (Z,U)$ be a partial Q-homomorphism into a complete STS. We define an extension f^+ of f to X^+ as follows. For every $u \in X^+ \setminus X$ there is a unique triple $\{x,y,n(x,y)\} \in T^+$ containing $u = n(x,y)$; either $f(x) = f(y) = z$ or there exists a unique $z \in Z$ with $\{f(x),f(y),z\} \in U$; in both cases set $f^+(n(x,y)) = z$. The mapping f^+ clearly is the only partial Q-homomorphism of S^+ into (Z,U) extending f.

For an arbitrary partial Steiner triple system S set $S_0 = S$ and $S_{k+1} = (S_k)^+$ for each $k \geq 0$. Since the inclusion $S_k \subseteq S_{k+1}$ is an S-homomorphism, we may define $X^* = \bigcup(X_k : k \geq 0)$, $T^* = \bigcup(T_k : k \geq 0)$, and $S^* = (X^*, T^*) = F(S)$. The set T^* clearly consists of three-element subsets of X^*. For every $x \in X^*$, define $\mathrm{rank}(x) = k$ if k is the least integer such that $x \in X_k$. Thus for any $t = \{x,y,z\} \in T^*$ either $t \subseteq X_0 = X$ or, say, $0 < k = \mathrm{rank}(z) > \mathrm{rank}(y), \mathrm{rank}(x)$. In the latter case $\{x,y,z\} \subseteq X_k$ and $z \in X_k \setminus X_{k-1}$, so that $\{x,y\} \subseteq X_{k-1}$ and $z = n(x,y)$. Therefore any triple $t \in T^*$ not contained in X_0 has a unique element of highest rank, and for every $z \in X^* \setminus X_0$ there is a single triple in T^* amongst whose elements z has the highest rank.

A Steiner quasigroup R is a <u>free extension</u> of a partial Steiner quasigroup Q if there is a partial Q-homomorphism $e : Q \to R$ such that for every partial Q-homomorphism $f : Q \to W$ into a complete Steiner quasigroup W there is a unique (complete) Q-homomorphism f^* such that $f = f^* \circ e$.

LEMMA 4.1. *If S is a PSTS corresponding to a partial Steiner quasigroup Q then $F(S)$ is a STS that corresponds to the free extension of Q.*

Proof. Clearly $X^* \subseteq \bigcup T^*$, that is, every vertex of $F(S)$ is contained in a triple. Let $x,y \in X^*$ be distinct and $k = \mathrm{rank}(x) < \mathrm{rank}(y) = n$; if $\{x,y\}$ is not contained in a triple $t \in T_n$ then there is a unique $z \in X_{n+1}$ with $\{x,y,z\} \in T_{n+1} \subseteq T^*$. In either case $\{x,y\}$ is contained in a triple of $F(S)$. Triples $\{x,y,z\}$, $\{x,y,v\}$ of $F(S)$ are triples of some PSTS S_m and hence $z = v$. Therefore $F(S)$ is a Steiner triple system.

For a PQ-homomorphism $f : S_0 \to W$ set $f_0 = f$ and $f_{k+1} = (f_k)^+$ when $k \geq 0$. In view of the remarks preceding this lemma it is easily seen that $f^* : F(S) \to W$ defined by $f^* = \bigcup(f_k : k \geq 0)$ is a unique Q-homomorphism extending f.

Free extensions of various combinatorial geometries have been investigated by M. Funk and A. Schleiermacher.

LEMMA 4.2. _If $P \subseteq F(S)$ is a finite PSTS whose every element belongs to at least two of its triples then_ $P \subseteq S$.

Proof. Let m be the least integer for which P is contained in X_m. If $m > 0$ then there is some $p \in P \cap (X_m \setminus X_{m-1})$. Furthermore, all triples of P are contained in T_m since any $t \in T^* \setminus T_m$ has an element of rank greater than m. There is, however, exactly one triple $s \in T_m$ of which p is a member, contrary to the hypothesis. Hence $m = 0$, that is, $P \subseteq S$.

We are now ready to prove Theorem 1.1. For any $G \in \mathbf{G}$ set $\Psi(G) = F(\Phi(G))$ where Φ is the functor defined in Section 3. Recall that $\Phi(g)$ is a partial Q-homomorphism for any compatible mapping $g : G \to G'$. Since the inclusion $\Phi(G') \subseteq F(\Phi(G')) = \Psi(G')$ also is a PQ-homomorphism, there exists a unique Q-homomorphism $\Psi(g) : \Psi(G) \to \Psi(G')$ extending $\Phi(g)$; it is easy to see that Ψ is a faithful functor and that $\Psi(g)$ is a nonconstant homomorphism for every compatible mapping g.

Consider a Q-homomorphism $f : \Psi(G) \to \Psi(G')$. If there exists a copy $\phi(E_r) \subseteq \Phi(G)$ of (A,B) on which the kernel of f is non-trivial then $f \upharpoonright \Phi(G)$ is a constant mapping and it is easy to see that the free extension f of $f \upharpoonright \Phi(G)$ is forced to be constant as well. Thus every restriction $f_r = f \upharpoonright \phi(E_r)$ is one-to-one for any nonconstant Q-homomorphism f. The range of every f_r must therefore be contained in $\Phi(G')$ by Lemma 4.2, that is, $f \upharpoonright \Phi(G)$ is a nonconstant partial Q-homomorphism into $\Phi(G')$; hence $f \upharpoonright \Phi(G) = \Phi(g)$ for some compatible $g : G \to G'$. From Lemma 4.1 it now follows that $f = F(\Phi(g)) = \Psi(g)$ as was to be shown. The almost universality of the variety of Steiner quasigroups is thus established.

Since S-homomorphisms are just one-to-one Q-homomorphisms, every S-homomorphism $\sigma : \Psi(G) \to \Psi(G')$ is of the form $\sigma = \Psi(g)$ for some compatible $g : G \to G'$. To investigate S-homomorphisms of this kind we need the concept of an _exact_ partial S-homomorphism $h : (X,T) \to (X',T')$: it is a PS-homomorphism satisfying

$$\{x,y\} \subseteq t \quad \text{for some} \quad t \in T \quad \text{whenever} \quad \{h(x),h(y)\} \subseteq t' \quad \text{for some} \quad t' \in T'. \tag{8}$$

To see that $\Phi(g)$ is exact for every one-to-one compatible g, recall that each triple t' of $\Phi(G')$ is of the form $t' = \{\phi'(a,s),\phi'(b,s),\phi'(c,s)\}$ with $\{a,b,c\} \in B$ and $s \in R'$, and that $\Phi(g)\phi(d,r) = \phi'(d,{}^2g(r))$. Let $\{\Phi(g)\phi(d,r),\Phi(g)\phi(e,q)\} \subseteq t'$. Since g is assumed to be one-to-one there is a unique $r \in R$ with ${}^2g(r) = s$. Thus $q = r$ and, for instance, $\Phi(g)\phi(d,r) = \phi'(a,s)$, $\Phi(g)\phi(e,r) = \phi'(b,s)$. Since ϕ' is one-to-one on E_s we conclude that $d = a$ and $e = b$. Consequently, the elements $\phi(d,r)$, $\phi(e,r)$ are contained in a triple $\{\phi(a,r),\phi(b,r),\phi(c,r)\}$ of $\Phi(G)$ as was to be shown.

To prove that $\Psi(g)$ is an S-homomorphism for each one-to-one compatible g we need the following claim.

LEMMA 4.3. $F(h) : F(S) \to F(S')$ *is an* S-*homomorphism for any exact partial* S-*homomorphism* $h : S \to S'$.

Proof. The exactness of h means that $\{x,y\} \in N(S)$ if and only if $\{h(x),h(y)\} \in N(S')$; therefore $h^+(n(x,y)) = n(h(x),h(y))$, so that the one-step extension h^+ of h is an exact partial S-homomorphism again. A simple induction based on this observation is all that is needed to conclude the proof.

When restricted to the category $\mathbf{G}_1$ of all one-to-one compatible mappings, $\Psi = F\circ\Phi$ is therefore a full embedding of $\mathbf{G}_1$ into the category of all S-homomorphisms. For any category $\mathbf{A}$ of algebras there is a full embedding $\Sigma : \mathbf{A} \to \mathbf{G}$ preserving one-to-one maps [7]; the result below then follows immediately.

THEOREM 4.4. *For any category* $\mathbf{A}_1$ *of all one-to-one homomorphisms of algebras of a given type there is a full embedding* Δ_1 *of* $\mathbf{A}_1$ *into the category of all* S-*homomorphisms of Steiner triple systems. In particular, there is a proper class of nonisomorphic* S-*rigid Steiner triple systems, and for every group* G *there exists a proper class* $(S_i : i \in I)$ *of nonisomorphic* STS *such that* $\mathrm{End}(S_i) \cong G \cong \mathrm{Aut}(S_i)$ *for all* $i \in I$.

The almost universality of Steiner quasigroups as established here applies to infinite quasigroups only. While every finite group appears as the full automorphism group of a finite STS - see E. Mendelsohn [6], representability of finite monoids by non-constant endomorphisms of finite Steiner quasigroups remains undecided. Since every finite monoid occurs as the endomorphism monoid of a finite graph, the existence of a locally finite almost universal variety **V** if Steiner quasigroups would answer the question provided a constructed full embedding $\Omega : \mathbf{G} \to \mathbf{V}$ maps finite graphs to finitely generated algebras. Alternately, is there an almost universal SQ variety generated by a single finite Steiner quasigroup?

REFERENCES

[1] M. E. Adams, V. Koubek, and J. Sichler, Endomorphisms of pseudocomplemented distributive lattices (with applications to Heyting algebras), to appear in Trans. Amer. math. Soc.

[2] G. Grätzer, H. Lakser, and R. Quackenbush, On the lattice of quasivarieties of distributive lattices with pseudocomplementation, Acta math. Acad. Sci. Hungar. 42 (1980), 257-263.

[3] J. Ježek, T. Kepka, and J. Němec, The category of totally symmetric quasigroups is binding, Acta Univ. Carolinae, Math. Phys. 19 (1978), 63-64.

[4] T.Katriňák, Über eine Konstruktion der distributiven pseudokomplementären Verbande, Math. Nachr. 53 (1971), 85-99.

[5] E. Mendelsohn, On the groups of automorphisms of Steiner triple and quadruple systems, J. combinat. Theory, Ser. A 25 (1978), 97-104.

[6] E. Mendelsohn, Every (finite) group is the automorphism group of a (finite) strongly regular graph, Ars comb. 6 (1978), 75-86.

[7] A. Pultr and V. Trnková, <u>Combinatorial, Algebraic and Topological Representations of Groups, Semigroups and Categories</u>, North-Holland Publishing Co., Amsterdam, 1980.

PRINCIPAL CONGRUENCE FORMULAS IN ARITHMETICAL VARIETIES

A. F. Pixley
Harvey Mudd College
Claremont, California 91711

1. INTRODUCTION AND SUMMARY. Following Baldwin and Berman [1] we define the congruence formulas of a variety V to be the positive existential 4-ary formulas $\varphi(u,v,x,y)$ of V such that $V \models \varphi(u,v,x,y) \rightarrow x = y$. Using this terminology Mal'cev's lemma ([6], page 54, Theorem 3) then asserts that for $u,v,x,y \in A$ where $\underset{\sim}{A}$ is an algebra in V,

$$x \equiv y \quad \theta(u,v) \quad \text{iff} \quad \underset{\sim}{A} \models \varphi(u,v,x,y) \tag{1.1}$$

for some congruence formula φ of V. (In fact Mal'cev's lemma asserts that φ may be taken to be of a special form $\varphi_1 := (\exists\bar{z})\varphi'(u,v,x,y,\bar{z})$ where φ' is a conjunction of equations of V.[1]) In general unrestricted varieties the particular syntactical form of congruence formulas, and the terms appearing in them, vary not only with the choice of algebra $\underset{\sim}{A} \in V$ but also with the individual quadruples of elements u,v,x,y in A. Consequently we are interested in the general problem of determining conditions on V which will insure some uniform structure for a set of congruence formulas sufficient for defining all principal congruences -- and thereby, of course, obtaining insight into the structure of V. In this context, when we say that a congruence formula φ defines principal congruences on a subset S of A, $\underset{\sim}{A} \in V$, we mean that (1.1) holds for all $u,v,x,y \in S$. The most uniform situation one might expect is when a single formula suffices to define principal congruences throughout V, i.e.: there is no dependence on $\underset{\sim}{A}$ or $u,v,x,y \in A$. In this case V is said to have definable principal congruences (DPC). This concept, introduced by Baldwin and Berman [1] initiated the general study of congruence formulas. The general problem of uniform syntactical structure is

[1] More accurately, Baldwin and Berman reserve the term "congruence formula" for the formulas of the form φ_1 and refer to our "congruence formulas" as "weak congruence formulas". Our use of the shorter terminology should cause no confusion.

considered by Baldwin and Berman in [2] and via a different approach in Fried, Grätzer, and Quackenbush [4].

The purpose of the present paper is to examine the nature of congruence formulas in arithmetical varieties. An immediate consequence (Corollary 4.2) of our most general result asserts that in such varieties V, principal congruences may be defined on an arbitrarily large finite subset of an algebra $\underset{\sim}{A} \in V$ by a single formula of the form

$$\varphi(u,v,x,y) := (\exists\bar{z})[t(u,v,x,\bar{z}) = t(u,v,y,\bar{z}) \wedge t(u,u,x,\bar{z}) = x \wedge t(u,u,y,\bar{z}) = y] \tag{1.2}$$

where $t = t_F$ is a term depending on the subset F. Our general result (Theorem 4.1) asserts that t_F can be chosen so that for suitable $\bar{a} = (a_1,\ldots,a_k)$, $a_i \in A$, the polynomial $t_F(u,v,x,\bar{a})$ satisfies the following two conditions for all $u,v,x,y \in F$:

$$\begin{aligned} &\text{a)}\quad t_F(u,u,x,\bar{a}) = x, \quad t_F(u,v,u,\bar{a}) = u, \quad t_F(u,v,v,\bar{a}) = u, \\ &\text{b)}\quad x \equiv y\ \theta(u,v) \Leftrightarrow t_F(u,v,x,\bar{a}) = t_F(u,v,y,\bar{a}). \end{aligned} \tag{1.3}$$

To understand the significance of (1.3) recall that V is arithmetical iff there is a single <u>term</u> $t(u,v,x)$ satisfying (1.3) a) identically in V. Such a term does not in general satisfy (1.3) b). An important case when it does is when V is a discriminator variety, which means simply that in addition to (1.3) a) for u,v,x in a subdirectly irreducible (SI) member of V, we have

$$u \neq v \Rightarrow t(u,v,x) = u. \tag{1.4}$$

In this case V has definable principal congruences and, in fact,

$$x \equiv y\ \theta(u,v) \Leftrightarrow \underset{\sim}{A} \models t(u,v,x) = t(u,v,y), \tag{1.5}$$

so that (1.2) simplifies to $\varphi(u,v,x,y) := t(u,v,x) = t(u,v,y)$. (See McKenzie [12].)

For a general arithmetical variety the polynomials $t_F(u,v,x,\bar{a})$ behave somewhat like the discriminator. In particular if $u,v,x \in F$ are also in the same monolith class of an SI member of V then (1.4) holds (Corollary 4.3). (The <u>monolith</u> of a non-trivial SI algebra is the unique congruence covering ω.) This fact is evidence for what we might call a general "principle" of arithmetical varieties: monolith classes of SI members of V behave "like" simple members of V (i.e.: like non-trivial SI members having a single monolith class). We will explore this "principle" a little further in Section 6 and show that if V is finitely generated arithmetical and $\mathrm{con}(\underset{\sim}{A})$ is a chain for each SI $\underset{\sim}{A} \in V$, then V has DPC (Theorem 6.5). E. Kiss [8] has recently given another proof of this.

In case V is locally finite, we have much stronger results. Here the polynomials $t_F(u,v,x,\bar{a})$ can, for all m-generated $(m \in \omega)$ members of V, be obtained uniformly from a single (m+3)-ary term $t(u,v,x,\bar{z})$ (Theorem 5.1). When specialized to finitely generated semi-simple varieties this fact coincides with an earlier result of S. Burris [3]. Moreover if V is finitely generated then for each finite $\underset{\sim}{A} \in V$ there is a unique polynomial satisfying (1.3) (with $\bar{a}$ generating $\underset{\sim}{A}$) iff V is semi-simple (and thus has DPC) (Theorem 5.1).

2. TERMINOLOGY AND BACKGROUND. For terminology we shall generally follow Grätzer [6] with the exception that we shall refer to his algebraic functions as <u>polynomial functions</u> and to his polynomials as <u>terms</u>. For an algebra $\underset{\sim}{A} = (A,F)$ with universe A and set of operations F, the congruence lattice of $\underset{\sim}{A}$ is denoted by $\mathrm{con}(\underset{\sim}{A})$. For $a \in A$ and $\varphi \in \mathrm{con}(\underset{\sim}{A})$, a/φ denotes the φ-class of a. We also use this notation for arbitrary equivalence relations on any set. For any set A a sublattice L of the lattice $\underset{\sim}{E}(A)$ of equivalence relations on A is an ω-ι sublattice if it contains both ω (identity) and ι (diversity). If L is such a sublattice, $\theta(a,b)$ denotes the principal equivalence determined by a,b (i.e.: the least relation in L collapsing a and b). If $\varphi \in L$, let L_φ denote the sublattice of $E(A/\varphi)$ which is naturally isomorphic to the interval $[\varphi,\iota]$ of L. We will use the following result due to Köhler and Pigozzi [9].

2.1 LEMMA. *Let* A *be a set and* L *an* ω-ι *sublattice of* $\underset{\sim}{E}(A)$. *Then for all* $a,b,c,d \in A$ *and* $\varphi \in L$, $\theta(c,d) \le \theta(a,b) \vee \varphi$ *in* L *iff* $\theta(c/\varphi,d/\varphi) \le \theta(a/\varphi,b/\varphi)$ *in* L_φ.

(Köhler and Pigozzi establish (2.1) for algebras but their proof applies just as well to the present formulation).

An algebra is <u>m-generated</u> if it is generated by a set of no more than m elements. A variety V is <u>locally finite</u> if each finitely generated member is finite. V is <u>(locally) semi-simple</u> if each (finitely generated) non-trivial SI member is simple. A sublattice L of $\underset{\sim}{E}(A)$ is <u>arithmetical</u> if it is distributive and the join of any pair of elements is their relation product. The following characerization is well known (see [14]).

2.2 LEMMA. *A sublattice* L *of* $\underset{\sim}{E}(A)$ *is arithmetical iff it satisfies the "Chinese remainder theorem":*

For each finite set $\theta_1,\ldots,\theta_n \in L$ *and* $a_1,\ldots,a_n \in A$, *the system*

$$x \equiv a_i(\theta_i) \qquad i = 1,\ldots,n$$

is solvable in A *iff for* $1 \le i < j \le n$,

$$a_i \equiv a_j \qquad (\theta_i \vee \theta_j).$$

An algebra $\underset{\sim}{A}$ is arithmetical if $con(\underset{\sim}{A})$ is arithmetical and a variety is arithmetical if all of its members are. Arithmetical varieties can be characterized in a number of interesting ways [15]. For the present paper we require only the following selection.

2.3 LEMMA. *For a variety* V *the following are equivalent:*

a) V *is arithmetical.*

b) For some term $t(x,y,z)$ *of* V, $t(x,x,z) = z$, $t(x,y,x) = x$, $t(x,y,y) = x$ *are equations of* V.

c) For any algebra $\underset{\sim}{A} \in V$, *integer* $k \ge 1$, *and finite partial function* f *in* $A^k \to A$, f *has an interpolating polynomial iff* f *is compatible with* $con(\underset{\sim}{A})$ *where defined. (*f *is a finite partial function means that domain* f *is a finite subset of* A^k; *a polynomial* p *interpolates such an* f *if the restriction* $p|_{\text{dom } f}$ *of* p *to the domain of* f *coincides with* f; f *is compatible with* $con(\underset{\sim}{A})$ *where defined means that if* $(x_1,\ldots,x_k)$ *and* $(y_1,\ldots,y_k)$ *are in domain* f *and* $x_i \equiv y_i(\theta)$, $\theta \in con(\underset{\sim}{A})$, $i = 1,\ldots,k$, *then* $f(x_1,\ldots,x_k) \equiv f(y_1,\ldots,y_k)(\theta)$.*)*

3. A FUNDAMENTAL LEMMA. Most of our results are applications of the following lemma (3.1) which we believe is fundamental in understanding arithmetical algebras and varieties. Notation: If $f:A^k \to A$ is a function which is compatible with an equivalence relation φ on A let f_φ denote the function $f_\varphi:(A/\varphi)^k \to A/\varphi$ induced by f, i.e.: defined by

$$f_\varphi(x_1/\varphi,\ldots,x_k/\varphi) = f(x_1,\ldots,x_k)/\varphi.$$

3.1 LEMMA. *Let* A *be a set and* L *a finite arithmetical* ω-ι *sub-*

lattice of $\underset{\sim}{E}(A)$. *Then there is a function* $f: A^3 \to A$ *having the following properties:*

a) f *is compatible with* L.

For all $\varphi \in L$ *and* $u,v,x,y \in A$,

b) $f_\varphi(u/\varphi,u/\varphi,v/\varphi) = v/\varphi$, $f_\varphi(u/\varphi,v/\varphi,u/\varphi) = u/\varphi$, $f_\varphi(u/\varphi,v/\varphi,v/\varphi) = u/\varphi$,

c) $x/\varphi \equiv y/\varphi\ \theta(u/\varphi,v/\varphi)$ *in* L_φ *iff* $f_\varphi(u/\varphi,v/\varphi,x/\varphi) = f_\varphi(u/\varphi,v/\varphi,y/\varphi)$.

d) There is only one function $f: A^3 \to A$ *satisfying a)-c) iff* L *if complemented (i.e.: is a Boolean lattice).*

A partial version of (3.1), containing only statements a) and b), was established in [14] (Lemma 3.1) and was formulated for algebras with $L = \mathrm{con}(\underset{\sim}{A})$. In that case as here, however, the proof depends only on L being an arithmetical sublattice of $\underset{\sim}{E}(A)$. The proof (below) of the present version is an elaboration of the proof of the partial version. I. Korec showed that the finiteness of L in the partial version can be dropped provided $|A| \le \omega$ and L is complete, and that, in general, the partial version is false if $|A| > \omega$. (See [10] and [11] for this and related results.) An alternative proof to Korec's extension appears in Kaarli [7]. In each of these proofs the result is obtained by representing A as the union of an expanding sequence of finite sets F_n, $n \in \omega$. It is then shown that for each F_n a finite partial function $f_n: F_n^3 \to A$ is definable which is compatible with L (where defined) and which satisfies statements a) and b) where defined. Moreover it is then shown that if $F_n \subset F_m$ then f_m can be taken as an extension of f_n to F_m^3 so that f can be defined as $f = \cup_n f_n$. In the present version of the lemma this proof strategy fails since satisfaction of statement c), in general, would require that each successively constructed f_m be redefined over its entire domain F_m and cannot be taken simply as an extension of the previous f_n.

Proof of Lemma 3.1. First notice that if f satisfies a) and b) then the "if" direction of c) is immediate; hence we need only prove the lemma for the "only if" direction of c). We do this as in [14] by induction on the height of L, fully defining $f_\varphi: (A/\varphi)^3 \to A/\varphi$ at successive levels of L, and begin at the top of L, defining f_φ, for maximal $\varphi \in L$, as the discriminator, i.e.:

$$
\begin{aligned}
f_\varphi(u/\varphi,v/\varphi,x/\varphi) &= x/\varphi \quad \text{if} \quad u/\varphi = v/\varphi, \\
&= u/\varphi \quad \text{if} \quad u/\varphi \neq v/\varphi.
\end{aligned}
$$

Notice that each such f_φ, φ maximal, satisfies both b) and c). Conversely, if φ is maximal and f_φ satisfies b) and c) then $u/\varphi \neq v/\varphi$ implies $f_\varphi(u/\varphi,v/\varphi,x/\varphi) = f_\varphi(u/\varphi,v/\varphi,u/\varphi) = u/\varphi$ for all x/φ so that f_φ is the discriminator. Hence for any f satisfying a), b) and c), f_φ is uniquely determined (as the discriminator) for maximal $\varphi \in L$. Note that, in particular, this definition establishes the lemma for the case height $L = 1$.

Now let $P(n)$ be the statement:

> For each φ of height $\geq n$ in L, there is a function $f_\varphi:(A/\varphi)^3 \to A/\varphi$ satisfying b) and c) on A/φ and
>
> $$\text{if } \varphi \leq \theta \text{ then for all } u,v,x \in A, \quad f_\varphi(u/\varphi,v/\varphi,x/\varphi) \subset f_\theta(u/\theta,v/\theta,x/\theta). \tag{3.2}$$

Assume $P(n)$, $0 < n <$ height L and let $\varphi \in L$ be of height $n-1$ and suppose $\varphi(1),\ldots,\varphi(k)$ are all of the elements of L (of height n) which cover φ.

<u>Case 1</u>. $k > 1$. (This is the only case which will ever occur iff L is complemented.) For each $u,v,x \in A$ pick $w_i \in f_{\varphi(i)}(u/\varphi(i), v/\varphi(i),x/\varphi(i))$. From (3.2) it follows that $w_i \equiv w_j(\varphi(i) \vee \varphi(j))$ for $1 \leq i < j \leq k$. Hence by Lemma 2.2 there is a $w \in A$ such that $w \equiv w_i(\varphi(i))$, $i = 1,\ldots,k$, and thus

$$w/\varphi \subset w/\varphi(i) = f_{\varphi(i)}(u/\varphi(i),v/\varphi(i),x/\varphi(i))$$

for $i = 1,\ldots,k$. Since the $\varphi(i)$ cover φ and $k > 1$, we have

$$w/\varphi = f_{\varphi(i)}(u/\varphi(i),v/\varphi(i),x/\varphi(i)) \cap f_{\varphi(j)}(u/\varphi(j),v/\varphi(j),x/\varphi(j)) \tag{3.3}$$

for any $i \neq j$. Hence we define

$$f_\varphi(u/\varphi,v/\varphi,x/\varphi) = w/\varphi$$

and, by $P(n)$, conclude that f_φ satisfies b) and $f_\varphi(u/\varphi,v/\varphi,x/\varphi) \subset f_{\varphi(i)}(u/\varphi(i),v/\varphi(i),x/\varphi(i))$ for each i, so that f_φ also satisfies (3.2). To establish c) use Lemma 2.1 to observe that

$$x/\varphi \equiv y/\varphi \quad \theta(u/\varphi,v/\varphi) \quad \text{in } L_\varphi$$

$$\Leftrightarrow \theta(x,y) \leq \theta(u,v) \vee \varphi \quad \text{in } L$$

$$\Leftrightarrow \theta(x,y) \leq \theta(u,v) \vee \varphi(i), \quad i = 1,\ldots,k$$

$\Rightarrow \theta(x/\varphi(i),y/\varphi(i)) \le \theta(u/\varphi(i),v/\varphi(i)), \quad i = 1,\dots,k$

$\Rightarrow f_{\varphi(i)}(u/\varphi(i),v/\varphi(i),x/\varphi(i)) = f_{\varphi(i)}(u/\varphi(i),v/\varphi(i),y/\varphi(i)),$

$i = 1,\dots,k, \quad \text{by} \quad P(n)$

$\Rightarrow f_\varphi(u/\varphi,v/\varphi,x/\varphi) = f_\varphi(u/\varphi,v/\varphi,y/\varphi) \quad \text{by} \quad (3.3).$

This establishes c).

Finally observe that if f satisfies a) then (3.2) must hold so that $k > 1$ implies that f_φ must be defined as above, i.e.: the $f_{\varphi(i)}$ uniquely determine f_φ. Moreover, if for some i there was a function $f'_{\varphi(i)} \neq f_{\varphi(i)}$, but still satisfying $P(n)$, then replacing $f_{\varphi(i)}$ by $f'_{\varphi(i)}$ in (3.3) would yield a different f_φ. This establishes the "if" direction of d) for f_φ, assuming a).

Case 2. $k = 1$. (This will occur at least once iff L is not complemented.) For each $u,v,x \in A$ we can obviously satisfy b) and (by $P(n)$) (3.2) be defining

$$f_\varphi(u/\varphi,v/\varphi,x/\varphi) = \begin{cases} \text{the value given by b) if any pair of } u/\varphi,\ v/\varphi,\ x/\varphi \text{ are equal,} \\ \text{any } \varphi\text{-class contained in } f_{\varphi(1)}(u/\varphi(1),v/\varphi(1), x/\varphi(1)) \text{ otherwise.} \end{cases} \tag{3.4}$$

Moreover, since φ is not maximal and is covered by $\varphi(1)$, there must be $u,v,x \in A$, all different and such that $u/\varphi(1) = x/\varphi(1) \neq v/\varphi(1)$ and $u/\varphi \neq x/\varphi$. Hence the second clause of (3.4) occurs at least once and at least two choices of $f_\varphi(u/\varphi,v/\varphi,x/\varphi)$ are possible in this case (e.g.: either u/φ or x/φ will do). Now if $x/\varphi \equiv y/\varphi\ \theta(u/\varphi,v/\varphi)$, then $x/\varphi(1) \equiv y/\varphi(1)\ \theta(u/\varphi(1),v/\varphi(1))$ by Lemma 2.1, as in case 1, and hence $f_{\varphi(1)}(u/\varphi(1),v/\varphi(1),x/\varphi(1)) = f_{\varphi(1)}(u/\varphi(1), v/\varphi(1), y/\varphi(1))$ by $P(n)$. Hence if $x/\varphi \equiv y/\varphi\ \theta(u/\varphi,v/\varphi)$ we choose $f_\varphi(u/\varphi,v/\varphi,y/\varphi) = f_\varphi(u/\varphi,v/\varphi,x/\varphi)$ and thus as a φ-class contained in $f_{\varphi(1)}(u/\varphi(1),v/\varphi(1),x/\varphi(1))$. In this way f_φ satisfies c) as well as b) and (3.2). Moreover for given $f_{\varphi(1)}$, f_φ can be defined in more than one way. Since L is complemented iff case 2 never occurs this establishes the "only if" direction of d) for f_φ and assuming a).

Repeating this construction for all φ of height $n-1$ we establish $P(n-1)$ and, by induction, $P(n)$ for all $n \ge 0$. Then define $f(u,v,x)$ to be the sole element in $f_\omega(u/\omega,v/\omega,x/\omega)$.

We must finally verify that f, so defined, satisfies a). Suppose $\theta \in L$ and $u \equiv u'$, $v \equiv v'$, $x \equiv x'$ (θ). Hence $f_\theta(u/\theta,v/\theta, x/\theta) = f_\theta(u'/\theta,v'/\theta,x'/\theta)$. Then

$$f(u,v,x) \in f_\omega(u/\omega,v/\omega,x/\omega) \subset f_\theta(u/\theta,v/\theta,x/\theta)$$

and

$$f(u',v',x') \in f_\omega(u'/\omega,v'/\omega,x'/\omega) \subset f_\theta(u'/\theta,v'/\theta,x'/\theta)$$

by P(n), so that

$$f(u,v,x)/\theta = f(u',v',x')/\theta.$$

Hence f is compatible with θ so a) is satisfied.

4. THE GENERAL CASE.

4.1 Theorem. *Let* V *be an arithmetical variety. For each* $\underset{\sim}{A} \in V$ *and finite subset* $F \subset A$ *there is an integer* m *and an* $(m+3)$*-ary term* $t_F(x,y,z,\bar{w})$ *such that for some* $\bar{a} \in A^m$ *the polynomial function* $p_F(x,y,z) = t_F(x,y,z,\bar{a})$ *satisfies the following conditions for all* $u,v,x,y \in F$:

a) $p_F(u,u,x) = x,\quad p_F(u,v,u) = u,\quad p_F(u,v,v) = u.$

b) $x \equiv y \quad \theta(u,v)$ *in* $\underset{\sim}{A}$ *iff* $p_F(u,v,x) = p_F(u,v,y)$.

Proof. Let $\underset{\sim}{A} \in V$ and let F be a finite subset of A. Let L be the sublattice of $\mathrm{con}(\underset{\sim}{A})$ generated by the set of congruences

$$\{\omega,\iota,\quad \theta(a,b):a,b \in F,\quad a \neq b\}.$$

Since L if a finitely generated distributive lattice it is finite and, as a sublattice of $\mathrm{con}(\underset{\sim}{A})$ it is arithmetical. Consequently there is a function $f:A^3 \to A$ which satisfies conditions a)-d) of Lemma 3.1. Let g be the restriction of f to F^3 so that g is a finite partial function in $A^3 \to A$. Let (x,y,z) and (x',y',z') be in F^3 and suppose that for some $\theta \in \mathrm{con}(\underset{\sim}{A})$

$$x \equiv x',\quad y \equiv y',\quad z \equiv z' \quad (\theta).$$

Then the same congruences hold modulo $\theta' = \theta(x,x') \vee \theta(y,y') \vee \theta(z,z')$ which is in L. Since $\theta' \leq \theta$, by the compatibility of f with L it follows that

$$g(x,y,z) \equiv g(x',y',z') \quad (\theta).$$

Therefore g is compatible with $\mathrm{con}(\underset{\sim}{A})$ where defined. By Lemma 2.3c g has an interpolating polynomial $p_F(x,y,z) = t_F(x,y,z,\bar{a})$ where

$t_F(x,y,z,\bar{w})$ is an (m+3)-ary term for some integer m and $\bar{a} \in A^m$. This establishes the theorem.

The following is immediate from 4.1.

4.2 COROLLARY. *If* V *is an arithmetical variety,* $\underset{\sim}{A} \in V$ *and* F *a finite subset of* A *then for a term* $t_F(x,y,z,\bar{w})$ *provided by Theorem 4.1, and all* $u,v,x,y \in F$,

$$x \equiv y \quad \theta(u,v) \quad \textit{in} \quad \underset{\sim}{A} \quad \textit{iff}$$

$$(\exists \bar{w})(t_F(u,v,x,\bar{w}) = t_F(u,v,y,\bar{w}) \wedge t_F(u,u,x,\bar{w}) = x \wedge t_F(u,u,y,\bar{w}) = y).$$

The polynomials $p_F(x,y,z) = t_F(x,y,z,\bar{a})$ not only satisfy the equations of Theorem 4.1, a) (which are formally the same as those satisfied by a term characterizing the arithmeticity of V -- see Lemma 2.3 b)) but they also equal the discriminator when restricted to monolith classes of SI members of V:

4.3 COROLLARY. *If* V *is arithmetical,* $\underset{\sim}{A} \in V$ *is SI, and* $F \subset A$ *and* $x,y,z \in F$ *are in the same monolith class of* $\underset{\sim}{A}$ *then*

$$p_F(x,y,z) = z \quad \textit{if} \quad x = y,$$
$$= x \quad \textit{if} \quad x \neq y.$$

Proof. If $x = y$ then $p_F(x,y,z) = z$ by a) of (4.1). If $x \neq y$ then $\theta(x,y)$ is the monolith of $\underset{\sim}{A}$ so $x \equiv z \quad \theta(x,y)$. Hence $p_F(x,y,z) = p_F(x,y,x) = x$.

Recall that Werner's characterization of functional completeness [18] asserts that for a finite set of F every function $f:F^n \to F$ is a composition of the discriminator, projections and constants of F. Corollary 4.3 then asserts that each monolith class of an SI $\underset{\sim}{A} \in V$ is "locally" functionally complete.

5. LOCALLY FINITE VARIETIES. For locally finite varieties we have a somewhat sharper version of (4.1); specifically, the polynomials $p_F(x,y,z)$ can be uniformly constructed and their uniqueness on all finite algebras is equivalent to local semi-simplicity:

5.1 THEOREM. *Let* V *be a locally finite arithmetical variety.*

1. *For each integer* $m \geq 0$ *there is an (m+3)-ary term* t_m *of* V *such that if* $\underset{\sim}{A} \in V$ *and* $g_1,\ldots,g_m$ *generate* A *then the polyno-*

nial

$$p_m(x,y,z) = t_m(x,y,z,g_1,\dots,g_m)$$

satisfies the following conditions for all $u,v,x,y \in A$:

a) $p_m(u,u,x) = x, \quad p_m(u,v,u) = u, \quad p_m(u,v,v) = u,$

b) $x \equiv y \quad \theta(u,v)$ *in* $\underset{\sim}{A}$ *iff* $p_m(u,v,x) = p_m(u,v,y)$.

2. V *is locally semi-simple iff for each finite* $\underset{\sim}{A} \in V$ *there is only one polynomial of* $\underset{\sim}{A}$ *which satisfies 1 a), b) for all* $u,v,$ $x,y \in A$. *Thus if* V *is finitely generated it is semi-simple iff for each finite* $\underset{\sim}{A} \in V$ *there is only one polynomial* p_m *satisfying 1 a), b).*

Proof. Statement 1: let $\underset{\sim}{F}(m)$ be the free V-algebra with free generators $v_1,\dots,v_m$. Then $\mathrm{con}(\underset{\sim}{F}(m))$ is finite so that (taking $L = \mathrm{con}(\underset{\sim}{F}(m))$) there is an $f:F(m)^3 \to F(m)$ satisfying a)-d) of Lemma 3.1. Since $F(m)$ is finite f is a polynomial, by Lemma 2.3 c. This means that for some (m+3)-ary term t_m of V

$$f(x,y,z) = t_m(x,y,z,v_1,\dots,v_m)$$

for all $x,y,z \in F(m)$. If $\underset{\sim}{A} \in V$ has generators $g_1,\dots,g_m$ then $v_i \to g_i$ induces a homomorphism of $\underset{\sim}{F}(m)$ onto $\underset{\sim}{A}$. If $\varphi \in \mathrm{con}(\underset{\sim}{F}(m))$ is the kernel then the natural isomorphism $\underset{\sim}{F}(m)/\varphi \cong A$ translates a)-c) of Lemma 3.1 into 1 a), b) of (5.1).

To prove statement 2 notice that each finitely generated SI in V is simple iff $\mathrm{con}(\underset{\sim}{F}(m)/\varphi)$ is complemented for all m and $\varphi \in \mathrm{con}(\underset{\sim}{F}(m))$. Hence statement 2 follows from 3.1 d).

Corresponding to (4.2) and (4.3) we have:

5.2 COROLLARY. *Let* V *be a locally finite arithmetical variety and for each* $m \geq 0$ *let* t_m *be an (m+3)-ary term provided by (5.1). Then*

1. The formula

$$\varphi_m(u,v,x,y) := (\exists \bar{w})[t_m(u,v,x,\bar{w}) = t_m(u,v,y,\bar{w})$$

$$\wedge\ t_m(u,u,x,\bar{w}) = x \wedge t_m(u,u,y,\bar{w}) = y]$$

defines principal congruences for all m-generated members of V.

2. If V *has DPC then some one of the formulas* φ_m *defines principal congruences throughout* V.

Statement 1 is immediate from (5.1) part 1. For statement 2 recall that if V has DPC then for some fixed m, if $x \equiv y \ \theta(u,v)$ in $\underset{\sim}{B} \in V$ then $x \equiv y \ \theta(u,v)$ in a subalgebra $\underset{\sim}{A} \leq \underset{\sim}{B}$ containing u,v,x,y and which is m-generated.

5.3 COROLLARY. *If* V *is locally finite and arithmetical and for each* $m \geq 0$ t_m *is an* $(m+3)$*-ary term provided by (5.1) then if* $\underset{\sim}{A} \in V$ *is SI with generators* $g_1,\ldots,g_m$, *and* x,y,z *are in the same monolith class of* $\underset{\sim}{A}$, *then*

$$t_m(x,y,z,g_1,\ldots,g_m) = z \quad \textit{if} \quad x = y,$$
$$= x \quad \textit{if} \quad x \neq y.$$

S. Burris [3] established (5.3) in the special case where V is finitely generated and semi-simple; specifically in this case there is evidently a single $(m+3)$-ary t_m, $m \leq \max\{|A| : \underset{\sim}{A}$ is simple in $V\}$, such that if $\underset{\sim}{A}$ is simple in V and $g_1,\ldots,g_m$ generate $\underset{\sim}{A}$ then $t_m(x,y,z,g_1,\ldots,g_m)$ is the discriminator on A. Of course such a variety is already known to have DPC (McKenzie [13]) but here we obtain the specific defining formula φ_m constructed from this t_m as in (5.2). In fact we have the following uniqueness result immediately from (5.1), 2:

5.4 COROLLARY. *If* V *is finitely generated arithmetical with DPC then* V *is semi-simple iff it has the following property:*

For any pair of terms t_m *and* t_n *such that each of the corresponding congruence formulas* φ_m *and* φ_n *defines principal congruences throughout* V, *then for any* $\underset{\sim}{A} \in V$ *and pair of generating sets* $\{g_1,\ldots,g_m\}$ *and* $\{h_1,\ldots,h_n\}$ *of* $\underset{\sim}{A}$,

$$t_m(x,y,z,g_1,\ldots,g_m) = t_n(x,y,z,h_1,\ldots,h_n)$$

for all $x,y,z \in \underset{\sim}{A}$.

6. VARIETIES WITH CHAIN-CONGRUENCE SUBDIRECTLY IRREDUCIBLES. In section 1 we alluded to what might be called a general "principle" of arithmetical varieties: monolith classes of SI algebras behave "like" simple algebras. Corollaries 4.3 and 5.3 offer some support for this principle. In the present section we shall show that if V is finitely generated and $\mathrm{con}(\underset{\sim}{A})$ is a chain for each SI $\underset{\sim}{A} \in V$, then V has DPC (Theorem 6.5). Our method of proof of this further illustrates

this "principle". This result can also be obtained as a consequence of a recent result of E. Kiss [8] (which actually gives an effective method for determining if a finitely generated congruence distributive variety has DPC) but his approach and ours are very different.

In part our proof of Theorem 6.5 is just an extension of McKenzie's proof that every directly representable variety has DPC [13]. McKenzie's proof depends essentially on the fact that if the algebra $\mathbf{B}$ is an arbitrary finite direct power of algebra $\mathbf{A}$ then there is a set of $N = |A|$ elements of B (namely the diagonal $\{(a,\ldots,a) : a \in A\}$) which projects onto each of the given direct factors of $\mathbf{B}$. The following lemma is an easy extension of McKenzie's observation. The important fact is that we can exhibit interesting varieties which are not directly representable where it applies.

[illegible].1 LEMMA. *Let* V *be a variety satisfying the following conditions:*

a) V *is locally finite and has only finitely many finite SI members.*

b) For each finite SI $\mathbf{A} \in V$ *there is a positive integer* $N = N(\mathbf{A})$ *such that for each finite subdirect power* $\mathbf{B}$ *of* $\mathbf{A}$ *there is a subset* $B_o \subset B$ *of size* $|B_o| \le N$ *such that* B_o *projects onto each of the given subdirect factors.*

Then V *has DPC.*

Proof. By a result of Quackenbush [16] V contains only finitely many SI algebras, all finite. Let these be $\mathbf{A}_1,\ldots,\mathbf{A}_k$ and put $N = N(\mathbf{A}_1) + \cdots + N(\mathbf{A}_k)$. Then if $\mathbf{B}$ is any finitely generated member of V, $\mathbf{B}$ may be taken as a subdirect product in $\mathbf{A}_1^{n(1)} \times\cdots\times \mathbf{A}_k^{n(k)}$. For each i, $1 \le i \le k$, pick $N(\mathbf{A}_i)$ elements of B which project onto each occurrence of A_i in the above subdirect factorization; hence, altogether, obtain a set $B_o = \{b^1,\ldots,b^N\}$ of N elements of B such that if $A(j)$ is any one of the above $n_1 + \cdots + n_k$ subdirect factors of B (each equal to some A_i) then $\{b^1_j,\ldots,b^N_j\} = A(j)$. Thus the hypothesis b) holds for all finite (not just all finite SI) members of V.

Next observe that by Mal'cev's lemma there is a single congruence formula $\varphi(u,v,x,y)$ such that for any of the $\mathbf{A}_i$ and $u_i,v_i,x_i,y_i \in A_i$,

$$x_i \equiv y_i \quad \theta(u_i,v_i) \quad \text{iff} \quad \mathbf{A}_i \models \varphi(u_i,v_i,x_i,y_i).$$

Moreover, we may assume that φ has the form

$$(\exists w_1)\cdots(\exists w_N)(\bigvee_{\varphi'\in\Phi} \varphi'(u,v,x,y,w_1,\ldots,w_N))$$

where Φ is a finite set and each $\varphi' \in \Phi$ is a conjunction of equations. Finally, by taking sufficiently many φ', we may assume that if $\underset{\sim}{A}(j)$ is any of the SI $\underset{\sim}{A}_i$ and $\{b_j^1,\ldots,b_j^N\} = \underset{\sim}{A}(j)$, then for $u_j,v_j,x_j,y_j \in A(j)$,

$$x_j \equiv y_j \quad \theta(u_j,v_j) \quad \text{iff}$$

$$\underset{\sim}{A}(j) \models \bigvee_{\varphi'\in\Phi} \varphi'(u_j,v_j,x_j,y_j,b_j^1,\ldots,b_j^N). \tag{6.2}$$

Now let $\underset{\sim}{C} \in V$ and $u,v,x,y \in C$ satisfy $x \equiv y \quad \theta(u,v)$. Then, again by Mal'cev's lemma, together with local finiteness, there is a finite subalgebra $\underset{\sim}{B}$ of $\underset{\sim}{C}$, containing u,v,x,y and such that $x \equiv y \quad \theta(u,v)$ in $\underset{\sim}{B}$. Represent $\underset{\sim}{B}$ as a subdirect product in $A_1^{n(1)} \times\cdots\times A_k^{n(k)}$. Then for each SI $\underset{\sim}{A}(j)$ occurring in this factorization, $x_j \equiv y_j \quad \theta(u_j,v_j)$ in $A(j)$ so that the right side of (6.2) holds where $\{b_j^1,\ldots,b_j^N\} = B_0$ is chosen for the subalgebra $\underset{\sim}{B}$ as described in the paragraph above. Hence $\underset{\sim}{B} \models \varphi(u,v,x,y)$ so $\underset{\sim}{C} \models \varphi(u,v,x,y)$ and thus V has DPC.

Now we turn more directly to the proof of Theorem 6.5. It is well known that if $\underset{\sim}{A}$ is a congruence permutable algebra and is an irredundant subdirect product of finitely many simple algebras $\underset{\sim}{A}_1,\ldots,\underset{\sim}{A}_n$ then $\underset{\sim}{A}$ is, in fact, the direct product $\underset{\sim}{A}_1 \times\cdots\times \underset{\sim}{A}_n$. (As usual "irredundant" means that no projection of A to a proper subset of the factors is an isomorphism.) This means that A is the cartesian product of classes of proper congruences of the factors (each A_i being such a class!). For an arithmetical algebra $\underset{\sim}{A}$ this fact generalizes to arbitrary, not necessarily simple, factors $\underset{\sim}{A}_i$: A becomes the <u>union</u> of cartesian products of classes of proper congruences. The following lemma can then be taken as support for our general "principle".

6.3 LEMMA. *Let $\underset{\sim}{A}$ be an arithmetical algebra which is an irredundant subdirect product in $\underset{\sim}{A}_1 \times\cdots\times \underset{\sim}{A}_n$. Then there are proper congruences $\rho_i \in \mathrm{con}(\underset{\sim}{A}_i)$, $\rho_i > \omega_i$, such that for each $a = (a_1,\ldots,a_n) \in \underset{\sim}{A}$,*

$$a/(\rho_1 \times\cdots\times \rho_n) = a_1/\rho_1 \times\cdots\times a_n/\rho_n \quad .$$

($\rho_1 \times\cdots\times \rho_n$ is the product congruence defined by $a \equiv b \quad (\rho_1 \times\cdots\times \rho_n)$ iff $a_i \equiv b_i \quad (\rho_i)$ for all $i = 1,\ldots,n$.) In particular if any $\underset{\sim}{A}_i$ is SI then ρ_i can be taken to be the monolith μ_i of $\underset{\sim}{A}_i$.

Proof. Let π_i be the kernel of the projection of $\underset{\sim}{A}$ onto $\underset{\sim}{A}_i$ and for each $i = 1,\ldots,n$, let

$$\begin{aligned}\rho_i' &= \pi_i \vee (\pi_1 \wedge \cdots \wedge \pi_{i-1} \wedge \pi_{i+1} \wedge \cdots \wedge \pi_n) \\ &= (\pi_i \vee \pi_1) \wedge \cdots \wedge (\pi_i \vee \pi_{i-1}) \wedge (\pi_i \vee \pi_{i+1}) \wedge \cdots \wedge (\pi_i \vee \pi_n).\end{aligned} \tag{6.4}$$

Under the natural isomorphism $\underset{\sim}{A}/\pi_i \cong \underset{\sim}{A}_i$ let $\rho_i \in \mathrm{con}(\underset{\sim}{A}_i)$ correspond to ρ_i'. Now $\rho_i' > \pi_i$, since $\pi_1 \wedge \cdots \wedge \pi_n = \omega$ and $\underset{\sim}{A}$ is an irredundant subdirect product of the $\underset{\sim}{A}_i$, so $\rho_i > \omega_i$ in $\mathrm{con}(\underset{\sim}{A}_i)$. Let $a = (a_1,\ldots,a_n) \in A$ and choose any $b = (b_1,\ldots,b_n) \in a_1/\rho_1 \times \cdots \times a_n/\rho_n$, i.e.: such that $b_i \equiv a_i \ (\rho_i)$. Then for all i there is $b^i \in A$ that $b_i^i = b_i$ and thus $b^i \equiv a \ (\rho_i')$ for $i = 1,\ldots,n$. But for $1 \le i < j \le n$, (6.4) implies $\rho_i' \vee \rho_j' \le \pi_i \vee \pi_j$, so that $b^i \equiv b^j \ (\pi_i \vee \pi_j)$. Hence by the Chinese remainder theorem (Lemma 2.2), we conclude that $b \in A$ and $b \in a/(\rho_1 \times \cdots \times \rho_n)$.

6.5 THEOREM. *Let* V *be a finitely generated arithmetical variety with SI members* $\underset{\sim}{A}_1,\ldots,\underset{\sim}{A}_n$, *and suppose that for each* i, $\mathrm{con}(\underset{\sim}{A}_i)$ *is a chain. Then* V *has DPC. In particular if* $\mathrm{con}(\underset{\sim}{A}_i)$ *has height* $h(i)$ *then the principal congruences of* V *are defined by one of the formulas* φ_m *(Corollary 5.2) with* $m \le |A_1|^{h(1)} + \cdots + |A_n|^{h(n)}$.

Proof. By Lemma 6.1 it is enough to establish the following assertion for each SI $\underset{\sim}{A} \in V$:

> If $\mathrm{con}(\underset{\sim}{A})$ is a chain of height h and $\underset{\sim}{B}$ is any finite subdirect power of $\underset{\sim}{A}$ then there is a subset $B_o \subset B$ with $|B_o| \le |A|^h$ such that B_o projects onto each of the given subdirect factors. (6.6)

To establish (6.6) suppose $\underset{\sim}{B} \le \underset{\sim}{A}^m$, m a positive integer. First we show that we may assume without loss of generality that $\underset{\sim}{B}$ is irredundant in $\underset{\sim}{A}^m$. Let p_i be the projection of $\underset{\sim}{B}$ onto the i-th component and $\pi_i = \ker p_i$, and suppose $\underset{\sim}{B}$ is redundant, i.e.: $\pi_1 \wedge \cdots \wedge \pi_k = \omega$ for some $k < m$. Then if $i > k$, $\pi_i \ge \pi_j$ for some $1 \le j \le k$, by congruence distributivity and the meet irreducibility of π_i. But since A is finite and $\underset{\sim}{B}/\pi_i \cong \underset{\sim}{A} \cong B/\pi_j$ we have $\pi_i = \pi_j$. Thus for all $x = (x_1,\ldots,x_m)$, $y = (y_1,\ldots,y_m)$ in B, $x_i = y_i \Leftrightarrow x_j = y_j$ and from this it follows that the correspondence $x_j \to x_i$ is an automorphism of $\underset{\sim}{A}$. Hence if $B_o \subset B$ is a subset such that $p_i(B_o) = A$ for $i = 1,\ldots,k$, then $p_i(B_o) = A$ for $i = k+1,\ldots,m$ as well, which means that we may assume $\underset{\sim}{B} \le \underset{\sim}{A}^m$ is irredundant.

We complete the proof of (6.6) by induction on the height h of con($\underset{\sim}{A}$).

h = 1. Then $\underset{\sim}{A}$ is simple so that by irredundancy and congruence permutability only, $\underset{\sim}{B} = \underset{\sim}{A}^m$. Hence let $B_o = \{(a,\dots,a) \in A^m : a \in A\}$, the diagonal. Then $|B_o| = |A|^h$ and $p_i(B_o) = A$ for all i.

Induction step. Suppose (6.6) is true for all $\underset{\sim}{A}$ with con($\underset{\sim}{A}$) a chain of height h-1. Let con($\underset{\sim}{A}$) have height $h > 1$ and suppose $\underset{\sim}{B} \leq \underset{\sim}{A}^m$ is an irredundant subdirect power. Let the monolith of $\underset{\sim}{A}$ be μ and let $\mu_o = \mu \times \cdots \times \mu$ (m factors), $\mu_o \in$ con($\underset{\sim}{B}$). Then $\underset{\sim}{B}/\mu_o$ is a subdirect power in $(\underset{\sim}{A}/\mu)^m$. Since con($\underset{\sim}{A}$) is a chain, $\underset{\sim}{A}/\mu$ is SI and con($\underset{\sim}{A}/\mu$) has height h-1, so that, by the induction hypothesis, there is a subset $\bar{B}$ of B/μ_o such that $|\bar{B}| \leq |A/\mu|^{h-1} < |A|^{h-1}$ and $p_i(\bar{B}) = A/\mu$ for i = 1,...,m. But, by Lemma 6.3, for each $b \in B$, $b/\mu_o = b_1/\mu \times \cdots \times b_m/\mu$ and thus there is a set of $\max\{|b_i/\mu| : 1 \leq i \leq m\}$ elements of b/μ_o which projects onto each B_i/μ. Then if we let $B_o = \cup\{b/\mu_o : b \in \bar{B}\}$, it follows that $|B_o| < |A|^{h-1} \cdot |A| = |A|^h$ and $p_i(B_o) = A$ for all i = 1,...,m. This proves (6.6) and hence Thoerem 6.5.

7. A PROBLEM. If V is a finitely generated congruence distributive variety with SI members $\underset{\sim}{A}_1,\dots,\underset{\sim}{A}_n$ then every algebra in V is a subdirect product in $\underset{\sim}{A}_1' \times \cdots \times \underset{\sim}{A}_n'$ where $\underset{\sim}{A}_i \in$ HSP($\underset{\sim}{A}_i$). Because of this it is easy to see that such a variety has DPC iff each of the subvarieties HSP($\underset{\sim}{A}_i$) has DPC. Now if A is a finite set and L is any arithmetical sublattice of the lattice of all equivalence relations on A such that both ι and ω are in L and ω is meet irreducible in L, then by a result of Quackenbush and Wolk [17] it is possible to construct an SI algebra $\underset{\sim}{A} = (A,F)$ on A with con($\underset{\sim}{A}$) = L. By Lemmas 3.1 and 2.3 we can do this so that the variety HSP($\underset{\sim}{A}$) is arithmetical. Also by adding nullary operations, one for each element of A, we can insure that each finite subdirect power of $\underset{\sim}{A}$ contains the diagonal. Hence by Lemma 6.1 HSP($\underset{\sim}{A}$) has DPC.

On the other hand E. Kiss [8] has recently given an example of a finitely generated arithmetical variety not having DPC. Hence if L is not a chain, one can sometimes construct $\underset{\sim}{A}$ so that HSP($\underset{\sim}{A}$) does not have DPC. The problem is whether this can always be done:

7.1 PROBLEM. If A and L are as given above and L is not a chain, can one always find an algebra $\underset{\sim}{A} = (A,F)$ with con($\underset{\sim}{A}$) = L, V = HSP($\underset{\sim}{A}$) arithmetical and V non-DPC?

ACKNOWLEDGEMENT. I am greatly indebted to J.D. Berman and W.J. Blok who carefully read an earlier version of this paper and kindly pointed out several errors.

REFERENCES

[1] Baldwin, J.T., and Berman, J., The number of subdirectly irreducible algebras in a vareity. Alg. Univ. 5 (1975), 379-389.

[2] __________, Definable principal congruence relations: kith and kin. Acta. Sci. Math. (Szeged) 44 (1982), 255-270.

[3] Burris, S., Arithmetical varieties and Boolean product representations. Univ. of Waterloo, manuscript 1978.

[4] Fried, E., Grätzer, G., and Quackenbush, R.W., Uniform congruence schemes. Alg. Univ. 10 (1980), 176-188.

[5] __________, and Kiss, E., Connections between congruence-lattices and polynomial properties. Alg. Univ. 17 (1983), 227-262.

[6] Grätzer, G., Universal Algebra, 2nd ed., Springer-Verlag, Berlin 1979.

[7] Kaarli, K., Compatible function extension property. Alg. Univ. 17 (1983), 200-207.

[8] Kiss, E., Definable principal congruences in congruence distributive varieties. Math. Institute of the Hungarian Academy of Sciences, manuscript 1984.

[9] Köhler, P., and Pigozzi, D., Varieties with equationally definable principal congruences. Alg. Univ. 11 (1980), 213-219.

[10] Korec, I., A ternary function for distributivity and permutability of an equivalence lattice. Proc. Amer. Math. Soc. 69 (1978), 8-10.

[11] __________, Concrete representation of some equivalence lattices. Math. Slovaca 31 (1981), 13-22.

[12] McKenzie, R., On spectra and the negative solution of the decision problem for identities having a finite non-trivial model. J. Symbolic Logic 40 (1975), 186-196.

[13] __________, Para-primal varieties: a study of finite axiomatizability and definable principal congruences in locally finite varieties. Alg. Univ. 8 (1978), 336-348.

[14] Pixley, A.F., Completeness in arithmetical algebras. Alg. Univ. 2 (1972), 179-196.

[15] __________, Characterizations of arithmetical varieties. Alg. Univ. 9 (1979) 87-98.

[16] Quackenbush, R.W., Equational classes generated by finite algebras. Alg. Univ. 1 (1971), 265-266.

[17] ________, and Wolk, B., Strong representation of congruence lattices. Alg. Univ. 1 (1971), 165-166.

[18] Werner, H., Eine Charakterisierung funktional vollständige Algebren. Arch. Math. (Basel) 21 (1970), 381-385.

FROM AFFINE TO PROJECTIVE GEOMETRY VIA CONVEXITY

Anna B. Romanowska
Warsaw Technical University
00661 Warsaw, Poland

and

Jonathan D. H. Smith
Iowa State University
Ames, Iowa 50011, U.S.A.

1. **INTRODUCTION** This paper sets out to study affine geometry, projective geometry, and the relationship between them. The study is undertaken from an algebraic standpoint subject to requirements of invariance and directness. "Invariance" means following Klein's Erlanger Programm for geometry [Kl,p.7] [Kn,p.463]:

> Suppose given a geometry and a group of transformations of the geometry; one should investigate features belonging to the geometry with regard to properties that are invariant under the transformations of the group.

The usual coordinatization of an affine geometry by a module over a commutative ring is not invariant under the affine group in this sense, since the zero element of the module is disturbed by affine translations. The algebraic descriptions of affine geometries to be used here are required to have affine groups as groups of automorphisms. The advantage of satisfying this requirement is that one may then identify the algebraic structure with the affine geometry, rather than having to maintain a duality between a description (the coordinatizing module) and the thing described (the affine geometry).

"Directness" is a requirement motivated by recent developments in applied mathematics [RS,p.103], [Sl,pp.388-9]. It means avoiding "secondary constructs" - features of a mathematical model introduced merely for pure mathematical convenience, and not corresponding to any phenomenon being modelled. In the present context, secondary constructs might take the form of ideal elements or points at infinity. Thus the aim of the paper is to describe affine geometry, projective geometry, and the passage between them purely algebraically, in an invariant way avoiding the setting up of auxiliary constructions.

2. **THE ALGEBRAIC FRAMEWORK** The algebraic approach to affine geometry here is that of [OS], [RS,2.5]. Let K be a field, and E a vector space over K. For each element k of K, define a binary operation

(2.1) $$\underline{k}: E \times E \to E;\ (x,t) \mapsto xy\underline{k} := x(1-k) + yk$$

on E, so that (E,K) becomes an algebra with the set K of binary operations. "Algebra" is meant in the sense of universal algebra: see [Co] or [RS,Chapter 1] for explanations of universal algebraic notions. On E, define the parallelogram-completion operation

$$(2.2) \qquad P: E \times E \times E \to E;\ (x,y,z) \mapsto x - y + z.$$

Then the algebra (E,K,P) with the ternary operation P and set K of binary operations has as its derived operations (those obtained from successive compositions of the basic operations P and <u>k</u> for k in K) precisely the affine combinations $x_1k_1 + \ldots + x_rk_r$ (with $k_1 + \ldots + k_r = 1$) of elements $x_1,\ldots,x_r$ of E. It follows that the algebra (E,K,P) has the affine group as its group of automorphisms, and may thus be identified as the affine geometry.

The corresponding projective geometry consists of the set L(E) of linear or vector subspaces of the vector space E, ordered by inclusion. This may be described algebraically as (L(E),+), where for subspaces U and V of E, U + V is the sum of U and V. The incidence or inclusion structure is recovered from (L(E),+) via $U \leq V$ iff $U + V = V$. Algebraically, (L(E),+) is a <u>semilattice</u> - the binary operation + is commutative, associative, and idempotent.

Both the algebras (L(E),+) describing the projective geometries and the algebras (E,K,P) describing the affine geometries have two special properties. They are <u>idempotent</u>, in the sense that each singleton is a subalgebra, and they are <u>entropic</u>, i.e. each operation, as a mapping from a direct power of the algebra into the algebra, is actually a homomorphism. Algebras with these two properties are called <u>modes</u>. They are studied in detail in [RS]. Given a mode (A,Ω), as a set A with a set Ω of operations $\omega: A^{\omega\tau} \to A$ on it, one may form the set $(A,\Omega)S$ or AS of non-empty subalgebras of (A,Ω). This set AS carries an Ω-algebra structure under the complex products

$$\omega: AS^{\omega\tau} \to AS;\ (X_1,\ldots,X_{\omega\tau}) \mapsto \{x_1\ldots x_{\omega\tau}\omega \mid x_i \in X_i\},$$

and it turns out that the algebra (AS,Ω) is again a mode, preserving many of the algebraic properties of (A,Ω) [RS,146]. The key idea of the current paper is to examine the algebras (E,K,P)S arising in this way from an affine space (E,K,P). The method is to obtain varieties nicely containing such (E,K,P)S, and to describe the structure of the (E,K,P)S within the varieties - see (3.7), (4.9), (5.13). It is the structure of the (E,K,P)S which yields the direct, invariant passage from the affine to the projective geometry.

The internal structure of the $(E,K,P)S$ is described using a construction known as the Płonka sum [Pł], [RS,236]. Let Ω be a non-empty domain of operations with an arity mapping $\tau:\Omega \to \{n \in \mathbf{N} | n > 1\}$. A semilattice $(H,+)$ may be considered as an Ω-algebra (H,Ω), a so-called Ω-semilattice, on defining $h_1 \ldots h_{\omega\tau}\omega = h_1 + \ldots + h_{\omega\tau}$ for h_i in H. The semilattice operation $+$ is then recovered as $h + k = hk\ldots k\omega$ for any ω from Ω. The semilattice $(H,+)$ may also be considered as a (small) category (H) with set H of objects, and with a unique morphism $h \to k$ precisely when $h + k = k$, i.e. $h \leq k$. The notion of Płonka sum depends on viewing the semilattice $(H,+)$ in this way both as a category and as an Ω-algebra. Let (Ω) denote the (concrete) category of Ω-algebras and homomorphisms between them. Let $F:(H) \to (\Omega)$ be a functor. Then the Płonka sum of the Ω-algebras (hF,Ω) (for h in H) over the semilattice $(H,+)$ by the functor F is the disjoint union $HF = \bigcup_{h \in H} hF$ of the underlying sets hF (h in H), equipped with the Ω-algebra structure given for an n-ary operation ω in Ω and $h_1,\ldots,h_n,k = h_1 + \ldots + h_n$ in H by

$$(2.3)\qquad \omega:h_1F\times\ldots\times h_nF \to kF;$$

$$(x_1,\ldots,x_n) \mapsto x_1(h_1\to k)F\ldots x_n(h_n\to k)F\omega.$$

The projection of the Płonka sum HF is the homomorphism $\pi_F:(HF,\Omega) \to (H,\Omega)$ with restrictions $\pi_F:hF \to \{h\}$. The subalgebras $(hF,\Omega) = (\pi_F^{-1}h,\Omega)$ of (HF,Ω) are referred to as the Płonka fibres.

The union $\{P\} \cup K$ forms a set of operations on $(E,K,P)S$, namely the set of corresponding complex products. For each of the three cases for K to be studied - namely characteristic 0, odd characteristic, and characteristic 2 - considerations of convexity suggest taking a certain subset Ω_K of $\{P\} \cup K$. The set Ω_K may be viewed as an algebraic analogue of the open unit interval in the field of rationals, and will actually be this interval in the characteristic zero case. The algebra $((E,K,P)S,\Omega_K)$ will then be shown to be a Płonka sum over the Ω_K-semilattice $(L(E),\Omega_K)$. The main result of this paper, the direct invariant passage from an affine to a projective geometry, follows:

THEOREM 2.4. *Let K be a field, and let (E,K,P) be an affine space over K. Then the projective geometry $(L(E),+)$ is the largest Ω_K-semilattice quotient of the algebra $((E,K,P)S,\Omega_K)$ of affine subspaces of (E,K,P).*

The three subsequent sections give the proof of the various cases of Theorem 2.4 in turn: Section 3 deals with K of characteristic 0, Section 4 with K of odd characteristic, and Section 5 with K of characteristic 2. In the course of describing the algebra $((E,K,P)S,\Omega_K)$ for the characteristic 2 case in Section 5, attention is drawn to some general results of Płonka theory. A method of determining a basis for the identities of the regularisation of a strongly irregular variety from a basis for the irregular variety is given. This method preserves finiteness of the basis. It is also observed that the automorphism group of a free algebra on a set in a regular variety is the permutation group of the set.

3. **CHARACTERISTIC ZERO** This section considers the case that the field K is of characteristic zero. Since 2 has the inverse 1/2 in K, the parallelogram operation P of (2.2) may be written as

$$xyzP = yxz\,\underline{1/2}\;\underline{2}, \tag{3.1}$$

the latter term being $y(1-2) + (x(1-\frac{1}{2}) + z\frac{1}{2})2 = -y + x + z$. This means that the algebraic structure (E,K,P) describing an affine geometry over K may be replaced by the structure (E,K). The class $\underline{K}$ consisting of these algebras (E,K) and the empty space (ϕ,K) forms a variety in the sense of universal algebra - the class of all algebras satisfying a given set of identities.

THEOREM 3.2 [OS,Satz 7] [RS,256]. *The class $\underline{K}$ of affine K-spaces is the variety of modes (E,K) of type $\tau:K \to \{2\}$ satisfying the identities:*

(a) $xy\underline{0} = x$;

(b) $xy\underline{1} = y$;

(c) $xy\underline{p}\;xy\underline{q}\underline{r} = xy(\underline{pq\underline{r}})$.

Given such an algebra (E,K), one may consider its reduct (E,I^o) obtained by admitting only those binary operations $\underline{k}$ as in (2.1) for which k lies in the open unit interval $I^o = \{x \in \mathbf{Q} | 0 < x < 1\}$ of the rationals $\mathbf{Q}$, the prime field of K. In the notation of Section 2, $I^o = \Omega_K$. The subalgebras (X,I^o) of (E,I^o) are precisely the $\mathbf{Q}$-convex subsets X of E. The class of all $\mathbf{Q}$-convex sets,i.e. the class of all such (X,I^o) for all such (E,I^o), does not itself form a variety, but the smallest variety containing the class, the variety of so-called rational barycentric algebras, is described as follows, where p' denotes $1 - p$.

THEOREM 3.3 [RS,214]. *The variety of rational barycentric algebras is the class of algebras of type $\tau:I^o \to \{2\}$ satisfying the identities:*

(a) $xx\underline{p} = x$

(b) $xy\underline{p} = yx\underline{p}'$

(c) $xy\underline{p}z\underline{q} = xyz(\underline{q/(p'q')'})(\underline{p'q'})'$.

(The reference [RS,214] actually treated the case of real barycentric algebras, with operations from the open unit interval in the reals, but the statement and proof for the rational case are completely analogous.) The identity (a) is just idempotence; (b) is a skew form of commutativity, while (c) is a skew form of associativity. Examples of rational barycentric algebras that are not convex sets are furnished by I^o-semilattices. For these, all the operations $\underline{p}$ with p in I^o coincide, so (b) becomes genuine commutativity and (c) becomes genuine associativity.

The choice of the set I^o of binary operations here is made for two reasons. Firstly, it leads to the readily available variety of rational barycentric algebras. Secondly, I^o avoids the elements 0 and 1 of K. The significance of this resides in the following general proposition.

PROPOSITION 3.4. *Let* K *be a field other than* $GF(2)$, *and let* k *be an element of* K *distinct from* 0 *and* 1. *Let* E *be a vector space over* K, *and* $(L(E),+)$ *the corresponding projective geometry.*

(i) *There is a homomorphism*

$$\pi:((E,K,P)S,\underline{k}) \rightarrow (L(E),+);\ x + U \mapsto U.$$

(ii) *For the functor* $F:L(E) \rightarrow (\{\underline{k}\})$ *with* $UF = \pi^{-1}(U)$ *and*

$$(U \rightarrow V)F;\ \pi^{-1}(U) \rightarrow \pi^{-1}(V);\ x + U \mapsto x + V,$$

the algebra $((E,K,P)S,\underline{k})$ *is the Płonka sum over* $(L(E),+)$ *by the functor* F.

Proof. For x,y in E and U,V in $L(E)$, $(x+U)(y+V)\underline{k} = \{(x+u)(y+v)\underline{k} \mid u \in U, v \in V\} = xy\underline{k} + UV\underline{k} = xy\underline{k} + U(1-k) + Vk = xy\underline{k} + (U+V)$. This shows that π is a homomorphism. Further, $xy\underline{k} + (U+V) = (x + (U+V))(y + (U+V))\underline{k} = ((x+U)(U\rightarrow U+V)F)((y+V)(V\rightarrow U+V)F)\underline{k}$, so this latter term is equal to $(x+U)(y+V)\underline{k}$, showing that $((E,K,P)S,\underline{k})$ is a Płonka sum as claimed.

In the case that K has characteristic zero, this proposition leads to the following theorem describing the structure of the I^o-algebra of affine subspaces of an affine K-space.

THEOREM 3.5. *For an affine* K-*space* (E,K) *in* $\underline{\underline{K}}$, *the* I^o-*algebra* $((E,K)S,I^o)$ *of affine subspaces of* (E,K) *is a Płonka sum of* $\mathbb{Q}$-*convex sets over the projective geometry* $(L(E),+)$ *by the functor* $F:(L(E)) \rightarrow (I^o)$ *with* $UF = \{x + U \mid x \in E\}$ *and* $(U\rightarrow V)F:UF \rightarrow VF;\ x + U \mapsto x + V$.

Proof. That $((E,K)S,I^{o})$ is the Płonka sum over $(L(E),+)$ by the functor F follows directly from Proposition 3.4. The Płonka fibres are the I^{o}-algebras $\{x + U | x \in E\}$ for fixed subspaces U, with $(x+U)(y+U)\underline{p} = xy\underline{p} + U$. Since these are the reducts $(E/U,I^{o})$ of the affine geometries $(E/U,K)$ coming from the quotient vector spaces E/U, they are $\mathbb{Q}$-convex sets.

COROLLARY 3.6. *The I^{o}-algebra $((E,K)S,I^{o})$ of affine subspaces of an affine K-space (E,K) is a rational barycentric algebra.*

Proof. The Płonka fibres $(E/U,I^{o})$, being $\mathbb{Q}$-convex sets, satisfy the identities (a)-(c) of Theorem 3.3. These identities are regular: in any of the identities, the same set of variables appears on each side of the identity. By a result of Płonka [Pł,Theorem I] [RS,238], it follows that the Płonka sum $((E,K)S,I^{o})$ of $\mathbb{Q}$-convex sets also satisfies the identities, and is thus a rational barycentric algebra.

Theorem 3.5 and Corollary 3.6 may be summarized as follows:

(3.7) The rational barycentric algebra of affine subspaces of affine K-space is a Płonka sum of $\mathbb{Q}$-convex sets over a projective geometry.

By Theorem 3.5, the projective geometry $(L(E),+)$ is an I^{o}-semilattice quotient of $((E,K)S,I^{o})$ by the projection π_F. To complete the proof of Theorem 2.4 for the characteristic zero case, it must be shown that the projective geometry is the largest semilattice quotient of $((E,K)S,I^{o})$. Since there is such a largest quotient, the so-called I^{o}-semilattice replica of $((E,K)S,I^{o})$ [Ma,11.3] [RS,1.5], and since projection onto this replica factorises the projection of $((E,K)S,I^{o})$ onto any semilattice quotient, it suffices to show that the Płonka fibres UF of Theorem 3.5 have no non-trivial I^{o}-semilattice quotient. Now these fibres are the I^{o}-reducts $(E/U,I^{o})$ of affine K-spaces $(E/U,K)$, so the proof of Theorem 2.4 in the characteristic zero case is concluded by the following result.

THEOREM 3.8. *For an affine K-space (E,K), the reduct (E,I^{o}) has no non-trivial I^{o}-semilattice quotient.*

Proof. If (E,I^{o}) has a non-trivial semilattice quotient, it has the two-element semilattice $\{0,1\}$ with $0 < 1$ as quotient, say by a surjective homomorphism $f:(E,I^{o}) \to (\{0,1\},I^{o})$. Then E is the disjoint union of non-empty subsets $f^{-1}(0)$ and $f^{-1}(1)$. Take x' in $f^{-1}(0)$ and y' in $f^{-1}(1)$. Consider the $\mathbb{Q}$-affine span of x' and y' in (E,K). This is a rational ffine line $(\mathbf{Q},\mathbf{Q})$. The homomorphism $f:(E,I^{o}) \to (\{0,1\},I^{o})$ restricts to a urjective homomorphism $g:(Q,I^{o}) \to (\{0,1\},I^{o})$, decomposing the rational affine ine as a disjoint union of non-empty fibres $g^{-1}(0)$ and $g^{-1}(1)$. These fibres re subalgebras of $(\mathbf{Q},I^{o})$, and so are convex subsets of $\mathbf{Q}$. Without loss of enerality, assume that elements of $g^{-1}(0)$ are less than elements of $g^{-1}(1)$ n the order $(\mathbf{Q},\leq)$. Then there are elements x,y of $g^{-1}(0)$ and z,t of $^{-1}(1)$ such that $x < y < z < t$. Take $p = (y-x)/(t-x)$ and $q = (z-x)/(t-x)$, o that $xt\underline{p} = y$ and $xt\underline{q} = z$. Then $0 = yg = xt\underline{p}g = xgtg\underline{p}$ and $1 = zg = xt\underline{q}g$ $xgtg\underline{q}$. But $xgtg\underline{p} = xgtg\underline{q}$ in an I^{o}-semilattice, a contradiction. It ollows that (E,I^{o}) has no non-trivial I^{o}-semilattice quotient.

. **ODD CHARACTERISTIC** In this section, the case that K has odd haracteristic is considered. As for the characteristic zero case, 2 is nvertible here, so the parallelogram operation P may be written in terms of he binary operations $\underline{1/2}$ and $\underline{2}$ using (3.1). It follows that the algebraic tructures (E,K,P) describing affine K-spaces may be replaced by the tructures (E,K). Then the set $(E,K)S$ of non-empty subalgebras of (E,K) is he set of affine K-subspaces of (E,K).

Let J denote the prime subfield of K. This subset J of K plays a ole analogous to that of the unit interval I in the rationals. The reduct $E,J)$ of (E,K) has as its subalgebras the affine J-subspaces of (E,K). hese may be viewed as analogues of the convex subsets of a rational affine pace. Consider the binary operation $\underline{1/2}$. Under this operation, $(E,\underline{1/2})$ is commutative binary mode, a reduct of (E,J).

ROPOSITION 4.1. *The binary derived operations of the algebra* $(E,\underline{1/2})$ *are he operations* $\underline{p}$ *with* p *from* J.

Proof. The binary derived operations of the algebra (E,J) are all of the orm $\underline{p}$ for p in J. Let X be the subset of J consisting of those p or which $\underline{p}$ is a binary derived operation of $(E,\underline{1/2})$. Certainly X ontains $1/2$. Now for p,q in K,

$$xxy\underline{p}\underline{q} = xy(\underline{pq}). \tag{4.2}$$

Thus X is a subsemigroup of the multiplicative group of $J - \{0\}$. In articular, 1 and 2 lie in X, and $2X$ is a subset of X. Since

$$xy\ \underline{2p}\ xy\ \underline{2q}\ \underline{1/2} = xy(\underline{p+q}) \tag{4.3}$$

for p,q in K, it follows that X is a subring of J. But X contains 1, and so is all of J.

In view of Proposition 4.1, Ω_K for K of odd characteristic will be taken to be the single binary operation $\underline{1/2}$, often written as a multiplication • or juxtaposition. Since (E,•) is a commutative binary mode, it follows [RS,146] that the set of affine subspaces of (E,K) forms a commutative binary mode ((E,K)S,•). The choice of Ω_K here is again made for two reasons similar to those involved in the choice of Ω_K for K of characteristic zero: firstly, Ω_K avoids 0 and 1, and secondly there is a readily available theory of commutative binary modes, due primarily to Ježek and Kepka [JK], [RS,Chapter 4].

The theory of commutative binary modes is based on the observation that the free commutative binary mode on the two-element set {0,1} may be realised as the unit interval D_1 in the set $\mathbf{D} = \{m2^{-n} | m,n \in \mathbf{Z}\}$ of dyadic rationals under the operation $\underline{1/2}$ [RS,424]. For an odd natural number m, let l(m) denote the least integer greater than $\log_2 m$, e.g. l(3) = 2. Then $m \cdot 2^{-l(m)}$ and $m \cdot 2^{-l(m)-1}$, as elements of $(D_1, \underline{1/2})$, represent words $w_m(0,1)$ and $w'_m(0,1)$ in {0,1} respectively. For example, 3/4 = 110 $\underline{1/2}$ $\underline{1/2}$ and 3/8 = 0110 $\underline{1/2}$ $\underline{1/2}$ $\underline{1/2}$, so $w_3(0,1)$ = 110 $\underline{1/2}$ $\underline{1/2}$ and $w'_3(0,1)$ = 0110 $\underline{1/2}$ $\underline{1/2}$ $\underline{1/2}$. Let $\underline{\underline{m}}$ denote the variety of commutative binary modes satisfying the identity $x = w_m(x,y)$, and $\tilde{\underline{\underline{m}}}$ the variety of those satisfying the identity $w_m(x,y) = w'_m(x,y)$. There is then the following classification theorem [JK,Theorem 4.9] [RS,454].

THEOREM 4.4. *Apart from the variety of all commutative binary modes, the varieties of commutative binary modes are the varieties* $\underline{\underline{m}}$ *and* $\tilde{\underline{\underline{m}}}$ *for odd natural numbers m. For each such m,* $\tilde{\underline{\underline{m}}}$ *is the variety of algebras satisfying the regular identities of* $\underline{\underline{m}}$.

Recall that a binary algebra (A,•) is said to be a quasigroup if there are derived binary operations / (called right division) and \ (called left division) on A such that the identities

$$(4.5) \qquad \begin{cases} (x \cdot y)/y = x, & (x/y) \cdot y = x, \\ y\backslash(y \cdot x) = x, & y \cdot (y\backslash x) = x \end{cases}$$

are satisfied. The commutative binary modes $(E,\cdot) = (E,\underline{1/2})$ coming from affine K-spaces (E,K) may then be described as follows.

PROPOSITION 4.6. *Let* u *be the multiplicative order of* 2 *in the field* K. *Then the reduct* $(E,\underline{1/2})$ *of an affine* K-*space* (E,K) *is a quasigroup in the variety* $\underline{\underline{m}}$ *for* $m = 2^u - 1$.

Proof For $m=2^u - 1$, $l(m) = u$. Then $m \cdot 2^{-l(m)} = 1 - (1/2^u) = 10(\underline{1/2^u}) = 1\ldots1\ 0\ \underline{1/2}\ldots\underline{1/2}$ with u applications of $\underline{1/2}$, the latter equality coming from (4.2). Thus $w_m(x,y) = y\ldots yx\ \underline{1/2}\ldots\underline{1/2}$. In $(E,\underline{1/2})$, $w_m(x,y) = y\ldots yx\ \underline{1/2}\ldots\underline{1/2} = yx(\underline{1/2^u}) = yx\ \underline{1} = x$, the second equality coming from (4.2). Thus the commutative binary mode $(E,\underline{1/2})$ lies in the variety $\underline{\underline{m}}$.

Consider the binary operations λ and ρ on E with $yx\lambda = yx(\underline{1/2^{u-1}})$ and $xy\rho = yx\lambda$. Using (4.2) and the commutativity of $\underline{1/2}$, the word $w_m(x,y) = y\ldots yx\ \underline{1/2}\ldots\underline{1/2}$ in $(E,\underline{1/2})$ may be written variously as $w_m(x,y) = y(yx\underline{1/2})\lambda = y(yx\lambda)\underline{1/2} = (xy\underline{1/2})y\rho = (xy\rho)y\underline{1/2}$. The identity $w_m(x,y) = x$ in $(E,\underline{1/2})$ then gives the quasigroup identities (4.5). Since λ and ρ are derived operations of $(E,\underline{1/2})$, it follows that $(E,\underline{1/2})$ is a quasigroup.

Propositions 3.4 and 4.6 may then be combined to give the following structural description of the commutative binary mode $((E,K)S,\cdot)$ of affine subspaces of the affine K-space (E,K).

THEOREM 4.7. *Let* K *be a field of odd characteristic* p. *Let* m *be the least integer multiple of* p *of the form* $2^u - 1$ *for a natural number* u. *Then for an affine* K-*space* (E,K), *the commutative binary mode* $((E,K)S,\cdot)$ *is a Płonka sum of quasigroups in the variety* $\underline{\underline{m}}$ *over the projective geometry* $(L(E),+)$ *by the functor* $F:(L(E)) \to (\{\cdot\})$ *with* $UF = \{x + U | x \in E\}$ *and* $(U \to V)F:UF \to VF; x + U \mapsto x + V$.

Proof. That $((E,K)S,\cdot)$ is a Płonka sum over $(L(E),+)$ by the functor F follows directly from Proposition 3.4 with $k = 1/2$. The Płonka fibres are the algebras $(\{x + U | x \in E\},\underline{1/2})$ for fixed subspaces U, with $(x+U)(y+U)\underline{1/2} = xy\underline{1/2} + U$. Since these are the reducts $(E/U,\underline{1/2})$ of the affine geometries $(E/U,K)$ coming from the quotient vector subspaces E/U, they are quasigroups in the variety $\underline{\underline{m}}$ by Proposition 4.6.

COROLLARY 4.8. *The commutative binary mode* $((E,K)S,\cdot)$ *lies in the variety* $\tilde{\underline{\underline{m}}}$.

Proof. By Theorem 4.7 and Płonka's result [Pł,Theorem I] [RS,238], the algebra $((E,K)S,\cdot)$ satisfies the regular identities of $\underline{\underline{m}}$ - the regular identities satisfied by each of the Płonka fibres of $((E,K)S,\cdot)$. Theorem 4.4 then shows that $((E,K)S,\cdot)$ lies in the variety $\tilde{\underline{\underline{m}}}$.

In analogy with (3.7), Theorem 4.7 and Corollary 4.8 may be summarized as:

$$(4.9) \quad \begin{cases} \text{The } \underline{\tilde{m}}\text{-algebra of affine subspaces of an affine } K\text{-space} \\ \text{is a Płonka sum of } \underline{m}\text{-quasigroups over a projective geometry.} \end{cases}$$

Just as for the characteristic zero case, the proof of Theorem 2.4 for the case that K has odd characteristic is completed by showing that the Płonka fibres UF of Theorem 4.7 have no non-trivial semilattice quotient. Now these Płonka fibres, lying in $\underline{m}$, satisfy the identity $w_m(x,y) = x$. In a semilattice quotient $(H,+)$ of such a fibre, this identity becomes $x + y = x$ or $y \leq x$. Thus for two elements h,k of H, one has $h \leq k$ and $k \leq h$, whence $h = k$ and the triviality of H.

5. **CHARACTERISTIC TWO** This section considers the case that the field K has characteristic 2. Since $2 = 0$ is no longer invertible, the ternary parallelogram operation P can no longer be made redundant by (3.1), and the full algebra structure (E,K,P) is needed to give the affine geometry. As in Section 4, J will denote the prime subfield $GF(2)$ of K. Note that every subset of E is a subalgebra of the reduct (E,J) of (E,K,P), since the binary operations J are just the projections $xy\underline{0} = x$ and $xy\underline{1} = y$. Let Ω_K in this case denote the singleton $\{P\}$ consisting of the ternary parallelogram operation (2.2). Thus the "convex subsets" of E will be taken to be the subalgebras of (E,P), the J-affine subspaces of (E,K,P). By [OS], [RS,255], the class of all J-affine spaces, together with the empty set, is the variety of all <u>minority modes</u> (A,P), algebras with a ternary operation P satisfying the entropic law

$$(5.1) \quad \begin{aligned} &x_{11}x_{12}x_{13}Px_{21}x_{22}x_{23}Px_{31}x_{32}x_{33}PP = \\ &x_{11}x_{21}x_{31}Px_{12}x_{22}x_{32}Px_{13}x_{23}x_{33}PP \end{aligned}$$

and the identities

$$(5.2) \quad yxyP = x, \quad xyyP = x, \quad yyxP = x.$$

The name comes from the observation that the value of the operation P in the identities (5.2) reduces to that one of its arguments, if any, that is in the minority. Note that idempotence is a consequence of (5.2), so minority modes really are modes.

By [RS,146], the set of affine subspaces of the affine K-space forms a ternary mode $((E,K,P)S,P)$. The structure of this algebra is given by the following theorem.

THEOREM 5.3. *Let K be a field of characteristic 2. Then for an affine K-space (E,K,P), the ternary mode ((E,K,P)S,P) of affine subspaces is a Płonka sum of minority modes over the projective geometry (L(E),+) by the functor F:(L(E)) → ({P}) with* $UF = \{x + U | x \in E\}$ *and* $(U \to V)F:UF \to VF:x + U \mapsto x + V$.

Proof. Since K is of characteristic 2, the operation P on E as in (2.2) becomes $xyzP = x + y + z$. For vector subspaces U,V,W, and $X = U + V + W$ of E, and for corresponding affine subspaces $x + U$, $y + V$, and $z + W$, one has $(x+U)(y+V)(z+W)P = x + U + y + V + z + W = xyzP + X = (x+X)(y+X)(z+X)P$ $= x(U\to X)Fy(V\to X)Fz(W\to X)FP$. Thus ((E,K,P)S,P) is a Płonka sum as claimed. The Płonka fibres UF = (E/U,P), as J-affine spaces, are minority modes.

In the characteristic zero case, Corollary 3.6 to the Structure Theorem 3.5 for the algebra of affine subspaces specified this algebra as lying in the variety of rational barycentric algebras, with identities given by Theorem 3.3. In the case that K had odd characteristic, Corollary 4.8 to the Structure Theorem 4.7 for the algebra of affine subspaces specified this algebra as lying in the variety $\tilde{\underline{\underline{m}}}$ of commutative binary modes satisfying the identity $w_m(x,y) = w'_m(x,y)$. It is thus of interest in the current case to find a variety nicely containing the algebra of affine subspaces, so that this algebra is described well as lying in the variety. By Theorem 5.3 and the result of Płonka quoted earlier [Pł,Theorem I] [RS,238], the algebra ((E,K,P)S,P) may be described as satisfying each regular identity satisfied by its Płonka fibres, i.e. each regular identity satisfied by each minority mode. Unfortunately, there are infinitely many such identities involving the single operation P, so that this description seriously lacks conciseness. The problem is to find a finite set of identities, a so-called finite basis, of which the set of all regular identities satisfied by all minority modes is the consequence.

A little universal algebra, essentially implicit in the work of Płonka, serves to solve the problem. A variety $\underline{V}$ of algebras (A,Ω) is called strongly irregular if there is a binary derived operation * such that $\underline{V}$-algebras may be characterised as the Ω-algebras satisfying some set of regular identities and the single irregular identity $x * y = x$. For example, the variety $\underline{m}$ of commutative binary modes is strongly irregular. Taking the binary derived operation $x * y = w_m(x,y)$, the variety $\underline{m}$ is specified by the regular commutative, idempotent, and entropic identities, together with the single irregular identity $w_m(x,y) = x$, i.e. $x * y = x$. In the present context, the variety of minority modes is strongly irregular. Define

$$x * y = yxyP. \tag{5.4}$$

Then the variety of minority modes is the variety of algebras (A,P) satisfying the regular identities of idempotence and entropicity (5.1), together with the three identities (5.2). The first of these is just $x * y = x$. When this obtains, the second and third of them, which appear to be irregular, may in fact be rewritten as the regular identities

$$xyyP = x * y \quad \text{and} \quad yyxP = x * y,$$

i.e. as

$$xyyP = yxyP \quad \text{and} \quad yyxP = yxyP. \tag{5.5}$$

In other words, minority modes are the ternary algebras (A,P) satisfying idempotence, entropicity (5.1), (5.5), and the irregular identity $x * y = x$ with $*$ as in (5.4).

A variety $\underline{V}$ of algebras (A,Ω) is called <u>irregular</u> if there is an irregular identity satisfied by each $\underline{V}$-algebra. The <u>regularised variety</u> or <u>regularisation</u> $\tilde{\underline{\underline{V}}}$ of such a variety $\underline{V}$ is the variety of algebras satisfying all the regular identities satisfied by all $\underline{V}$-algebras. The current task is to specify the regularisation of a strongly irregular variety. Now a binary operation $*$ on an algebra (A,Ω) is said to be a <u>partition operation</u> on A if (A,Ω) satisfies the following identities: $(A,*)$ is a <u>left normal band</u>, i.e.

$$\begin{cases} x * x = x, \\ (x * y) * z = x * (y * z), \quad \text{and} \\ x * y * z = x * z * y; \end{cases} \tag{5.6}$$

$*$ <u>distributes from the right</u> over ω in Ω, i.e.

$$x_1 \ldots x_n\omega * y = (x_1 * y)\ldots(x_n * y)\omega; \tag{5.7}$$

and $*$ <u>breaks</u> ω <u>from the left</u>, i.e.

$$y * (x_1 \ldots x_n\omega) = y * x_1 * \ldots * x_n. \tag{5.8}$$

(Note that no bracketing is necessary in the right hand side of (5.8) once (5.6) holds.) The significance of partition operations comes from the following result of Płonka.

PROPOSITION 5.9. [Pł] [RS,237]. *An algebra* (A,Ω) *is a Płonka sum iff there is a partition operation* $*$ *on* A*. If these conditions obtain, the identity* $x * y = x$ *is satisfied by each fibre.*

Using this result, the following characterisation of the regularisation of a strongly irregular variety may be given.

THEOREM 5.10. *Let* $\mathbf{V}$ *be a strongly irregular variety, specified by* $x * y = x$ *and a set* R *of regular identities. Then the regularisation* $\tilde{\mathbf{V}}$ *is specified by* R *and the identities* (5.6), (5.7), (5.8).

Proof. Let $\mathbf{W}$ be the variety of algebras, of the same type as $\mathbf{V}$, satisfying R and (5.6), (5.7), (5.8). Let (A,Ω) be an algebra in $\mathbf{W}$. By (5.6), (5.7), 5.8), the derived binary operation $*$ is a partition operation on (A,Ω). Proposition 5.9 then shows that (A,Ω) is a Płonka sum of algebras satisfying the identity $x * y = x$. Since the Płonka fibres, as subalgebras of (A,Ω), also satisfy the identities in R, it follows that the fibres lie in $\mathbf{V}$. Consequently [Pł] [RS238] (A,Ω) satisfies the regular identities of $\mathbf{V}$, and so is in the regularisation $\tilde{\mathbf{V}}$. This shows that $\tilde{\mathbf{V}}$ contains $\mathbf{W}$.

Conversely, consider a $\mathbf{V}$-algebra (B,Ω). Since $x * y = x$ on B, the identities (5.6), (5.7), (5.8) are all satisfied by (B,Ω). As a $\mathbf{V}$-algebra, (B,Ω) also satisfies the identities R. Thus (B,Ω) lies in $\mathbf{W}$, and $\mathbf{W}$ contains $\mathbf{V}$. But since the identities specifying $\mathbf{W}$ are all regular, $\mathbf{W}$ also contains $\tilde{\mathbf{V}}$. The equality of $\tilde{\mathbf{V}}$ with $\mathbf{W}$, and the theorem, follow.

COROLLARY 5.11. *If a strongly irregular variety* $\mathbf{V}$ *has a finite basis for its identities, then so does its regularisation* $\tilde{\mathbf{V}}$.

Define an algebra (A,P) with a single ternary operation P to be a <u>regularised minority mode</u> if it satisfies the identities of idempotence, entropicity (5.1), (5.5), the <u>associative law</u> zyxyPzP = zyzPxzyzPP, the <u>left normal law</u> zyxyPzP = yzxzPyP, and the <u>left breaking law</u> xyzPtxyzPP = zyxtxPyPzP. Writing the derived operation $*$ as in (5.4), the idempotence, associative, and left normal laws show that (5.6) hold for regularised minority modes. The distributive law (5.7) follows from the idempotence and entropicity, while (5.8) follows from the left breaking law. Theorem 5.10 then shows that the regularisation of the variety of minority modes may thus be concisely described as the finitely based variety of regularised minority modes.

Regularised minority modes appear to have some interesting properties worthy of further investigation. There is a result of Płonka stating that the free algebra over a set X in the regularisation of a strongly irregular variety is a Płonka sum over the join semilattice of non-empty subsets of X, by the free algebra functor for the strongly irregular variety [Po] [RS,273]. As a consequence of this, it turns out that the free regularised minority mode on $n + 1$ elements has the cardinality of n-dimensional projective space over GF(3). As the following general result implies, the two structures have different automorphism groups, but it would nevertheless be useful to set up some correspondence between them in order to facilitate manipulation of the identities for regularised minority modes.

THEOREM 5.12. *Let* $\underline{\underline{W}}$ *be a variety specified by regular identities. For a set* X, *let* $X\underline{\underline{W}}$ *denote the free* $\underline{\underline{W}}$*-algebra on* X. *Then the automorphism group of* $X\underline{\underline{W}}$ *is isomorphic to the group of permutations of* X.

Proof. Since the identities of $\underline{\underline{W}}$ are regular, an element x of X can only lie in the subalgebra of $X\underline{\underline{W}}$ generated by a subset Y of $X\underline{\underline{W}}$ if x actually appears as an element of Y. If f is an automorphism of $X\underline{\underline{W}}$, this forces Xf to be equal to X. Conversely, knowledge of the restriction of f to the generating set X determines f uniquely. Thus restriction to X provides an isomorphism from the automorphism group of $X\underline{\underline{W}}$ to the permutation group of X.

Returning to the algebra of affine subspaces of an affine K-space, it is now possible to formulate the following corollary to Theorem 5.3.

COROLLARY 5.13. *The ternary algebra* ((E,K,P)S,P) *of affine subspaces of an affine* K*-space* (E,K,P) *is a regularised minority mode.*

Proof. By Theorem 5.3, ((E,K,P)S,P) is a Płonka sum of minority modes. Thus [Pł,Theorem I] [RS,238] it satisfies the regular identities satisfied by minority modes. Theorem 5.9 then shows that it is a regularised minority mode.

Theorem 5.3 and Corollary 5.12 summarize as:

(5.13) $\left\{\begin{array}{l}\text{The regularised minority mode of affine subspaces}\\ \text{of an affine K-space is a Płonka sum of}\\ \text{minority modes over a projective geometry.}\end{array}\right.$

To complete the proof of Theorem 2.4 for the case that K has characteristic 2, note that the Płonka fibres UF of Theorem 5.3 satisfy the irregular identity $x * y = x$. An argument identical to that given in the odd characteristic case then shows that these fibres have no non-trivial semilattice quotient, so the projective geometry (L(E),+) is the largest such quotient of ((E,K,P)S,P).

ACKNOWLEDGEMENTS

We are grateful to the mathematics departments at Iowa State University, the Université de Montréal, and Temple University (Philadelphia) for support and facilities during the preparation of this manuscript. Particular thanks are due to Ivo Rosenberg in Montréal and Hala Pflugfelder in Philadelphia.

REFERENCES

[Co] P. M. COHN, "Universal Algebra", Harper and Row, New York, 1965.

[JK] J. JEŽEK and T. KEPKA, The lattice of varieties of commutative abelian distributive groupoids, Alg. Univ. 5(1975),225-237.

[Kl] F. KLEIN, "Le Programme d'Erlangen", Gauthier-Villars, Paris 1974.

[Kn] F. KLEIN, "Gesammelte Mathematische Abhandlungen Bd.I", Springer, Berlin 1973.

[Ma] A. I. MAL'CEV (tr. A. P. DOOHOVSKOY and B. D. SECKLER), "Algebraic Systems", Springer, Berlin 1973.

[OS] F. OSTERMANN and J. SCHMIDT, Baryzentrischer Kalkül als axiomatische Grundlage der affinen Geometrie, J. reine angew. Math. 224(1966),44-57.

[Pł] J. PŁONKA, On a method of contruction of abstract algebras, Fund. Math. 61(1967),183-189.

[Po] J. PŁONKA, On free algebras and algebraic decompositions of algebras from some equational classes defined by regular equations, Alg. Univ. 1(1971),261-267.

[RS] A. B. ROMANOWSKA and J. D. H. SMITH, "Modal Theory - An Algebraic Approach to Order, Geometry, and Convexity", Heldermann-Verlag, Berlin, to appear.

[Sl] D. SLEPIAN, Some comments on Fourier analysis, uncertainty, and modeling, SIAM Review 25(1983),379-393.

MORE CONDITIONS EQUIVALENT TO CONGRUENCE MODULARITY

Steven T. Tschantz
Vanderbilt University
Nashville, Tennessee 37235

1. INTRODUCTION AND GENERAL THEORY. Mal'cev conditions equivalent to congruence modularity have been given by A. Day [1] and by H. P. Gumm [3]. In this paper we build a broad theory of similar equivalent conditions generalizing these results and develop techniques for dealing with such conditions effectively. We are able to find a single condition equivalent to congruence modularity that has both Day's condition and Gumm's condition as special cases.

If σ and τ are terms in variables α, β, γ...(standing for congruences) and binary operations $\circ$ and $\cdot$ (for composition of binary relations and intersection), then the condition $\sigma \leq \tau$ (set inclusion), holding for all algebras A of a variety and all α, β, $\gamma\ldots$ congruences on A, is equivalent to a certain strong Mal'cev condition holding in the variety [7]. The procedure for translating from σ and τ to the identities defining the corresponding strong Mal'cev condition is straight-forward, easily verified in any particular case, and best explained by example (see proof of lemma 3). If τ also involves the binary operation $+$ (for join), then the condition $\sigma \leq \tau$ (for all algebras, congruences) is equivalent to the Mal'cev condition defined by the union of the strong Mal'cev conditions corresponding to $\sigma \leq \tau_n$ where τ_n is τ with the joins replaced by n compositions $(1 \leq n < \omega)$. (If σ involves joins then we get corresponding weak Mal'cev conditions as intersections of simpler Mal'cev conditions.) For example, we have easily

LEMMA 1 [1]. *The following are equivalent for any variety V.*

a) for all $A \in V$ and α, β, $\gamma \in Con(A)$, $\gamma\cdot(\beta \circ \gamma\alpha \circ \beta) \leq \gamma\alpha + \gamma\beta$

b) (Day terms) there exists an n and 4-ary terms $m_0, m_1, \ldots m_n$ such that V satisfies

i) $m_0(xyzu) = x$ and $m_n(xyzu) = u$

ii) $m_i(xyyx) = x$ for all $i \leq n$

iii) $m_i(xxzz) = m_{i+1}(xxzz)$ for all even $i < n$

iv) $m_i(xyyu) = m_{i+1}(xyyu)$ for all odd $i < n$.

LEMMA 2 [4,6]. *The following are equivalent for any variety V.*

a) for all $A \in V$ and $\alpha, \beta, \gamma \in Con(A)$, $\gamma \cdot (\alpha \circ \beta) \leq \gamma \cdot (\beta \circ \alpha) \circ (\gamma\alpha + \gamma\beta)$

b) (Gumm terms) there exists an n and 3-ary terms $p, q_1, \ldots q_n$ such that V satisfies

i) $p(xzz) = x$ and $q_n(xyz) = z$

ii) $p(xxz) = q_1(xxz)$

iii) $q_i(xyx) = x$ for all $i \leq n$

iv) $q_i(xxz) = q_{i+1}(xxz)$ for all even $i < n$

v) $q_i(xzz) = q_{i+1}(xzz)$ for all odd $i < n$.

We add to these results the following lemmas.

LEMMA 3. *The following are equivalent for any variety V.*

a) for all algebras $A \in V$ and $\alpha, \beta, \gamma \in Con(A)$,

$$\gamma\beta \circ \gamma \cdot (\alpha \circ \beta) \leq \gamma \cdot (\alpha \circ \beta) \circ (\gamma\alpha + \gamma\beta)$$

b) there exists an n and 4-ary terms $p', q'_1, \ldots q'_n$ such that V satisfies

i) $p'(xyyu) = x$ and $q'_n(xyzu) = u$

ii) $p'(xxzz) = q'_1(xxzz)$

iii) $q'_i(xxzx) = x$ for all $i \leq n$

iv) $q'_i(xxzz) = q'_{i+1}(xxzz)$ for all even $i < n$

v) $q'_i(xyyu) = q'_{i+1}(xyyu)$ for all odd $i < n$.

LEMMA 4. *The following are equivalent for any variety V.*

a) for all algebras $A \in V$ and $\alpha, \beta, \gamma \in Con(A)$,

$$\gamma \cdot (\beta \circ \alpha \circ \beta) \leq \gamma \cdot (\alpha \circ \beta) \circ (\gamma\alpha + \gamma\beta)$$

b) there exists an n and 4-ary terms $p'', q''_1, \ldots q''_n$ such that V satisfies

i) $p''(xyyu) = x$ and $q''_n(xyzu) = u$

ii) $p''(xxzz) = q''_1(xxzz)$

iii) $q''_i(xyzx) = x$ for all $i \leq n$

iv) $q''_i(xxzz) = q''_{i+1}(xxzz)$ for all even $i < n$

v) $q''_i(xyyu) = q''_{i+1}(xyyu)$ for all odd $i < n$.

We give a detailed proof of lemma 3. The other lemmas can all be proved by similar reasoning.

Proof of 3. Assuming a) holds, we take $A = F_V(4)$ generated by x, y, z and u and take $\alpha = Cg(\{(y,z)\})$, $\beta = Cg(\{(x,y),(z,u)\})$, and $\gamma = Cg(\{(x,y),(x,u)\})$. Then

$$(x,u) \in \gamma\beta \circ \gamma\cdot(\alpha \circ \beta) \leq \gamma\cdot(\alpha \circ \beta) \circ (\gamma\alpha + \gamma\beta)$$

so for some n

$$(x,u) \in \gamma\cdot(\alpha \circ \beta) \circ \underbrace{(\gamma\alpha \circ \gamma\beta \circ \gamma\alpha \circ \ldots)}_{n-1 \text{ terms}}$$

so there exist $p', q_1', \ldots q_n'$, elements of A, with

i) $p' \; \alpha \; x$ and $q_n' = u$

ii) $p' \; \beta \; q_1'$

iii) $q_i' \; \gamma \; x$ for all $i \leq n$

iv) $q_i' \; \beta \; q_{i+1}'$ for all even $i < n$

v) $q_i' \; \alpha \; q_{i+1}'$ for all odd $i < n$.

Taking 4-ary terms in x, y, z and u representing these elements of A we get the 5 parts of b) (since e.g. α is the kernel of the map fixing x, y, and u, and taking z to y). Conversely, assuming b) holds and given $A \in V$, $\alpha, \beta, \gamma \in \mathrm{Con}(A)$, suppose $(x,u) \in \gamma\beta \circ \gamma\cdot(\alpha \circ \beta)$ and take y and z showing this, i.e. $x \; \gamma\beta \; y \; \alpha \; z \; \beta \; y$ and $y \; \gamma \; u$. Then $p'(xyzu)$, $q_1'(xyzu), \ldots q_n'(xyzu)$ are elements showing $(x,u) \in \gamma\cdot(\alpha \circ \beta) \circ (\gamma\alpha + \gamma\beta)$.

We will show that all of the above conditions (and more) are in fact equivalent. Thus, any variety with terms satisfying part b) of one of conditions 1-4 always has terms for each of these. In particular, a variety defined by the equations in one of 1b)-4b) will satisfy all of conditions 1-4. We conclude therefore, that there must be a way of writing the terms in each of these condtions as expressions in the terms for each of the other conditions.

The underlying method in the following analyses is to establish implications between Mal'cev conditions by simply exhibiting the construction of terms for one Mal'cev condition from terms for another. For example, terms satisfying 4b) immediately satisfy 3b) and by identifying variables will yield 2b). By taking $m_1(xyzu) = x$ we also have the terms in 4b) giving us terms for 1b). Thus, in some sense, conditions 1-3 are contained in condition 4 as special cases. Of course, explicit constructions relating Mal'cev conditions are not usually this trivial. The construction of terms for 1b) from terms for 2b) is easy but not obvious, while the construction of terms for 2b) from terms for 1b) is neither easy nor obvious [4,6].

Another approach to showing implications between these conditions is to use part a) of conditions 1-4. Since for congruences on any algebra,

$$\gamma\beta \circ \gamma\cdot(\alpha \circ \beta) = \gamma\cdot(\gamma\beta \circ \alpha \circ \beta) \leq \gamma\cdot(\beta \circ \alpha \circ \beta) \quad \text{and} \quad \gamma\cdot(\beta \circ \alpha) \leq \gamma\cdot(\beta \circ \alpha \circ \beta),$$

we have that (4a) $\gamma\cdot(\beta \circ \alpha \circ \beta) \leq \gamma\cdot(\alpha \circ \beta) \circ (\gamma\alpha + \gamma\beta)$ implies both

$$\gamma\beta \circ \gamma\cdot(\alpha \circ \beta) \leq \gamma\cdot(\alpha \circ \beta) \circ (\gamma\alpha + \gamma\beta) \quad \text{and} \quad \gamma\cdot(\beta \circ \alpha) \leq \gamma\cdot(\alpha \circ \beta) \circ (\gamma\alpha + \gamma\beta),$$

(3a and 2a rewritten). For any algebra, if $\gamma\cdot(\beta \circ \alpha \circ \beta) \leq \gamma\cdot(\alpha \circ \beta) \circ (\gamma\alpha + \gamma\beta)$ holds for all congruences α, β, γ then (1a)

$$\gamma\cdot(\beta \circ \gamma\alpha \circ \beta) \leq \gamma\cdot(\gamma\alpha \circ \beta) \circ (\gamma\alpha + \gamma\beta) = \gamma\alpha + \gamma\beta.$$

These derivations reflect the correspondence between the terms in 4b) and 1b)-3b) noted above and are considerably easier to check and understand. Unfortunately, such an easy derivation can be expected only when the implication holds for every algebra and not just for algebras generating varieties in which every algebra satisfies the conditions. (Thus we see the sense in which 4 is a stronger condition than 1-3 even though we will show that, for varieties, they are all equivalent.)

There is a useful generalization of this last approach suggested by the following incorrect proof that 2a) implies 1a). If $\gamma\cdot(\alpha \circ \beta) \leq \gamma\cdot(\beta \circ \alpha) \circ (\gamma\alpha + \gamma\beta)$ then taking $\beta \circ \gamma\alpha$ in place of α we get $\gamma\cdot(\beta \circ \gamma\alpha \circ \beta) \leq \gamma\cdot(\beta \circ \beta \circ \gamma\alpha) \circ (\gamma\cdot(\beta \circ \gamma\alpha) + \gamma\beta) = \gamma\alpha + \gamma\beta$. The problem is that we have 2a) holding for congruences but that, in general, $\beta \circ \gamma\alpha$ will not be a congruence. There is a corresponding correct proof involving the terms of 2b). What we need is a generalization of 2a) to arbitrary binary reflexive admissible relations (subalgebras of the square of the algebra containing the identity relation) such as $\beta \circ \gamma\alpha$.

If α and β are reflexive admissible relations (hereafter r.a.r.s) we will write $\breve{\alpha}$ for the converse of α, $\alpha \circ \beta$ for relational composition, $\alpha\cdot\beta$ for intersection, $\alpha + \beta$ for the union of the relations $\alpha \circ \beta \circ \alpha \circ \ldots$ taking compostions of n terms, $n \geq 2$, (note that $\alpha + \alpha$ is then the transitive closure of α), and define $\overline{\alpha} = (\alpha \circ \breve{\alpha})(\breve{\alpha} \circ \alpha)$. Thus, an r.a.r. α is a congruence iff $\alpha = \breve{\alpha} = \alpha \circ \alpha = \alpha + \alpha$. The important advantage over restricting ourselves to congruences is that if α and β are r.a.r.s, then $\breve{\alpha}, \alpha \circ \beta, \alpha\cdot\beta$, and $\alpha + \beta$ all are also. If σ and τ are terms in variables $\alpha, \beta, \gamma \ldots$ standing for r.a.r.s instead of congruences and involving $\breve{\ }$, $\cdot$, $\circ (+)$

then $\sigma \leq \tau$ holding in a variety is equivalent to a strong Mal'cev condition (Mal'cev or weak Mal'cev condition) by a procedure similar to that for congruences [5,7].

Our strategy for proving the following results and establishing new equivalent conditions can now be outlined. Starting from an identity $\sigma \leq \tau$ holding for congruences (or r.a.r.s) on algebras in a variety, we form the corresponding Mal'cev condition. Using direct constructions from these terms, we find generalizations of the form $\sigma' \leq \tau'$ which hold for arbitrary r.a.r.s on algebras in the variety. Special cases of this new identity should yield the original $\sigma \leq \tau$ and more candidate conditions. The process is repeated in hopes of chaining together a large number of equivalent conditions. Conditions can be combined to establish new conditions for comparison.

The difficult steps in this process are in choosing appropriate and useful generalizations and in eliminating special cases that are too weak to possibly chain back to the original condition. In principle, all of these steps can be replaced by direct constructions relating the corresponding Mal'cev conditions. In practice, it has been much easier to organize and use the conditions on congruences (and r.a.r.s).

2. MAIN RESULTS. We are now ready to state and prove our main results, lisiting conditions equivalent to congruence modularity. In each case we give an identity which is to hold for all algebras in the variety and for all $\alpha, \beta, \gamma \ldots$. congruences of the algebra (or for all $\alpha, \beta, \gamma \ldots$r.a.r.s of the algebra). In each case there is also a corresponding equivalent (possibly weak) Mal'cev condition. The Mal'cev conditions corresponding to the important parts of our main theorem have been given in the last section and the other Mal'cev conditions can be easily derived. The most interesting conditions below are 0-5.

THEOREM 5. In 6 and 7 below we take $\alpha, \beta, \gamma \ldots$ to be reflexive admissible relations (r.a.r.s) and otherwise we take $\alpha, \beta, \gamma \ldots$ to be congruences. For any variety V, the following conditions holding for all algebras $A \in V$ and all $\alpha, \beta, \gamma \ldots$ on A are equivalent.

0a) $\gamma \cdot (\gamma\alpha + \beta) \leq \gamma\alpha + \gamma\beta$

0b) $\gamma \cdot (\gamma\alpha + \gamma\beta) = \gamma\alpha + \gamma\beta$ (congruence modularity)

1) $\gamma \cdot (\beta \circ \gamma\alpha \circ \beta) \leq \gamma\alpha + \gamma\beta$

2) $\gamma \cdot (\alpha \circ \beta) \leq \gamma \cdot (\beta \circ \alpha) \circ (\gamma\alpha + \gamma\beta)$

3) $\gamma\beta \circ \gamma \cdot (\alpha \circ \beta) \leq \gamma \cdot (\alpha \circ \beta) \circ (\gamma\alpha + \gamma\beta)$

4) $\gamma \cdot (\beta \circ \alpha \circ \beta) \leq \gamma \cdot (\alpha \circ \beta) \circ (\gamma\alpha + \gamma\beta)$

5a) $\gamma \cdot (\alpha + \beta) \leq \gamma \cdot (\alpha \circ \beta) \circ (\gamma\alpha + \gamma\beta)$

5b) $\gamma\cdot(\alpha+\beta)=\gamma\cdot(\alpha\circ\beta)\circ(\gamma\alpha+\gamma\beta)$

6) for r.a.r.s, $\delta_1\circ\delta_2\circ\gamma\cdot(\alpha\circ\beta)\le(\delta_1\circ\gamma\gamma^{\smile})(\beta\beta^{\smile}\circ\delta_1\circ\alpha^{\smile})(\alpha\alpha^{\smile}\circ\delta_1\circ\beta)$
$\circ\,\delta_2\circ(\overline{\gamma}\alpha+\overline{\gamma}\alpha^{\smile}+\overline{\gamma}\beta+\overline{\gamma}\beta^{\smile})$

7) for r.a.r.s, $\gamma\cdot(\alpha\circ\beta)\le\gamma\gamma^{\smile}\cdot(\beta\beta^{\smile}\circ\alpha\alpha^{\smile})\circ(\overline{\gamma}\alpha+\overline{\gamma}\alpha^{\smile}+\overline{\gamma}\beta+\overline{\gamma}\beta^{\smile})$

8) $\gamma\cdot(\alpha\circ\beta)\circ\gamma\cdot(\alpha\circ\beta)\le\gamma\cdot(\alpha\circ\beta)\circ(\gamma\alpha+\gamma\beta)$

9) $\delta\circ\gamma\cdot(\alpha\circ\beta)\le\gamma\cdot(\alpha\circ\beta)\circ\delta\circ(\gamma\alpha+\gamma\beta)$

10) $\gamma\cdot(\alpha\circ\beta)+(\gamma\alpha+\gamma\beta)\le\gamma\cdot(\alpha\circ\beta)\circ(\gamma\alpha+\gamma\beta)$

11) $\delta\circ\gamma\cdot(\alpha\circ\beta)\le(\delta\circ\gamma)(\alpha\circ\delta\circ\beta)\circ(\gamma\alpha+\gamma\beta)$

12) $\gamma\cdot(\beta\circ\alpha\circ\beta)\le\gamma\cdot(\alpha\circ\beta\circ\alpha)\circ(\gamma\alpha+\gamma\beta)$

13) $\gamma\cdot(\beta\circ\alpha\circ\beta)\le\gamma\cdot((\alpha+\gamma\beta)\circ\beta\circ(\alpha+\gamma\beta))$

14) $\gamma\cdot(\alpha+\beta)\le\gamma\cdot(\alpha\circ\beta\circ\alpha)\circ(\gamma\alpha+\gamma\beta)$

Proof. We have that 0a) and 0b) are equivalent since $\gamma\alpha+\gamma\beta\le\gamma\cdot(\gamma\alpha+\beta)$ and 5a) and 5b) are equivalent since $\gamma\cdot(\alpha\circ\beta)\circ(\gamma\alpha+\gamma\beta)\le\gamma\cdot(\alpha+\beta)$ always. The equivalence of 0, 1, and 2 is just a restatement of the results of [1] and [3]. Our chain of reasoning here gives an alternate proof of this using the techniques described in the previous section with the major steps being the proofs that 1 implies 2, 3 implies 2, and 2 implies 6. These steps employ the terms from lemmas 1-3 but then we will be able to complete the chain of equivalent conditions without further resorting to this approach. For the proof along these lines that 1 implies 2 the reader is referred to [4,6].

Proof of 3 implies 2. By Lemma 3 we have terms as in 3b). Assume α,β, and γ are congruences on an algebra in our variety and take $(x,z)\in\gamma\cdot(\alpha\circ\beta)$. Hence, there is a y such that $x\ \alpha\ y\ \beta\ z$. Then we have

$x=p'(xzzz)\ \beta\ p'(xyzz)\ \alpha\ p'(xxzz)=q_1'(xxzz)$,

$x=q_1'(xxzx)\ \gamma\ q_1'(xxzz)$,

$q_1'(xxzz)\ \beta\ q_1'(xxyz)$,

$q_i'(xxyz)\ \gamma\ q_i'(xxyx)=x$ for all $i\le n$,

$q_i'(xxyz)\ \alpha\ q_i'(xxxz)=q_{i+1}'(xxxz)\ \alpha\ q_{i+1}'(xxyz)$ for all odd $i<n$,

$q_i'(xxyz)\ \beta\ q_i'(xxzz)=q_{i+1}'(xxzz)\ \beta\ q_{i+1}'(xxyz)$ for all even $i<n$, and

$q_n'(xxyz)=z$.

Hence we have $(x,z)\in\gamma\cdot(\beta\circ\alpha)\circ(\gamma\alpha+\gamma\beta)$, giving us 2.

Proof of 2 implies 6. By lemma 2 we have Gumm terms as in 2b). Now let $\delta_1, \delta_2, \alpha, \beta, \gamma$ be reflexive admissible relations on an algebra in our variety, $(v, z) \in \delta_1 \circ \delta_2 \circ \gamma \cdot (\alpha \circ \beta)$ and take w, x, and y so that $v \; \delta_1 \; w \; \delta_2 \; x \; \alpha \; y \; \beta \; z$ and $x \; \gamma \; z$. Then

$$v \; \delta_1 \; w = p(wxx) \; \gamma \; p(wxz), \quad \text{and} \quad w = p(wzz) \check{\gamma} \, p(wxz),$$

so we have $(v, p(wxz)) \in \delta_1 \circ \gamma\check{\gamma}$. Similarly,

$$v = p(vyy) \; \beta \; p(vyz), \quad \text{and} \quad v = p(vzz) \; \check{\beta} \; p(vyz) \; \delta_1 \; p(wyz) \; \check{\alpha} \; p(wxz),$$

so $(v, p(wxz)) \in \beta\check{\beta} \circ \delta_1 \circ \check{\alpha}$. And again,

$$v = p(vxx) \; \alpha \; p(vxy), \quad \text{and} \quad v = p(vyy) \; \check{\alpha} \; p(vxy) \; \delta_1 \; p(wxy) \; \beta \; p(wxz),$$

so $(v, p(wxz)) \in \alpha\check{\alpha} \circ \delta_1 \circ \beta$. Finally,

$$\begin{aligned}
&p(wxz) \; \delta_2 \; p(xxz) = q_1(xxz), \\
&q_i(xxz) \; \alpha \; q_i(xyz) \; \beta \; q_i(xzz) = q_{i+1}(xzz) \quad \text{for all odd } i < n, \\
&q_i(xzz) \; \check{\beta} \; q_i(xyz) \; \check{\alpha} \; q_i(xxz) = q_{i+1}(xxz) \quad \text{for all even } i < n, \\
&x = q_i(xxx) \; \gamma \; q_i(xxz) \; \gamma \; q_i(zxz) = z \quad \text{for all } i \leq n, \\
&x = q_i(xyx) \; \gamma \; q_i(xyz) \; \gamma \; q_i(zyz) = z \quad \text{for all } i \leq n, \\
&x = q_i(xzx) \; \gamma \; q_i(xzz) \; \gamma \; q_i(zzz) = z \quad \text{for all } i \leq n,
\end{aligned}$$

so for odd $i < n$ we have

$$q_i(xxz) \; \overline{\gamma}\alpha \; q_i(xyz) \; \overline{\gamma}\beta \; q_i(xzz) = q_{i+1}(xzz) \; \overline{\gamma}\check{\beta} \; q_{i+1}(xyz) \; \overline{\gamma}\check{\alpha} \; q_{i+1}(xxz) = q_{i+2}(xxz).$$

But $q_n(xyz) = z$, so we have 6 since

$$(v, z) \in (\delta_1 \circ \gamma\check{\gamma})(\beta\check{\beta} \circ \delta_1 \circ \check{\alpha})(\alpha\check{\alpha} \circ \delta_1 \circ \beta) \circ \delta_2 \circ (\overline{\gamma}\alpha + \overline{\gamma}\check{\alpha} + \overline{\gamma}\beta + \overline{\gamma}\check{\beta}).$$

We now use 6 to derive 7-9 each of which in turn implies 2 or 3, thus completing the proof of the equivalence of these conditions. With δ_1 and δ_2 equal to the identity relation, we have 6 implies $\gamma \cdot (\alpha \circ \beta) \leq \gamma \cdot (\alpha\check{\alpha} \circ \beta) \circ (\overline{\gamma}\alpha + \overline{\gamma}\check{\alpha} + \overline{\gamma}\beta + \overline{\gamma}\check{\beta})$. Applying 6 again to $\gamma \cdot (\alpha\check{\alpha} \circ \beta)$ on the right hand side we get

$$\begin{aligned}
\gamma \cdot (\alpha \circ \beta) &\leq \gamma \cdot (\alpha\check{\alpha} \circ \beta) \circ (\overline{\gamma}\alpha + \overline{\gamma}\check{\alpha} + \overline{\gamma}\beta + \overline{\gamma}\check{\beta}) \\
&\leq \gamma\check{\gamma} \cdot (\beta\check{\beta} \circ (\alpha\check{\alpha})\check{\ }) \circ (\overline{\gamma}\alpha + \overline{\gamma}\check{\alpha} + \overline{\gamma}\beta + \overline{\gamma}\check{\beta}) \\
&= \gamma\check{\gamma} \cdot (\beta\check{\beta} \circ \alpha\check{\alpha}) \circ (\overline{\gamma}\alpha + \overline{\gamma}\check{\alpha} + \overline{\gamma}\beta + \overline{\gamma}\check{\beta}).
\end{aligned}$$

This is 7 and if we take α, β, γ to be congruences we have 7 implies 2. Next, with $\delta_1 = \gamma\cdot(\alpha \circ \beta)$, δ_2 the identity relation, and α, β, γ congruences, we have 6 implies

$$\gamma\cdot(\alpha \circ \beta) \circ \gamma\cdot(\alpha \circ \beta) \leq \gamma\cdot(\alpha \circ \gamma\cdot(\alpha \circ \beta) \circ \beta) \circ (\gamma\alpha + \gamma\beta) = \gamma\cdot(\alpha \circ \beta) \circ (\gamma\alpha + \gamma\beta).$$

This is 8 and since $\gamma\beta \circ \gamma\cdot(\alpha \circ \beta) \leq \gamma\cdot(\alpha \circ \beta) \circ \gamma\cdot(\alpha \circ \beta)$ this implies 3. With δ_1 the identity and $\delta_2 = \delta$, for congruences we have 6 implies 9. With $\delta = \gamma\beta$, we have 9 implies 3.

Since 8 and 9 are equivalent now, we use these together to establish 10. It suffices to show that for every n,

$$\underbrace{\gamma\cdot(\alpha \circ \beta) \circ (\gamma\alpha + \gamma\beta) \circ \gamma\cdot(\alpha \circ \beta) \circ \ldots}_{n \text{ terms}} \leq \gamma\cdot(\alpha \circ \beta) \circ (\gamma\alpha + \gamma\beta).$$

We proceed by induction on n. If $n < 3$ the conclusion is trivial. If $n \geq 3$ and we assume the result for fewer compositions then we have by induction hypothesis applied to the last $n - 2$ terms on the left hand side,

$$\begin{aligned} &\gamma(\alpha \circ \beta) \circ (\gamma\alpha + \gamma\beta) \circ \underbrace{(\gamma\cdot(\alpha \circ \beta) \circ \cdots)}_{n-2 \text{ terms}} \\ &\qquad \leq \gamma\cdot(\alpha \circ \beta) \circ (\gamma\alpha + \gamma\beta) \circ (\gamma(\alpha \circ \beta) \circ (\gamma\alpha + \gamma\beta)) \\ &\qquad \leq \gamma\cdot(\alpha \circ \beta) \circ \gamma\cdot(\alpha \circ \beta) \circ (\gamma\alpha + \gamma\beta) \quad \text{by } 9 \\ &\qquad \leq \gamma\cdot(\alpha \circ \beta) \circ (\gamma\alpha + \gamma\beta) \text{ by } 8 \end{aligned}$$

so we have the conclusion for n terms on the left. Thus, we have 10 by induction. Since $\gamma\beta \circ \gamma\cdot(\alpha \circ \beta) \leq \gamma\cdot(\alpha \circ \beta) + (\gamma\alpha + \gamma\beta)$, 10 implies 3 showing that 10 is equivalent to the previously established conditions.

Now we can use 2, 7, and 10 to derive 5. It suffices to show that for every n,

$$\gamma\cdot\underbrace{(\alpha \circ \beta \circ \alpha \circ \ldots)}_{n \text{ terms}} \leq \gamma\cdot(\alpha \circ \beta) \circ (\gamma\alpha + \gamma\beta).$$

We proceed by induction on n. If $n < 3$ then the conclusion is trivial. If $n \geq 3$ and we assume the result for fewer compositions then for even n using 7

$$\begin{aligned} &\gamma\cdot((\alpha \circ \beta) \circ \underbrace{(\alpha \circ \cdots \circ \beta)}_{n-2 \text{ terms}} \\ &\quad \leq \gamma\cdot(\underbrace{(\alpha \circ \cdots \circ \beta)}_{n-2 \text{ terms}} \circ (\beta \circ \alpha)) \circ (\gamma\cdot(\alpha \circ \beta) + \gamma\cdot(\beta \circ \alpha) + \gamma\cdot\underbrace{(\alpha \circ \ldots \circ \beta)}_{n-2 \text{ terms}} + \gamma\cdot\underbrace{(\beta \circ \ldots \circ \alpha)}_{n-2 \text{ terms}}) \\ &\quad \leq \gamma\cdot\underbrace{(\alpha \circ \ldots \circ \beta \circ \alpha)}_{n-1 \text{ terms}} \circ (\gamma\cdot(\alpha \circ \beta) + \gamma\cdot(\beta \circ \alpha) + \gamma\cdot\underbrace{(\alpha \circ \cdot \circ \beta \circ \alpha)}_{n-1 \text{ terms}} + \gamma\cdot\underbrace{(\alpha \circ \beta \cdots \circ \alpha)}_{n-1 \text{ terms}}) \end{aligned}$$

so by induction hypothesis

$$\leq \gamma\cdot(\alpha\circ\beta)\circ(\gamma\alpha+\gamma\beta)\circ(\gamma\cdot(\alpha\circ\beta)+\gamma\cdot(\beta\circ\alpha)+(\gamma\cdot(\alpha\circ\beta)\circ(\gamma\alpha\circ\gamma\beta))$$
$$\leq \gamma\cdot(\alpha\circ\beta)+(\gamma\alpha+\gamma\beta)\leq\gamma\cdot(\alpha\circ\beta)\circ(\gamma\alpha+\gamma\beta)$$

by 2 and 10. For odd n we start with

$$\gamma\cdot((\alpha\circ\beta)\circ\underbrace{(\alpha\circ\ldots\circ\alpha)}_{n-2\text{ terms}})\leq\gamma\cdot(\underbrace{(\alpha\circ\ldots\circ\alpha)}_{n-2\text{ terms}}\circ(\alpha\circ\beta))\circ(\gamma\cdot(\alpha\circ\beta)+\gamma\cdot\underbrace{(\alpha\circ\ldots\circ\alpha)}_{n-2\text{ terms}})$$

and proceed as before. Hence, 5 follows by induction.

Taking $\gamma\alpha$ in place of α, 5 implies 0, and 0 implies 1, since $\gamma\cdot(\beta\circ\gamma\alpha\circ\beta)\leq\gamma\cdot(\gamma\alpha+\beta)$. Since $\gamma\cdot(\beta\circ\alpha\circ\beta)\leq\gamma\cdot(\alpha+\beta)$, 5 implies 4, which in turn implies 3. The remaining conditions (and many others) are implied by either 4 or 6, and in turn imply 1, 2, or 3 by similar reasoning, completing the proof of the theorem.

3. FURTHER GENERALIZATIONS. Condition 5 of the last theorem suggests further generalizations, e.g. the condition $\gamma\cdot(\alpha_0+\alpha_1+\alpha_2)=\gamma\cdot(\alpha_0\circ\alpha_1\circ\alpha_2)\circ(\gamma\alpha_0+\gamma\alpha_1+\gamma\alpha_2)$. Several partial results can be derived without returning to constructions with terms. We instead produce a construction giving us generalizations encompassing all of these results.

THEOREM 6. For any variety V and any $m\geq 2$, congruence modularity for V is equivalent to each of the following conditions.

15) For all $A\in V$, δ any r.a.r., and γ_j,α_k congruences on A,

$$\delta\cdot(\gamma_0\circ\gamma_1\circ\gamma_2\circ\ldots\gamma_{m-1})(\alpha_0\circ\alpha_1\circ\alpha_2\circ\ldots\circ\alpha_{m-1})$$
$$\leq\delta\cdot(\gamma_0\circ\gamma_1\circ\gamma_2\circ\ldots\circ\gamma_{m-1})(\gamma_1\circ\gamma_0\circ\gamma_2\circ\ldots\circ\gamma_{m-1})(\gamma_1\circ\gamma_2\circ\ldots\circ\gamma_{m-1}\circ\gamma_0)$$
$$\cdot(\alpha_0\circ\alpha_1\circ\alpha_2\circ\ldots\circ\alpha_{m-1})(\alpha_1\circ\alpha_0\circ\alpha_2\circ\ldots\circ\alpha_{m-1})(\alpha_1\circ\alpha_2\circ\ldots\circ\alpha_{m-1}\circ\alpha_0)$$
$$\circ(\textstyle\sum\{\gamma_j\alpha_k:0\leq j,k<m\}).$$

16) For all $A\in V$, and γ_j,α_k congruences on A,

$$(\gamma_0\circ\gamma_1\circ\gamma_2\circ\ldots\circ\gamma_{m-1})(\alpha_0\circ\alpha_1\alpha_2\circ\ldots\circ\alpha_{m-1})$$
$$\leq\Pi\{(\gamma_{\pi(0)}\circ\gamma_{\pi(1)}\circ\gamma_{\pi(2)}\circ\ldots\circ\gamma_{\pi(m-1)}):\pi\in\mathrm{Sym}(m)\}$$
$$\cdot\Pi\{(\alpha_{\pi(0)}\circ\alpha_{\pi(1)}\circ\alpha_{\pi(2)}\circ\ldots\circ\alpha_{\pi(m-1)}):\pi\in\mathrm{Sym}(m)\}$$
$$\circ(\textstyle\sum\{\gamma_j\alpha_k:0\leq j,k<m\}).$$

17) For all $A \in V$, and γ_j, α_k congruences, on A,

$$(\sum\{\gamma_j : 0 \leq j < m\})(\sum\{\alpha_k : 0 \leq k < m\})$$
$$\leq (\gamma_0 \circ \gamma_1 \circ \ldots \circ \gamma_{m-1})(\alpha_0 \circ \alpha_1 \circ \ldots \circ \alpha_{m-1}) \circ (\sum\{\gamma_j \alpha_k : 0 \leq j, k < m\}).$$

Proof. We prove 15 using modularity and the terms in lemma 2. Taking γ_j, α_k congruences on $A \in V$ a variety satisfying conditions 1 and 2 of theorem 5, suppose

$$(x, z) \in (\gamma_0 \circ \gamma_1 \circ \gamma_2 \circ \ldots \circ \gamma_{m-1})(\alpha_0 \circ \alpha_1 \circ \alpha_1 \circ \alpha_2 \circ \ldots \circ \alpha_{m-1}).$$

Take $x_j \in A$, $0 \leq j \leq m$ so $x_0 = x$, $x_m = z$ and $x_j \; \gamma_j \; x_{j+1}$ and take $y_k \in A$, $0 \leq k \leq m$ so $y_0 = x$, $y_m = z$ and $y_k \; \alpha_k \; y_{k+1}$. Then

$$x = p(xxx_0), \quad p(xxx_j) \; \gamma_j \; p(xxx_{j+1}), \quad \text{and} \quad p(xxx_m) = p(xxz),$$

so $x \, (\gamma_0 \circ \gamma_1 \circ \gamma_2 \circ \ldots \gamma_{m-1}) \, p(xxz)$. Likewise,

$$x = p(xx_1x_1) \; \gamma_1 \; p(xx_1x_2) \; \gamma_0 \; p(xx_0x_2) = p(xxx_2) \, (\gamma_2 \circ \ldots \circ \gamma_{m-1}) \, p(xxz), \quad \text{and}$$
$$x = p(xx_1x_1) \; \gamma_1 \; p(xx_1x_2) \, (\gamma_2 \circ \ldots \circ \gamma_{m-1}) \, p(xx_1x_m) \; \gamma_0 \; p(xxz).$$

Similarly for the α's

$$x = p(xxy_0) \; \alpha_0 \; p(xxy_1) \; \alpha_1 \; p(xxy_2) \, (\alpha_2 \circ \ldots \circ \alpha_{m-1}) \, p(xxy_m) = p(xxz),$$
$$x = p(xy_1y_1) \; \alpha_1 \; p(xy_1y_2) \; \alpha_0 \; p(xxy_2) \, (\alpha_2 \circ \ldots \circ \alpha_{m-1}) \, p(xxz), \quad \text{and}$$
$$x = p(xy_1y_1) \; \alpha_1 \; p(xy_1y_2) \, (\alpha_2 \circ \ldots \circ \alpha_{m-1}) \, p(xy_1y_m) \; \alpha_0 \; p(xxz).$$

Finally $x = p(xxx) \; \delta \; p(xxz) = q_1(xxz)$, hence

$$(x, q_1(xxz)) \in \delta \cdot (\gamma_0 \circ \gamma_1 \circ \ldots \circ \gamma_{m-1})(\gamma_1 \circ \gamma_0 \circ \gamma_2 \circ \ldots \circ \gamma_{m-1})(\gamma_1 \circ \ldots \circ \gamma_{m-1} \circ \gamma_0)$$
$$\cdot (\alpha_0 \circ \alpha_1 \circ \ldots \circ \alpha_{m-1})(\alpha_1 \circ \alpha_0 \circ \alpha_2 \circ \ldots \circ \alpha_{m-1})(\alpha_1 \circ \ldots \circ \alpha_{m-1} \circ \alpha_0).$$

Now for each $i \leq n$ and $k < m$ we show by induction that for all $j < m$,

$$q_i(xy_kx_j) \; (\alpha_k\gamma_0 + \alpha_k\gamma_1 + \ldots + \alpha_k\gamma_{m-1}) \; q_i(xy_{k+1}x_j).$$

For $j = 0$, we have immediately $q_i(xy_kx_0) = x = q_i(xy_{k+1}x_0)$. Assuming the statement is true for $j < m$ we have

$$q_i(xy_kx_{j+1}) \; \gamma_j \; q_i(xy_kx_j) \, (\alpha_k\gamma_0 + \alpha_k\gamma_1 + \ldots + \alpha_k\gamma_{m-1}) \; q_i(xy_{k+1}x_j) \; \gamma_j \; q_i(xy_{k+1}x_{j+1}).$$

But $q_i(xy_kx_{j+1})\,\alpha_k\,q_i(xy_{k+1}x_{j+1})$ also, so we have

$$\begin{aligned}(q_i(xy_kx_{j+1}), q_i(xy_{k+1}x_{j+1})) &\in \alpha_k(\gamma_j \circ (\alpha_k\gamma_0 + \alpha_k\gamma_1 + \ldots + \alpha_k\gamma_{m-1}) \circ \gamma_j) \\ &\leq (\alpha_k\gamma_0 + \alpha_k\gamma_1 + \ldots + \alpha_k\gamma_{m-1})\end{aligned}$$

(by 1 of Th.5) establishing the induction step and, by induction, the claim. Then

$$q_i(xxz)\,(\alpha_0\gamma_0 + \alpha_0\gamma_1 + \ldots)\,q_i(xy_1z)\,(\alpha_1\gamma_0 + \alpha_1\gamma_1 + \ldots)\,q_i(xy_2z)\ldots q_i(xzz)$$

and since $q_i(xxz) = q_{i+1}(xxz)$ if i even, and $q_i(xzz) = q_{i+1}(xzz)$ if i odd, we have

$$q_1(xxz)\;(\textstyle\sum\{\gamma_j\alpha_k : 0 \leq j, k < m\})\;q_n(xxz) = z$$

giving us 15.

We next show that 16 follows from 15. If in 16 we have instead a product over some subset of permutations of $\{0, 1, \ldots m-1\}$ appearing, we can apply 15 to some pair of factors $(\gamma_{\pi(0)} \circ \gamma_{\pi(1)} \circ \gamma_{\pi(2)} \circ \ldots \circ \gamma_{\pi(m-1)})(\alpha_{\pi(0)} \circ \alpha_{\pi(1)} \circ \alpha_{\pi(2)} \circ \ldots \alpha_{\pi(m-1)})$ from this product, taking δ to be the remaining factors, and obtain a new condition involving the previous permutations together with the products of π with the permutations in 15. Since each permutation can be expressed as a product of the transposition and the m -cycle represented in 15, by an induction argument we can get all permutations eventually, thus establishing 16. Now for any $m \geq 2$, 16 implies 2 as we can take $\gamma_j = \gamma$, $\alpha_0 = \alpha$, $\alpha_k = \beta$ for $k \geq 1$, and choose an appropriate permutation.

To show 17, it suffices to show for all M,

$$\begin{aligned}&((\gamma_0 \circ \gamma_1 \circ \ldots \circ \gamma_{m-1}) \circ (\gamma_0 \circ \ldots) \ldots \circ \gamma_r)((\alpha_0 \circ \alpha_1 \circ \ldots \circ \alpha_{m-1}) \circ (\alpha_0 \circ \ldots) \ldots \circ \alpha_r) \\ &\leq (\gamma_0 \circ \gamma_1 \circ \ldots \gamma_{m-1})(\alpha_0 \circ \alpha_1 \circ \ldots \alpha_{m-1}) \circ (\textstyle\sum\{\gamma_j\alpha_k : 0 \leq j, k < m\gamma_j\alpha k\})\end{aligned}$$

where M terms appear in each factor of the left hand side. We apply 16 with M terms and choose the permutation that arranges all of the γ_0's first then the γ_1's etc. Thus

$$\begin{aligned}&((\gamma_0 \circ \gamma_1 \circ \ldots \gamma_{m-1}) \circ (\gamma_0 \circ \ldots) \ldots \circ \gamma_r)((\alpha_0 \circ \alpha_1 \circ \ldots \alpha_{m-1}) \circ (\alpha_0 \circ \ldots) \ldots \circ \alpha_r) \\ &\leq ((\gamma_0 \circ \gamma_0 \circ \ldots) \circ (\gamma_1 \circ \gamma_1 \circ \ldots) \ldots)((\alpha_0 \circ \alpha_0 \circ \ldots) \circ (\alpha_1 \circ \alpha_1 \circ \ldots) \ldots) \\ &\qquad \circ (\textstyle\sum\{\gamma_j\alpha_k : 0 \leq j, k < m\}) \\ &= (\gamma_0 \circ \gamma_1 \circ_2 \circ \ldots \gamma_{m-1})(\alpha_0 \circ \alpha_1 \circ \alpha_2 \circ \ldots \gamma_{m-1}) \circ (\textstyle\sum\{\gamma_j\alpha_k : 0 \leq j, k < m\})\end{aligned}$$

Finally, for any $m \geq 2$, 17 implies 5, thus completing the loop and showing that all the conditions are equivalent to those of Theorem 5.

4. CONCLUSION. Many more conditions equivalent to congruence modularity can easily be found. Any condition which follows from **6** or **17** and implies any one of **1-3** can be added to our list, e.g. taking any single permutation of the γ's and any non-identity permutation of the α's in **16**. In some sense, **6** and **17** are the locally strongest conditions and **1-3** are the locally weakest conditions we have derived. To improve and expand upon our understanding of conditions equivalent to congruence modularity we can look not only for stronger, more general conditions, but also for weaker and simpler conditions. The techniques employed here should prove useful in establishing more results of this type. We conclude by mentioning three conditions which arose during this investigation but were never linked back into our chain of equivalent conditions.

$$A)\quad \gamma\cdot(\alpha\circ\beta) \le (\gamma\alpha+\gamma\beta)\circ\gamma\cdot(\beta\circ\alpha)\circ(\gamma\alpha+\gamma\beta)$$

$$B)\quad \gamma\cdot(\alpha\circ\beta) \le \gamma\cdot(\beta\circ(\gamma\alpha+\gamma\beta)\circ\alpha)$$

$$C)\quad (\gamma_0\circ\gamma_1)(\beta_0\circ\beta_1)(\alpha_0\circ\alpha_1) \le (\gamma_0\circ\gamma_1)(\beta_0\circ\beta_1)(\alpha_1\circ\alpha_0)\circ\left(\sum\{\gamma_i\beta_j\alpha_k : i,j,k<2\}\right)$$

Condition A is implied by **2** but then the question is does $\gamma\cdot(\alpha\circ\beta) \le \gamma\alpha\circ\gamma\cdot(\beta\circ\alpha)\circ\gamma\beta$ even imply congruence modularity? Condition B is implied indirectly by **2** but does $\gamma\cdot(\alpha\circ\beta) \le \gamma\cdot(\beta\circ\gamma\alpha\circ\gamma\beta\circ\alpha)$ even imply congruence modularity? Condition C is an attempt to generalize **15-17**. It implies congruence modularity but is it implied by congruence modularity? What we need to go with our techniques of showing equivalences is a complementary technique for deciding when such conditions are inequivalent.

Part of the motivation for this investigation is the question of whether two Mal'cev conditions together can imply congruence modularity without one or the other alone implying congruence modularity (see [2]). If an example could be found using conditions expressed as we have done here, the proof that together they imply congruence modularity may be considerably simplified by the techniques and results we have presented. On the other hand, we have shown that a wide variety of conditions similar to congruence modularity are in fact equivalent to congruence modularity. This suggests the likelihood that an example with neither condition alone equivalent to congruence modularity will be hard to find. If no such simple example exists, then there may be some abstract property of the conditions we have been exploring that explains this fact. Whether this would help solve the problem in the general case is doubtful, but it seems clear that if no example exists the proof of this fact will be difficult by any approach.

REFERENCES

1. Day. A, *A characterization of modularity for congruence lattices of algebras*, Canad. Math. Bull. **12**(1969), 167-173.

2. Garcia, O. C., and W. Taylor, *The Lattice of Interpretability Types*, Memoirs of the AMS # 305, 1984

3. Gumm, H. P., *Congruence modularity is permutability composed with distributivity*, Arch. Math. (Basel) **36**(1981), 569-576.

4. Lakser, H., W. Taylor, and S. Tschantz, *A new proof of Gumm's theorem* Alg. Univ. (to appear.)

5. Pixley, A. F., *Local Mal'cev conditions*, Canad. Math. Bull. **15** (1972), 559-568.

6. Tschantz, S., *Constructions in Clone Theory*, Ph.D. thesis, Univ. of Calif., Berkeley, 1983

7. Wille, R., Kongruenzklassengeometrien, Lecture Notes in Mathematics, Vol. 113, Springer-Verlag, Berlin-New York.

ɔl. 1008: Algebraic Geometry. Proceedings, 1981. Edited by J. Dolachev. V, 138 pages. 1983.

ɔl. 1009: T. A. Chapman, Controlled Simple Homotopy Theory and pplications. III, 94 pages. 1983.

ɔl. 1010: J.-E. Dies, Chaînes de Markov sur les permutations. IX, 26 pages. 1983.

ɔl. 1011: J. M. Sigal. Scattering Theory for Many-Body Quantum echanical Systems. IV, 132 pages. 1983.

ɔl. 1012: S. Kantorovitz, Spectral Theory of Banach Space Operors. V, 179 pages. 1983.

ɔl. 1013: Complex Analysis – Fifth Romanian-Finnish Seminar. art 1. Proceedings, 1981. Edited by C. Andreian Cazacu, N. Boboc, Jurchescu and I. Suciu. XX, 393 pages. 1983.

ɔl. 1014: Complex Analysis – Fifth Romanian-Finnish Seminar. art 2. Proceedings, 1981. Edited by C. Andreian Cazacu, N. Boboc, Jurchescu and I. Suciu. XX, 334 pages. 1983.

l. 1015: Equations différentielles et systèmes de Pfaff dans le amp complexe – II. Seminar. Edited by R. Gérard et J. P. Ramis. 411 pages. 1983.

l. 1016: Algebraic Geometry. Proceedings, 1982. Edited by M. ynaud and T. Shioda. VIII, 528 pages. 1983.

l. 1017: Equadiff 82. Proceedings, 1982. Edited by H. W. Knobloch d K. Schmitt. XXIII, 666 pages. 1983.

l. 1018: Graph Theory, Łagów 1981. Proceedings, 1981. Edited by Borowiecki, J. W. Kennedy and M. M. Sysło. X, 289 pages. 1983.

l. 1019: Cabal Seminar 79–81. Proceedings, 1979–81. Edited by S. Kechris, D. A. Martin and Y. N. Moschovakis. V, 284 pages. 1983.

l. 1020: Non Commutative Harmonic Analysis and Lie Groups. oceedings, 1982. Edited by J. Carmona and M. Vergne. V, 187 ges. 1983.

l. 1021: Probability Theory and Mathematical Statistics. Proedings, 1982. Edited by K. Itô and J.V. Prokhorov. VIII, 747 pages. 83.

l. 1022: G. Gentili, S. Salamon and J.-P. Vigué. Geometry Seminar ıigi Bianchi", 1982. Edited by E. Vesentini. VI, 177 pages. 1983.

l. 1023: S. McAdam, Asymptotic Prime Divisors. IX, 118 pages. 83.

l. 1024: Lie Group Representations I. Proceedings, 1982–1983. ited by R. Herb, R. Lipsman and J. Rosenberg. IX, 369 pages. 1983.

. 1025: D. Tanré, Homotopie Rationnelle: Modèles de Chen, illen, Sullivan. X, 211 pages. 1983.

l. 1026: W. Plesken, Group Rings of Finite Groups Over p-adic egers. V, 151 pages. 1983.

. 1027: M. Hasumi, Hardy Classes on Infinitely Connected Rienn Surfaces. XII, 280 pages. 1983.

l. 1028: Séminaire d'Analyse P. Lelong – P. Dolbeault – H. Skoda. nées 1981/1983. Edité par P. Lelong, P. Dolbeault et H. Skoda. I, 328 pages. 1983.

. 1029: Séminaire d'Algèbre Paul Dubreil et Marie-Paule Malliavin. ɔceedings, 1982. Edité par M.-P. Malliavin. V, 339 pages. 1983.

. 1030: U. Christian, Selberg's Zeta-, L-, and Eisensteinseries. 196 pages. 1983.

. 1031: Dynamics and Processes. Proceedings, 1981. Edited by Blanchard and L. Streit. IX, 213 pages. 1983.

. 1032: Ordinary Differential Equations and Operators. Proedings, 1982. Edited by W. N. Everitt and R. T. Lewis. XV, 521 pages. 33.

. 1033: Measure Theory and its Applications. Proceedings, 1982. ted by J. M. Belley, J. Dubois and P. Morales. XV, 317 pages. 1983.

Vol. 1034: J. Musielak, Orlicz Spaces and Modular Spaces. V, 222 pages. 1983.

Vol. 1035: The Mathematics and Physics of Disordered Media. Proceedings, 1983. Edited by B. D. Hughes and B. W. Ninham. VII, 432 pages. 1983.

Vol. 1036: Combinatorial Mathematics X. Proceedings, 1982. Edited by L. R. A. Casse. XI, 419 pages. 1983.

Vol. 1037: Non-linear Partial Differential Operators and Quantization Procedures. Proceedings, 1981. Edited by S. I. Andersson and H.-D. Doebner. VII, 334 pages. 1983.

Vol. 1038: F. Borceux, G. Van den Bossche, Algebra in a Localic Topos with Applications to Ring Theory. IX, 240 pages. 1983.

Vol. 1039: Analytic Functions, Błażejewko 1982. Proceedings. Edited by J. Ławrynowicz. X, 494 pages. 1983

Vol. 1040: A. Good, Local Analysis of Selberg's Trace Formula. III, 128 pages. 1983.

Vol. 1041: Lie Group Representations II. Proceedings 1982–1983. Edited by R. Herb, S. Kudla, R. Lipsman and J. Rosenberg. IX, 340 pages. 1984.

Vol. 1042: A. Gut, K. D. Schmidt, Amarts and Set Function Processes. III, 258 pages. 1983.

Vol. 1043: Linear and Complex Analysis Problem Book. Edited by V. P. Havin, S. V. Hruščëv and N. K. Nikol'skii. XVIII, 721 pages. 1984.

Vol. 1044: E. Gekeler, Discretization Methods for Stable Initial Value Problems. VIII, 201 pages. 1984.

Vol. 1045: Differential Geometry. Proceedings, 1982. Edited by A. M. Naveira. VIII, 194 pages. 1984.

Vol. 1046: Algebraic K-Theory, Number Theory, Geometry and Analysis. Proceedings, 1982. Edited by A. Bak. IX, 464 pages. 1984.

Vol. 1047: Fluid Dynamics. Seminar, 1982. Edited by H. Beirão da Veiga. VII, 193 pages. 1984.

Vol. 1048: Kinetic Theories and the Boltzmann Equation. Seminar, 1981. Edited by C. Cercignani. VII, 248 pages. 1984.

Vol. 1049: B. Iochum, Cônes autopolaires et algèbres de Jordan. VI, 247 pages. 1984.

Vol. 1050: A. Prestel, P. Roquette, Formally p-adic Fields. V, 167 pages. 1984.

Vol. 1051: Algebraic Topology, Aarhus 1982. Proceedings. Edited by I. Madsen and B. Oliver. X, 665 pages. 1984.

Vol. 1052: Number Theory. Seminar, 1982. Edited by D. V. Chudnovsky, G. V. Chudnovsky, H. Cohn and M. B. Nathanson. V, 309 pages. 1984.

Vol. 1053: P. Hilton, Nilpotente Gruppen und nilpotente Räume. V, 221 pages. 1984.

Vol. 1054: V. Thomée, Galerkin Finite Element Methods for Parabolic Problems. VII, 237 pages. 1984.

Vol. 1055: Quantum Probability and Applications to the Quantum Theory of Irreversible Processes. Proceedings, 1982. Edited by L. Accardi, A. Frigerio and V. Gorini. VI, 411 pages. 1984.

Vol. 1056: Algebraic Geometry. Bucharest 1982. Proceedings, 1982. Edited by L. Bădescu and D. Popescu. VII, 380 pages. 1984.

Vol. 1057: Bifurcation Theory and Applications. Seminar, 1983. Edited by L. Salvadori. VII, 233 pages. 1984.

Vol. 1058: B. Aulbach, Continuous and Discrete Dynamics near Manifolds of Equilibria. IX, 142 pages. 1984.

Vol. 1059: Séminaire de Probabilités XVIII, 1982/83. Proceedings. Edité par J. Azéma et M. Yor. IV, 518 pages. 1984.

Vol. 1060: Topology. Proceedings, 1982. Edited by L. D. Faddeev and A. A. Mal'cev. VI, 389 pages. 1984.

Vol. 1061: Séminaire de Théorie du Potentiel. Paris, No. 7. Proceedings. Directeurs: M. Brelot, G. Choquet et J. Deny. Rédacteurs: F. Hirsch et G. Mokobodzki. IV, 281 pages. 1984.

Vol. 1062: J. Jost, Harmonic Maps Between Surfaces. X, 133 pages. 1984.

Vol. 1063: Orienting Polymers. Proceedings, 1983. Edited by J.L. Ericksen. VII, 166 pages. 1984.

Vol. 1064: Probability Measures on Groups VII. Proceedings, 1983. Edited by H. Heyer. X, 588 pages. 1984.

Vol. 1065: A. Cuyt, Padé Approximants for Operators: Theory and Applications. IX, 138 pages. 1984.

Vol. 1066: Numerical Analysis. Proceedings, 1983. Edited by D.F. Griffiths. XI, 275 pages. 1984.

Vol. 1067: Yasuo Okuyama, Absolute Summability of Fourier Series and Orthogonal Series. VI, 118 pages. 1984.

Vol. 1068: Number Theory, Noordwijkerhout 1983. Proceedings. Edited by H. Jager. V, 296 pages. 1984.

Vol. 1069: M. Kreck, Bordism of Diffeomorphisms and Related Topics. III, 144 pages. 1984.

Vol. 1070: Interpolation Spaces and Allied Topics in Analysis. Proceedings, 1983. Edited by M. Cwikel and J. Peetre. III, 239 pages. 1984.

Vol. 1071: Padé Approximation and its Applications, Bad Honnef 1983. Prodeedings. Edited by H. Werner and H.J. Bünger. VI, 264 pages. 1984.

Vol. 1072: F. Rothe, Global Solutions of Reaction-Diffusion Systems. V, 216 pages. 1984.

Vol. 1073: Graph Theory, Singapore 1983. Proceedings. Edited by K.M. Koh and H.P. Yap. XIII, 335 pages. 1984.

Vol. 1074: E.W. Stredulinsky, Weighted Inequalities and Degenerate Elliptic Partial Differential Equations. III, 143 pages. 1984.

Vol. 1075: H. Majima, Asymptotic Analysis for Integrable Connections with Irregular Singular Points. IX, 159 pages. 1984.

Vol. 1076: Infinite-Dimensional Systems. Proceedings, 1983. Edited by F. Kappel and W. Schappacher. VII, 278 pages. 1984.

Vol. 1077: Lie Group Representations III. Proceedings, 1982–1983. Edited by R. Herb, R. Johnson, R. Lipsman, J. Rosenberg. XI, 454 pages. 1984.

Vol. 1078: A.J.E.M. Janssen, P. van der Steen, Integration Theory. V, 224 pages. 1984.

Vol. 1079: W. Ruppert. Compact Semitopological Semigroups: An Intrinsic Theory. V, 260 pages. 1984

Vol. 1080: Probability Theory on Vector Spaces III. Proceedings, 1983. Edited by D. Szynal and A. Weron. V, 373 pages. 1984.

Vol. 1081: D. Benson, Modular Representation Theory: New Trends and Methods. XI, 231 pages. 1984.

Vol. 1082: C.-G. Schmidt, Arithmetik Abelscher Varietäten mit komplexer Multiplikation. X, 96 Seiten. 1984.

Vol. 1083: D. Bump, Automorphic Forms on GL (3,IR). XI, 184 pages. 1984.

Vol. 1084: D. Kletzing, Structure and Representations of Q-Groups. VI, 290 pages. 1984.

Vol. 1085: G.K. Immink, Asymptotics of Analytic Difference Equations. V, 134 pages. 1984.

Vol. 1086: Sensitivity of Functionals with Applications to Engineering Sciences. Proceedings, 1983. Edited by V. Komkov. V, 130 pages. 1984

Vol. 1087: W. Narkiewicz, Uniform Distribution of Sequences of Integers in Residue Classes. VIII, 125 pages. 1984.

Vol. 1088: A.V. Kakosyan, L.B. Klebanov, J.A. Melamed, Characterization of Distributions by the Method of Intensively Monotone Operators. X, 175 pages. 1984.

Vol. 1089: Measure Theory, Oberwolfach 1983. Proceedings. Edited by D. Kölzow and D. Maharam-Stone. XIII, 327 pages. 1984.

Vol. 1090: Differential Geometry of Submanifolds. Proceedin 1984. Edited by K. Kenmotsu. VI, 132 pages. 1984.

Vol. 1091: Multifunctions and Integrands. Proceedings, 1983. Edi by G. Salinetti. V, 234 pages. 1984.

Vol. 1092: Complete Intersections. Seminar, 1983. Edited by S. Gre and R. Strano. VII, 299 pages. 1984.

Vol. 1093: A. Prestel, Lectures on Formally Real Fields. XI, 125 pag 1984.

Vol. 1094: Analyse Complexe. Proceedings, 1983. Edité par E. An R. Gay et Nguyen Thanh Van. IX, 184 pages. 1984.

Vol. 1095: Stochastic Analysis and Applications. Proceedings, 19 Edited by A. Truman and D. Williams. V, 199 pages. 1984.

Vol. 1096: Théorie du Potentiel. Proceedings, 1983. Edité par Mokobodzki et D. Pinchon. IX, 601 pages. 1984.

Vol. 1097: R.M. Dudley, H. Kunita, F. Ledrappier, École d'Éte Probabilités de Saint-Flour XII – 1982. Edité par P.L. Hennequin 396 pages. 1984.

Vol. 1098: Groups – Korea 1983. Proceedings. Edited by A.C. K and B.H. Neumann. VII, 183 pages. 1984.

Vol. 1099: C.M. Ringel, Tame Algebras and Integral Quadratic Forr XIII, 376 pages. 1984.

Vol. 1100: V. Ivrii, Precise Spectral Asymptotics for Elliptic Operat Acting in Fiberings over Manifolds with Boundary. V, 237 pag 1984.

Vol. 1101: V. Cossart, J. Giraud, U. Orbanz, Resolution of Surfa Singularities. VII, 132 pages. 1984.

Vol. 1102: A. Verona, Stratified Mappings – Structure and Trian lability. IX, 160 pages. 1984.

Vol. 1103: Models and Sets. Proceedings, Logic Colloquium, 19 Part I. Edited by G.H. Müller and M.M. Richter. VIII, 484 pages. 19

Vol. 1104: Computation and Proof Theory. Proceedings, Lo Colloquium, 1983, Part II. Edited by M.M. Richter, E. Börger, W. Ob schelp, B. Schinzel and W. Thomas. VIII, 475 pages. 1984.

Vol. 1105: Rational Approximation and Interpolation. Proceedin 1983. Edited by P.R. Graves-Morris, E.B. Saff and R.S. Var XII, 528 pages. 1984.

Vol. 1106: C.T. Chong, Techniques of Admissible Recursion The IX, 214 pages. 1984.

Vol. 1107: Nonlinear Analysis and Optimization. Proceedings, 19 Edited by C. Vinti. V, 224 pages. 1984.

Vol. 1108: Global Analysis – Studies and Applications I. Edited Yu.G. Borisovich and Yu.E. Gliklikh. V, 301 pages. 1984.

Vol. 1109: Stochastic Aspects of Classical and Quantum Syste Proceedings, 1983. Edited by S. Albeverio, P. Combe and M. Sirug Collin. IX, 227 pages. 1985.

Vol. 1110: R. Jajte, Strong Limit Theorems in Non-Commuta Probability. VI, 152 pages. 1985.

Vol. 1111: Arbeitstagung Bonn 1984. Proceedings. Edited by F. Hi bruch, J. Schwermer and S. Suter. V, 481 pages. 1985.

Vol. 1112: Products of Conjugacy Classes in Groups. Edited Z. Arad and M. Herzog. V, 244 pages. 1985.

Vol. 1113: P. Antosik, C. Swartz, Matrix Methods in Analysis. 114 pages. 1985.

Vol. 1114: Zahlentheoretische Analysis. Seminar. Herausgege von E. Hlawka. V, 157 Seiten. 1985.

Vol. 1115: J. Moulin Ollagnier, Ergodic Theory and Statistical Mech ics. VI, 147 pages. 1985.

Vol. 1116: S. Stolz, Hochzusammenhängende Mannigfaltigke und ihre Ränder. XXIII, 134 Seiten. 1985.